Approximation Algorithms

Springer
Berlin
Heidelberg
New York
Hong Kong
London
Milan
Paris
Tokyo

Vijay V. Vazirani

Approximation Algorithms

Springer

Vijay V. Vazirani
Georgia Institute of Technology
College of Computing
801 Atlantic Avenue
Atlanta, GA 30332-0280
USA

vazirani@cc.gatech.edu
http://www.cc.gatech.edu/fac/Vijay.Vazirani

Corrected Second Printing 2003

Library of Congress Cataloging-in-Publication Data
Vazirani, Vijay V.
Approximation algorithms / Vijay V. Vazirani.
p.cm.
Includes bibliographical references and index.

1. Computer algorithms. 2. Mathematical optimization. I. Title.
QA76.9.A43 V39 2001
005-1–dc21

ACM Computing Classification (1998): F.1–2, G.1.2, G.1.6, G2–4

AMS Mathematics Classification (2000): 68-01; 68W05, 20, 25, 35,40; 68Q05–17, 25; 68R05, 10; 90-08; 90C05, 08, 10, 22, 27, 35, 46, 47, 59, 90; 05A05; 05C05, 12, 20, 38, 40, 45, 69, 70, 85, 90; 11H06; 15A03, 15, 18, 39, 48

ISBN 978-3-642-08469-0

Springer-Verlag Berlin Heidelberg New York
a member of Springer Science+Business Media
http://www.springer.de

Printed in Germany

Cover Design: KünkelLopka, Heidelberg

Printed on acid-free paper

To my parents

Preface

Although this may seem a paradox, all exact science is dominated by the idea of approximation.
Bertrand Russell (1872–1970)

Most natural optimization problems, including those arising in important application areas, are **NP**-hard. Therefore, under the widely believed conjecture that $\mathbf{P} \neq \mathbf{NP}$, their exact solution is prohibitively time consuming. Charting the landscape of approximability of these problems, via polynomial time algorithms, therefore becomes a compelling subject of scientific inquiry in computer science and mathematics. This book presents the theory of approximation algorithms as it stands today. It is reasonable to expect the picture to change with time.

This book is divided into three parts. In Part I we cover combinatorial algorithms for a number of important problems, using a wide variety of algorithm design techniques. The latter may give Part I a non-cohesive appearance. However, this is to be expected – nature is very rich, and we cannot expect a few tricks to help solve the diverse collection of **NP**-hard problems. Indeed, in this part, we have purposely refrained from tightly categorizing algorithmic techniques so as not to trivialize matters. Instead, we have attempted to capture, as accurately as possible, the individual character of each problem, and point out connections between problems and algorithms for solving them.

In Part II, we present linear programming based algorithms. These are categorized under two fundamental techniques: rounding and the primal–dual schema. But once again, the exact approximation guarantee obtainable depends on the specific LP-relaxation used, and there is no fixed recipe for discovering good relaxations, just as there is no fixed recipe for proving a theorem in mathematics (readers familiar with complexity theory will recognize this as the philosophical point behind the $\mathbf{P} \neq \mathbf{NP}$ question).

Part III covers four important topics. The first is the problem of finding a shortest vector in a lattice which, for several reasons, deserves individual treatment (see Chapter 27).

The second topic is the approximability of counting, as opposed to optimization, problems (counting the number of solutions to a given instance). The counting versions of almost all known **NP**-complete problems are #**P**-complete. Interestingly enough, other than a handful of exceptions, this is true of problems in **P** as well. An impressive theory has been built for obtaining efficient approximate counting algorithms for this latter class of problems. Most of these algorithms are based on the Markov chain Monte Carlo (MCMC) method, a topic that deserves a book by itself and is therefore not treated here. In Chapter 28 we present combinatorial algorithms, not using the MCMC method, for two fundamental counting problems.

The third topic is centered around recent breakthrough results, establishing hardness of approximation for many key problems, and giving new legitimacy to approximation algorithms as a deep theory. An overview of these results is presented in Chapter 29, assuming the main technical theorem, the PCP Theorem. The latter theorem, unfortunately, does not have a simple proof at present.

The fourth topic consists of the numerous open problems of this young field. The list presented should by no means be considered exhaustive, and is moreover centered around problems and issues currently in vogue. Exact algorithms have been studied intensively for over four decades, and yet basic insights are still being obtained. Considering the fact that among natural computational problems, polynomial time solvability is the exception rather than the rule, it is only reasonable to expect the theory of approximation algorithms to grow considerably over the years.

The set cover problem occupies a special place, not only in the theory of approximation algorithms, but also in this book. It offers a particularly simple setting for introducing key concepts as well as some of the basic algorithm design techniques of Part I and Part II. In order to give a complete treatment for this central problem, in Part III we give a hardness result for it, even though the proof is quite elaborate. The hardness result essentially matches the guarantee of the best algorithm known – this being another reason for presenting this rather difficult proof.

Our philosophy on the design and exposition of algorithms is nicely illustrated by the following analogy with an aspect of Michelangelo's art. A major part of his effort involved looking for interesting pieces of stone in the quarry and staring at them for long hours to determine the form they naturally wanted to take. The chisel work exposed, in a minimalistic manner, this form. By analogy, we would like to start with a clean, simply stated problem (perhaps a simplified version of the problem we actually want to solve in practice). Most of the algorithm design effort actually goes into understanding the algorithmically relevant combinatorial structure of the problem. The algorithm exploits this structure in a minimalistic manner. The exposition of algorithms in this book will also follow this analogy, with emphasis on stating the structure offered by problems, and keeping the algorithms minimalistic.

An attempt has been made to keep individual chapters short and simple, often presenting only the key result. Generalizations and related results are relegated to exercises. The exercises also cover other important results which could not be covered in detail due to logistic constraints. Hints have been provided for some of the exercises; however, there is no correlation between the degree of difficulty of an exercise and whether a hint is provided for it.

This book is suitable for use in advanced undergraduate and graduate level courses on approximation algorithms. It has more than twice the material that can be covered in a semester long course, thereby leaving plenty of room for an instructor to choose topics. An undergraduate course in algorithms and the theory of **NP**-completeness should suffice as a prerequisite for most of the chapters. For completeness, we have provided background information on several topics: complexity theory in Appendix A, probability theory in Appendix B, linear programming in Chapter 12, semidefinite programming in Chapter 26, and lattices in Chapter 27. (A disproportionate amount of space has been devoted to the notion of self-reducibility in Appendix A because this notion has been quite sparsely treated in other sources.) This book can also be used as supplementary text in basic undergraduate and graduate algorithms courses. The first few chapters of Part I and Part II are suitable for this purpose. The ordering of chapters in both these parts is roughly by increasing difficulty.

In anticipation of this wide audience, we decided not to publish this book in any of Springer's series – even its prestigious Yellow Series. (However, we could not resist spattering a patch of yellow on the cover!) The following translations are currently planned: French by Claire Kenyon, Japanese by Takao Asano, and Romanian by Ion Măndoiu. Corrections and comments from readers are welcome. We have set up a special email address for this purpose: approx@cc.gatech.edu.

Finally, a word about practical impact. With practitioners looking for high performance algorithms having error within 2% or 5% of the optimal, what good are algorithms that come within a factor of 2, or even worse, $O(\log n)$, of the optimal? Further, by this token, what is the usefulness of improving the approximation guarantee from, say, factor 2 to 3/2?

Let us address both issues and point out some fallacies in these assertions. The approximation guarantee only reflects the performance of the algorithm on the most pathological instances. Perhaps it is more appropriate to view the approximation guarantee as a measure that forces us to explore deeper into the combinatorial structure of the problem and discover more powerful tools for exploiting this structure. It has been observed that the difficulty of constructing tight examples increases considerably as one obtains algorithms with better guarantees. Indeed, for some recent algorithms, obtaining a tight example has been a paper by itself (e.g., see Section 26.7). Experiments have confirmed that these and other sophisticated algorithms do have error bounds of the desired magnitude, 2% to 5%, on typical instances, even though their worst case error bounds are much higher. Additionally, the theoretically proven algorithm should be viewed as a core algorithmic idea that needs to be fine tuned to the types of instances arising in specific applications.

We hope that this book will serve as a catalyst in helping this theory grow and have practical impact.

Acknowledgments

This book is based on courses taught at the Indian Institute of Technology, Delhi in Spring 1992 and Spring 1993, at Georgia Tech in Spring 1997, Spring 1999, and Spring 2000, and at DIMACS in Fall 1998. The Spring 1992 course resulted in the first set of class notes on this topic. It is interesting to note that more than half of this book is based on subsequent research results.

Numerous friends – and family members – have helped make this book a reality. First, I would like to thank Naveen Garg, Kamal Jain, Ion Măndoiu, Sridhar Rajagopalan, Huzur Saran, and Mihalis Yannakakis – my extensive collaborations with them helped shape many of the ideas presented in this book. I was fortunate to get Ion Măndoiu's help and advice on numerous matters – his elegant eye for layout and figures helped shape the presentation. A special thanks, Ion!

I would like to express my gratitude to numerous experts in the field for generous help on tasks ranging all the way from deciding the contents and its organization, providing feedback on the writeup, ensuring correctness and completeness of references to designing exercises and helping list open problems. Thanks to Sanjeev Arora, Alan Frieze, Naveen Garg, Michel Goemans, Mark Jerrum, Claire Kenyon, Samir Khuller, Daniele Micciancio, Yuval Rabani, Sridhar Rajagopalan, Dana Randall, Tim Roughgarden, Amin Saberi, Leonard Schulman, Amin Shokrollahi, and Mihalis Yannakakis, with special thanks to Kamal Jain, Éva Tardos, and Luca Trevisan.

Numerous other people helped with valuable comments and discussions. In particular, I would like to thank Sarmad Abbasi, Cristina Bazgan, Rogerio Brito, Gruia Calinescu, Amit Chakrabarti, Mosses Charikar, Joseph Cheriyan, Vasek Chvátal, Uri Feige, Cristina Fernandes, Ashish Goel, Parikshit Gopalan, Mike Grigoriadis, Sudipto Guha, Dorit Hochbaum, Howard Karloff, Leonid Khachian, Stavros Kolliopoulos, Jan van Leeuwen, Nati Lenial, George Leuker, Vangelis Markakis, Aranyak Mehta, Rajeev Motwani, Prabhakar Raghavan, Satish Rao, Miklos Santha, Jiri Sgall, David Shmoys, Alistair Sinclair, Prasad Tetali, Pete Veinott, Ramarathnam Venkatesan, Nisheeth Vishnoi, and David Williamson. I am sure I am missing several names – my apologies and thanks to these people as well. A special role was played by the numerous students who took my courses on this topic and scribed notes. It will be impossible to individually remember their names. I would like to express my gratitude collectively to them.

I would like to thank IIT Delhi – with special thanks to Shachin Maheshwari – Georgia Tech, and DIMACS for providing pleasant, supportive and academically rich environments. Thanks to NSF for support under grants CCR-9627308 and CCR-9820896.

It was a pleasure to work with Hans Wössner on editorial matters. The personal care with which he handled all such matters and his sensitivity to an author's unique point of view were especially impressive. Thanks also to Frank Holzwarth for sharing his expertise with LaTeX.

A project of this magnitude would be hard to pull off without wholehearted support from family members. Fortunately, in my case, some of them are also fellow researchers – my wife, Milena Mihail, and my brother, Umesh Vazirani. Little Michel's arrival, halfway through this project, brought new joys and energies, though made the end even more challenging! Above all, I would like to thank my parents for their unwavering support and inspiration – my father, a distinguished author of several Civil Engineering books, and my mother, with her deep understanding of Indian Classical Music. This book is dedicated to them.

Atlanta, Georgia, May 2001 Vijay Vazirani

Table of Contents

1 **Introduction** 1
1.1 Lower bounding OPT 2
1.1.1 An approximation algorithm for cardinality vertex cover 3
1.1.2 Can the approximation guarantee be improved? 3
1.2 Well-characterized problems and min–max relations 5
1.3 Exercises 7
1.4 Notes 10

Part I. Combinatorial Algorithms

2 **Set Cover** 15
2.1 The greedy algorithm 16
2.2 Layering 17
2.3 Application to shortest superstring 19
2.4 Exercises 22
2.5 Notes 26

3 **Steiner Tree and TSP** 27
3.1 Metric Steiner tree 27
3.1.1 MST-based algorithm 28
3.2 Metric TSP 30
3.2.1 A simple factor 2 algorithm 31
3.2.2 Improving the factor to 3/2 32
3.3 Exercises 33
3.4 Notes 37

4 **Multiway Cut and k-Cut** 38
4.1 The multiway cut problem 38
4.2 The minimum k-cut problem 40
4.3 Exercises 44
4.4 Notes 46

5 k-Center 47
5.1 Parametric pruning applied to metric k-center 47
5.2 The weighted version 50
5.3 Exercises 52
5.4 Notes 53

6 Feedback Vertex Set 54
6.1 Cyclomatic weighted graphs 54
6.2 Layering applied to feedback vertex set 57
6.3 Exercises 60
6.4 Notes 60

7 Shortest Superstring 61
7.1 A factor 4 algorithm 61
7.2 Improving to factor 3 64
7.2.1 Achieving half the optimal compression 66
7.3 Exercises 66
7.4 Notes 67

8 Knapsack 68
8.1 A pseudo-polynomial time algorithm for knapsack 69
8.2 An FPTAS for knapsack 69
8.3 Strong **NP**-hardness and the existence of FPTAS's 71
8.3.1 Is an FPTAS the most desirable approximation algorithm? 72
8.4 Exercises 72
8.5 Notes 73

9 Bin Packing 74
9.1 An asymptotic PTAS 74
9.2 Exercises 77
9.3 Notes 78

10 Minimum Makespan Scheduling 79
10.1 Factor 2 algorithm 79
10.2 A PTAS for minimum makespan 80
10.2.1 Bin packing with fixed number of object sizes 81
10.2.2 Reducing makespan to restricted bin packing 81
10.3 Exercises 83
10.4 Notes 83

11 Euclidean TSP 84
11.1 The algorithm 84
11.2 Proof of correctness 87
11.3 Exercises 89
11.4 Notes 89

Part II. LP-Based Algorithms

12 Introduction to LP-Duality ... 93
12.1 The LP-duality theorem ... 93
12.2 Min–max relations and LP-duality ... 97
12.3 Two fundamental algorithm design techniques ... 100
12.3.1 A comparison of the techniques and the notion of integrality gap ... 101
12.4 Exercises ... 103
12.5 Notes ... 107

13 Set Cover via Dual Fitting ... 108
13.1 Dual-fitting-based analysis for the greedy set cover algorithm 108
13.1.1 Can the approximation guarantee be improved? ... 111
13.2 Generalizations of set cover ... 112
13.2.1 Dual fitting applied to constrained set multicover ... 112
13.3 Exercises ... 116
13.4 Notes ... 117

14 Rounding Applied to Set Cover ... 118
14.1 A simple rounding algorithm ... 118
14.2 Randomized rounding ... 119
14.3 Half-integrality of vertex cover ... 121
14.4 Exercises ... 122
14.5 Notes ... 123

15 Set Cover via the Primal–Dual Schema ... 124
15.1 Overview of the schema ... 124
15.2 Primal–dual schema applied to set cover ... 126
15.3 Exercises ... 128
15.4 Notes ... 129

16 Maximum Satisfiability ... 130
16.1 Dealing with large clauses ... 131
16.2 Derandomizing via the method of conditional expectation ... 131
16.3 Dealing with small clauses via LP-rounding ... 133
16.4 A 3/4 factor algorithm ... 135
16.5 Exercises ... 136
16.6 Notes ... 138

17 Scheduling on Unrelated Parallel Machines ... 139
17.1 Parametric pruning in an LP setting ... 139
17.2 Properties of extreme point solutions ... 140
17.3 The algorithm ... 141

17.4 Additional properties of extreme point solutions 142
17.5 Exercises 143
17.6 Notes 144

18 Multicut and Integer Multicommodity Flow in Trees 145
18.1 The problems and their LP-relaxations 145
18.2 Primal–dual schema based algorithm 148
18.3 Exercises 151
18.4 Notes 153

19 Multiway Cut 154
19.1 An interesting LP-relaxation 154
19.2 Randomized rounding algorithm 156
19.3 Half-integrality of node multiway cut 159
19.4 Exercises 162
19.5 Notes 166

20 Multicut in General Graphs 167
20.1 Sum multicommodity flow 167
20.2 LP-rounding-based algorithm 169
20.2.1 Growing a region: the continuous process 170
20.2.2 The discrete process 171
20.2.3 Finding successive regions 172
20.3 A tight example 174
20.4 Some applications of multicut 175
20.5 Exercises 176
20.6 Notes 178

21 Sparsest Cut 179
21.1 Demands multicommodity flow 179
21.2 Linear programming formulation 180
21.3 Metrics, cut packings, and ℓ_1-embeddability 182
21.3.1 Cut packings for metrics 182
21.3.2 ℓ_1-embeddability of metrics 184
21.4 Low distortion ℓ_1-embeddings for metrics 185
21.4.1 Ensuring that a single edge is not overshrunk 186
21.4.2 Ensuring that no edge is overshrunk 189
21.5 LP-rounding-based algorithm 190
21.6 Applications 191
21.6.1 Edge expansion 191
21.6.2 Conductance 191
21.6.3 Balanced cut 192
21.6.4 Minimum cut linear arrangement 193
21.7 Exercises 194
21.8 Notes 196

22 Steiner Forest 197
22.1 LP-relaxation and dual 197
22.2 Primal–dual schema with synchronization 198
22.3 Analysis 203
22.4 Exercises 206
22.5 Notes 211

23 Steiner Network 212
23.1 LP-relaxation and half-integrality 212
23.2 The technique of iterated rounding 216
23.3 Characterizing extreme point solutions 218
23.4 A counting argument 220
23.5 Exercises 223
23.6 Notes 230

24 Facility Location 231
24.1 An intuitive understanding of the dual 232
24.2 Relaxing primal complementary slackness conditions 233
24.3 Primal–dual schema based algorithm 234
24.4 Analysis 235
24.4.1 Running time 237
24.4.2 Tight example 237
24.5 Exercises 238
24.6 Notes 241

25 *k*-Median 242
25.1 LP-relaxation and dual 242
25.2 The high-level idea 243
25.3 Randomized rounding 246
25.3.1 Derandomization 247
25.3.2 Running time 248
25.3.3 Tight example 248
25.3.4 Integrality gap 249
25.4 A Lagrangian relaxation technique for approximation algorithms 249
25.5 Exercises 250
25.6 Notes 253

26 Semidefinite Programming 255
26.1 Strict quadratic programs and vector programs 255
26.2 Properties of positive semidefinite matrices 257
26.3 The semidefinite programming problem 258
26.4 Randomized rounding algorithm 260
26.5 Improving the guarantee for MAX-2SAT 263
26.6 Exercises 265
26.7 Notes 268

Part III. Other Topics

27 Shortest Vector 273
27.1 Bases, determinants, and orthogonality defect 274
27.2 The algorithms of Euclid and Gauss 276
27.3 Lower bounding OPT using Gram–Schmidt orthogonalization 278
27.4 Extension to n dimensions 280
27.5 The dual lattice and its algorithmic use 284
27.6 Exercises 288
27.7 Notes 292

28 Counting Problems 294
28.1 Counting DNF solutions 295
28.2 Network reliability 297
28.2.1 Upperbounding the number of near-minimum cuts 298
28.2.2 Analysis 300
28.3 Exercises 302
28.4 Notes 305

29 Hardness of Approximation 306
29.1 Reductions, gaps, and hardness factors 306
29.2 The PCP theorem 309
29.3 Hardness of MAX-3SAT 311
29.4 Hardness of MAX-3SAT with bounded occurrence of variables 313
29.5 Hardness of vertex cover and Steiner tree 316
29.6 Hardness of clique 318
29.7 Hardness of set cover 322
29.7.1 The two-prover one-round characterization of **NP** 322
29.7.2 The gadget 324
29.7.3 Reducing error probability by parallel repetition 325
29.7.4 The reduction 326
29.8 Exercises 329
29.9 Notes 332

30 Open Problems 334
30.1 Problems having constant factor algorithms 334
30.2 Other optimization problems 336
30.3 Counting problems 338
30.4 Notes 343

Appendix

A An Overview of Complexity Theory for the Algorithm Designer ... 344
A.1 Certificates and the class **NP** ... 344
A.2 Reductions and **NP**-completeness ... 345
A.3 **NP**-optimization problems and approximation algorithms ... 346
A.3.1 Approximation factor preserving reductions ... 348
A.4 Randomized complexity classes ... 348
A.5 Self-reducibility ... 349
A.6 Notes ... 352

B Basic Facts from Probability Theory ... 353
B.1 Expectation and moments ... 353
B.2 Deviations from the mean ... 354
B.3 Basic distributions ... 355
B.4 Notes ... 355

References ... 357

Problem Index ... 373

Subject Index ... 377

1 Introduction

NP-hard optimization problems exhibit a rich set of possibilities, all the way from allowing approximability to any required degree, to essentially not allowing approximability at all. Despite this diversity, underlying the process of design of approximation algorithms are some common principles. We will explore these in the current chapter.

An optimization problem is polynomial time solvable only if it has the algorithmically relevant combinatorial structure that can be used as "footholds" to efficiently home in on an optimal solution. The process of designing an exact polynomial time algorithm is a two-pronged attack: unraveling this structure in the problem and finding algorithmic techniques that can exploit this structure.

Although **NP**-hard optimization problems do not offer footholds for finding optimal solutions efficiently, they may still offer footholds for finding near-optimal solutions efficiently. So, at a high level, the process of design of approximation algorithms is not very different from that of design of exact algorithms. It still involves unraveling the relevant structure and finding algorithmic techniques to exploit it. Typically, the structure turns out to be more elaborate, and often the algorithmic techniques result from generalizing and extending some of the powerful algorithmic tools developed in the study of exact algorithms.

On the other hand, looking at the process of designing approximation algorithms a little more closely, one can see that it has its own general principles. We illustrate some of these principles in Section 1.1, using the following simple setting.

Problem 1.1 (Vertex cover) Given an undirected graph $G = (V, E)$, and a cost function on vertices $c : V \to \mathbf{Q}^+$, find a minimum cost *vertex cover*, i.e., a set $V' \subseteq V$ such that every edge has at least one endpoint incident at V'. The special case, in which all vertices are of unit cost, will be called the *cardinality vertex cover problem.*

Since the design of an approximation algorithm involves delicately attacking **NP**-hardness and salvaging from it an efficient approximate solution, it will be useful for the reader to review some key concepts from complexity theory. Appendix A and some exercises in Section 1.3 have been provided for this purpose.

It is important to give precise definitions of an **NP**-optimization problem and an approximation algorithm for it (e.g., see Exercises 1.9 and 1.10). Since these definitions are quite technical, we have moved them to Appendix A. We provide essentials below to quickly get started.

An **NP**-optimization problem Π is either a minimization or a maximization problem. Each valid instance I of Π comes with a nonempty set of feasible solutions, each of which is assigned a nonnegative rational number called its objective function value. There exist polynomial time algorithms for determining validity, feasibility, and the objective function value. A feasible solution that achieves the optimal objective function value is called an optimal solution. $\mathrm{OPT}_\Pi(I)$ will denote the objective function value of an optimal solution to instance I. We will shorten this to OPT when there is no ambiguity. For the problems studied in this book, computing $\mathrm{OPT}_\Pi(I)$ is **NP**-hard.

For example, valid instances of the vertex cover problem consist of an undirected graph $G = (V, E)$ and a cost function on vertices. A feasible solution is a set $S \subseteq V$ that is a cover for G. Its objective function value is the sum of costs of all vertices in S. A feasible solution of minimum cost is an optimal solution.

An approximation algorithm, $\mathcal{A}$, for Π produces, in polynomial time, a feasible solution whose objective function value is "close" to the optimal; by "close" we mean within a guaranteed factor of the optimal. In the next section, we will present a factor 2 approximation algorithm for the cardinality vertex cover problem, i.e., an algorithm that finds a cover of cost $\leq 2 \cdot \mathrm{OPT}$ in time polynomial in $|V|$.

1.1 Lower bounding OPT

When designing an approximation algorithm for an **NP**-hard **NP**-optimization problem, one is immediately faced with the following dilemma. In order to establish the approximation guarantee, the cost of the solution produced by the algorithm needs to be compared with the cost of an optimal solution. However, for such problems, not only is it **NP**-hard to find an optimal solution, but it is also **NP**-hard to compute the cost of an optimal solution (see Appendix A). In fact, in Section A.5 we show that computing the cost of an optimal solution (or even solving its decision version) is precisely the difficult core of such problems. So, how do we establish the approximation guarantee? Interestingly enough, the answer to this question provides a key step in the design of approximation algorithms.

Let us demonstrate this in the context of the cardinality vertex cover problem. We will get around the difficulty mentioned above by coming up with a "good" polynomial time computable *lower bound* on the size of the optimal cover.

1.1.1 An approximation algorithm for cardinality vertex cover

We provide some definitions first. Given a graph $H = (U, F)$, a subset of the edges $M \subseteq F$ is said to be a *matching* if no two edges of M share an endpoint. A matching of maximum cardinality in H is called a *maximum matching*, and a matching that is maximal under inclusion is called a *maximal matching*. A maximal matching can clearly be computed in polynomial time by simply greedily picking edges and removing endpoints of picked edges. More sophisticated means lead to polynomial time algorithms for finding a maximum matching as well.

Let us observe that the size of a maximal matching in G provides a lower bound. This is so because *any* vertex cover has to pick at least one endpoint of each matched edge. This lower bounding scheme immediately suggests the following simple algorithm:

Algorithm 1.2 (Cardinality vertex cover)
Find a maximal matching in G and output the set of matched vertices.

Theorem 1.3 *Algorithm 1.2 is a factor 2 approximation algorithm for the cardinality vertex cover problem.*

Proof: No edge can be left uncovered by the set of vertices picked – otherwise such an edge could have been added to the matching, contradicting its maximality. Let M be the matching picked. As argued above, $|M| \leq \text{OPT}$. The approximation factor follows from the observation that the cover picked by the algorithm has cardinality $2|M|$, which is at most $2 \cdot \text{OPT}$. □

Observe that the approximation algorithm for vertex cover was very much related to, and followed naturally from, the lower bounding scheme. This is in fact typical in the design of approximation algorithms. In Part II of this book, we show how linear programming provides a unified way of obtaining lower bounds for several fundamental problems. The algorithm itself is designed around the LP that provides the lower bound.

1.1.2 Can the approximation guarantee be improved?

The following questions arise in the context of improving the approximation guarantee for cardinality vertex cover:

1. Can the approximation guarantee of Algorithm 1.2 be improved by a better analysis?

2. Can an approximation algorithm with a better guarantee be designed using the lower bounding scheme of Algorithm 1.2, i.e., size of a maximal matching in G?
3. Is there some other lower bounding method that can lead to an improved approximation guarantee for vertex cover?

Example 1.4 shows that the answer to the first question is "no", i.e., the analysis presented above for Algorithm 1.2 is tight. It gives an infinite family of instances in which the solution produced by Algorithm 1.2 is twice the optimal. An infinite family of instances of this kind, showing that the analysis of an approximation algorithm is tight, will be referred to as a *tight example*. The importance of finding tight examples for an approximation algorithm one has designed cannot be overemphasized. They give critical insight into the functioning of the algorithm and have often led to ideas for obtaining algorithms with improved guarantees. (The reader is advised to run algorithms on the tight examples presented in this book.)

Example 1.4 Consider the infinite family of instances given by the complete bipartite graphs $K_{n,n}$.

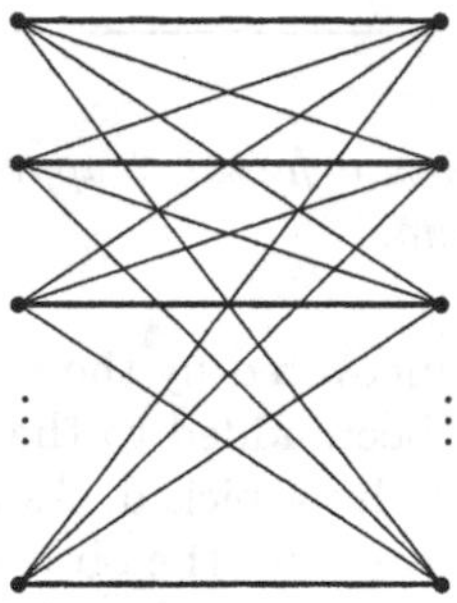

When run on $K_{n,n}$, Algorithm 1.2 will pick all $2n$ vertices, whereas picking one side of the bipartition gives a cover of size n. □

Let us assume that we will establish the approximation factor for an algorithm by simply comparing the cost of the solution it finds with the lower bound. Indeed, almost all known approximation algorithms operate in this manner. Under this assumption, the answer to the second question is also "no". This is established in Example 1.5, which gives an infinite family of instances on which the lower bound, of size of a maximal matching, is in fact half the size of an optimal vertex cover. In the case of linear-programming-based approximation algorithms, the analogous question will be answered by determining a fundamental quantity associated with the linear programming relaxation – its integrality gap (see Chapter 12).

The third question, of improving the approximation guarantee for vertex cover, is currently a central open problem in the field of approximation algorithms (see Section 30.1).

Example 1.5 The lower bound, of size of a maximal matching, is half the size of an optimal vertex cover for the following infinite family of instances. Consider the complete graph K_n, where n is odd. The size of any maximal matching is $(n-1)/2$, whereas the size of an optimal cover is $n-1$. □

1.2 Well-characterized problems and min–max relations

Consider decision versions of the cardinality vertex cover and maximum matching problems.

- Is the size of the minimum vertex cover in G at most k?
- Is the size of the maximum matching in G at least l?

Both these decision problems are in **NP** and therefore have Yes certificates (see Appendix A for definitions). Do these problems also have No certificates? We have already observed that the size of a maximum matching is a lower bound on the size of a minimum vertex cover. If G is bipartite, then in fact equality holds; this is the classic König-Egerváry theorem.

Theorem 1.6 *In any bipartite graph,*

$$\max_{\textit{matching } M} |M| = \min_{\textit{vertex cover } U} |U|.$$

Therefore, if the answer to the first decision problem is "no", there must be a matching of size $k+1$ in G that suffices as a certificate. Similarly, a vertex cover of size $l-1$ must exist in G if the answer to the second decision problem is "no". Hence, when restricted to bipartite graphs, both vertex cover and maximum matching problems have No certificates and are in co-**NP**. In fact, both problems are also in **P** under this restriction. It is easy to see that any problem in **P** trivially has Yes as well as No certificates (the empty string suffices). This is equivalent to the statement that $\mathbf{P} \subseteq \mathbf{NP} \cap \text{co-}\mathbf{NP}$. It is widely believed that the containment is strict; the conjectured status of these classes is depicted below.

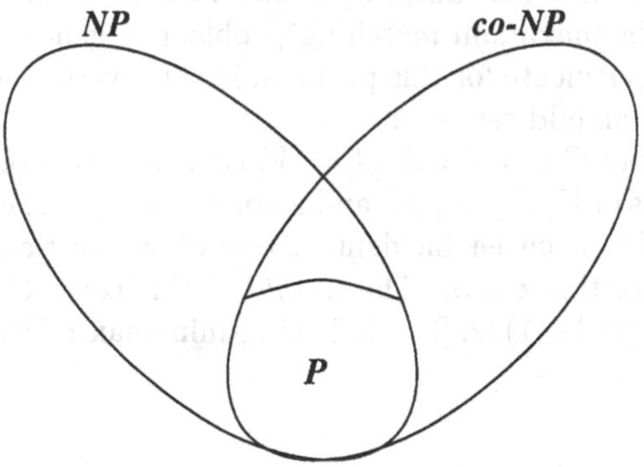

Problems that have Yes and No certificates, i.e., are in **NP** ∩ co-**NP**, are said to be *well-characterized*. The importance of this notion can be gauged from the fact that the quest for a polynomial time algorithm for matching started with the observation that it is well-characterized.

Min–max relations of the kind given above provide proof that a problem is well-characterized. Such relations are some of the most powerful and beautiful results in combinatorics, and some of the most fundamental polynomial time algorithms (exact) have been designed around such relations. Most of these min–max relations are actually special cases of the LP-duality theorem (see Section 12.2). As pointed out above, LP-duality theory plays a vital role in the design of approximation algorithms as well.

What if G is not restricted to be bipartite? In this case, a maximum matching may be strictly smaller than a minimum vertex cover. For instance, if G is simply an odd length cycle on $2p+1$ vertices, then the size of a maximum matching is p, whereas the size of a minimum vertex cover is $p+1$. This may happen even for graphs having a perfect matching, for instance, the Petersen graph:

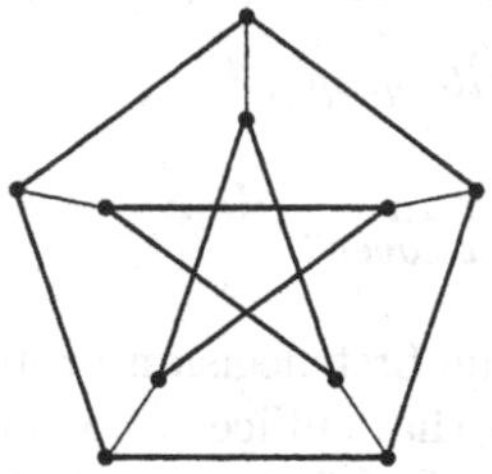

This graph has a perfect matching of cardinality 5; however, the minimum vertex cover has cardinality 6. One can show that there is no vertex cover of size 5 by observing that any vertex cover must pick at least $p+1$ vertices from an odd cycle of length $2p+1$, just to cover all the edges of the cycle, and the Petersen graph has two disjoint cycles of length 5.

Under the widely believed assumption that **NP** $\neq$ co-**NP**, **NP**-hard problems do not have No certificates. Thus, the minimum vertex cover problem in general graphs, which is **NP**-hard, does not have a No certificate, assuming **NP** $\neq$ co-**NP**. The maximum matching problem in general graphs is in **P**. However, the No certificate for this problem is not a vertex cover, but a more general structure: an odd set cover.

An *odd set cover* C in a graph $G=(V,E)$ is a collection of disjoint odd cardinality subsets of V, $S_1,\ldots,S_k$, and a collection $v_1,\ldots,v_l$ of vertices such that each edge of G is either incident at one of the vertices v_i or has both endpoints in one of the sets S_i. The *weight* of this cover C is defined to be $w(C)=l+\sum_{i=1}^{k}(|S_i|-1)/2$. The following min–max relation holds.

Theorem 1.7 *In any graph,* $$\max_{\text{matching } M} |M| = \min_{\text{odd set cover } C} w(C).$$

As shown above, in general graphs a maximum matching can be smaller than a minimum vertex cover. Can it be arbitrarily smaller? The answer is "no". A corollary of Theorem 1.3 is that in any graph, the size of a maximum matching is at least half the size of a minimum vertex cover. More precisely, Theorem 1.3 gives, as a corollary, the following approximate min–max relation. Approximation algorithms frequently yield such approximate min–max relations, which are of independent interest.

Corollary 1.8 *In any graph,*

$$\max_{\text{matching } M} |M| \leq \min_{\text{vertex cover } U} |U| \leq 2 \cdot \left(\max_{\text{matching } M} |M| \right).$$

Although the vertex cover problem does not have No certificate under the assumption $\mathbf{NP} \neq \text{co-}\mathbf{NP}$, surely there ought to be a way of certifying that (G, k) is a "no" instance for small enough values of k. Algorithm 1.2 (more precisely, the lower bounding scheme behind this approximation algorithm) provides such a method. Let $\mathcal{A}(G)$ denote the size of vertex cover output by Algorithm 1.2. Then, $\text{OPT}(G) \leq \mathcal{A}(G) \leq 2 \cdot \text{OPT}(G)$. If $k < \mathcal{A}(G)/2$ then $k < \text{OPT}(G)$, and therefore (G, k) must be a "no" instance. Furthermore, if $k < \text{OPT}(G)/2$ then $k < \mathcal{A}(G)/2$. Hence, Algorithm 1.2 provides a No certificate for all instances (G, k) such that $k < \text{OPT}(G)/2$.

A No certificate for instances (I, B) of a minimization problem Π satisfying $B < \text{OPT}(I)/\alpha$ is called a *factor α approximate No certificate.* As in the case of normal Yes and No certificates, we do not require that this certificate be polynomial time computable. An α factor approximation algorithm $\mathcal{A}$ for Π provides such a certificate. Since $\mathcal{A}$ has polynomial running time, this certificate is polynomial time computable. In Chapter 27 we will show an intriguing result – that the shortest vector problem has a factor n approximate No certificate; however, a polynomial time algorithm for constructing such a certificate is not known.

1.3 Exercises

1.1 Give a factor 1/2 algorithm for the following.

Problem 1.9 (Acyclic subgraph) Given a directed graph $G = (V, E)$, pick a maximum cardinality set of edges from E so that the resulting subgraph is acyclic.
Hint: Arbitrarily number the vertices and pick the bigger of the two sets, the forward-going edges and the backward-going edges. What scheme are you using for upper bounding OPT?

1.2 Design a factor 2 approximation algorithm for the problem of finding a minimum cardinality maximal matching in an undirected graph.
Hint: Use the fact that any maximal matching is at least half the maximum matching.

1.3 (R. Bar-Yehuda) Consider the following factor 2 approximation algorithm for the cardinality vertex cover problem. Find a depth first search tree in the given graph, G, and output the set, say S, of all the nonleaf vertices of this tree. Show that S is indeed a vertex cover for G and $|S| \leq 2 \cdot \text{OPT}$.
Hint: Show that G has a matching of size $\lceil |S|/2 \rceil$.

1.4 Perhaps the first strategy one tries when designing an algorithm for an optimization problem is the greedy strategy. For the vertex cover problem, this would involve iteratively picking a maximum degree vertex and removing it, together with edges incident at it, until there are no edges left. Show that this algorithm achieves an approximation guarantee of $O(\log n)$. Give a tight example for this algorithm.
Hint: The analysis is similar to that in Theorem 2.4.

1.5 A maximal matching can be found via a greedy algorithm: pick an edge, remove its two endpoints, and iterate until there are no edges left. Does this make Algorithm 1.2 a greedy algorithm?

1.6 Give a lower bounding scheme for the arbitrary cost version of the vertex cover problem.
Hint: Not easy if you don't use LP-duality.

1.7 Let $A = \{a_1, \ldots, a_n\}$ be a finite set, and let "$\leq$" be a relation on A that is reflexive, antisymmetric, and transitive. Such a relation is called a *partial ordering of* A. Two elements $a_i, a_j \in A$ are said to be *comparable* if $a_i \leq a_j$ or $a_j \leq a_i$. Two elements that are not comparable are said to be *incomparable*. A subset $S \subseteq A$ is a *chain* if its elements are pairwise comparable. If the elements of S are pairwise incomparable, then it is an *antichain*. A *chain (antichain) cover* is a collection of chains (antichains) that are pairwise disjoint and cover A. The size of such a cover is the number of chains (antichains) in it. Prove the following min–max result. The size of a longest chain equals the size of a smallest antichain cover.
Hint: Let the size of the longest chain be m. For $a \in A$, let $\phi(a)$ denote the size of the longest chain in which a is the smallest element. Now, consider the partition of A, $A_i = \{a \in A \mid \phi(a) = i\}$, for $1 \leq i \leq m$.

1.8 (Dilworth's theorem, see [202]) Prove that in any finite partial order, the size of a largest antichain equals the size of a smallest chain cover.
Hint: Derive from the König-Egerváry Theorem. Given a partial order on n-element set A, consider the bipartite graph $G = (U, V, E)$ with $|U| = |V| = n$ and $(u_i, v_j) \in E$ iff $a_i \leq a_j$.

The next ten exercises are based on Appendix A.

1.9 Is the following an **NP**-optimization problem? Given an undirected graph $G = (V, E)$, a cost function on vertices $c : V \to \mathbf{Q}^+$, and a positive integer k, find a minimum cost vertex cover for G containing at most k vertices.
Hint: Can valid instances be recognized in polynomial time (such an instance must have at least one feasible solution)?

1.10 Let $\mathcal{A}$ be an algorithm for a minimization **NP**-optimization problem Π such that the expected cost of the solution produced by $\mathcal{A}$ is $\leq \alpha\text{OPT}$, for a constant $\alpha > 1$. What is the best approximation guarantee you can establish for Π using algorithm $\mathcal{A}$?
Hint: A guarantee of $2\alpha - 1$ follows easily. For guarantees arbitrarily close to α, run the algorithm polynomially many times and pick the best solution. Apply Chernoff's bound.

1.11 Show that if SAT has been proven **NP**-hard, and SAT has been reduced, via a polynomial time reduction, to the decision version of vertex cover, then the latter is also **NP**-hard.
Hint: Show that the composition of two polynomial time reductions is also a polynomial time reduction.

1.12 Show that if the vertex cover problem is in co-**NP**, then **NP** = co-**NP**.

1.13 (Pratt [230]) Let L be the language consisting of all prime numbers. Show that $L \in \mathbf{NP}$.
Hint: Consider the multiplicative group mod$n$, $Z_n^* = \{a \in \mathbf{Z}^+ \mid 1 \leq a < n \text{ and } (a, n) = 1\}$. Clearly, $|Z_n^*| \leq n - 1$. Use the fact that $|Z_n^*| = n - 1$ iff n is prime, and that Z_n^* is cyclic if n is prime. The Yes certificate consists of a primitive root of Z_n^*, the prime factorization of $n - 1$, and, recursively, similar information about each prime factor of $n - 1$.

1.14 Give proofs of self-reducibility for the optimization problems discussed later in this book, in particular, maximum matching, MAX-SAT (Problem 16.1), clique (Problem 29.15), shortest superstring (Problem 2.9), and Minimum makespan scheduling (Problem 10.1).
Hint: For clique, consider two possibilities, that v is or isn't in the optimal clique. Correspondingly, either restrict G to v and its neighbors, or remove v from G. For shortest superstring, remove two strings and replace them by a legal overlap (may even be a simple concatenation). If the length of the optimal superstring remains unchanged, work with this smaller instance. Generalize the scheduling problem a bit – assume that you are also given the number of time units already scheduled on each machine as part of the instance.

1.15 Give a suitable definition of self-reducibility for problems in **NP**, i.e., decision problems and not optimization problems, which enables you to obtain a polynomial time algorithm for finding a feasible solution given an oracle for the decision version, and moreover, yields a self-reducibility tree for instances.
Hint: Impose an arbitrary order among the atoms of a solution, e.g., for SAT, this was achieved by arbitrarily ordering the n variables.

1.16 Let Π_1 and Π_2 be two minimization problems such that there is an approximation factor preserving reduction from Π_1 to Π_2. Show that if there is an α factor approximation algorithm for Π_2 then there is also an α factor approximation algorithm for Π_1.
Hint: First prove that if the reduction transforms instance I_1 of Π_1 to instance I_2 of Π_2 then $\mathrm{OPT}_{\Pi_1}(I_1) = \mathrm{OPT}_{\Pi_2}(I_2)$.

1.17 Show that

$$L \in \mathbf{ZPP} \text{ iff } L \in (\mathbf{RP} \cap \text{co-}\mathbf{RP}).$$

1.18 Show that if $\mathbf{NP} \subseteq \text{co-}\mathbf{RP}$ then $\mathbf{NP} \subseteq \mathbf{ZPP}$.
Hint: If SAT instance ϕ is satisfiable, a satisfying truth assignment for ϕ can be found, with high probability, using self-reducibility and the co-**RP** machine for SAT. If ϕ is not satisfiable, a "no" answer from the co-**RP** machine for SAT confirms this; the machine will output such an answer with high probability.

1.4 Notes

The notion of well-characterized problems was given by Edmonds [75] and was precisely formulated by Cook [53]. In the same paper, Cook initiated the theory of **NP**-completeness. Independently, this discovery was also made by Levin [193]. It gained its true importance with the work of Karp [171], showing **NP**-completeness of a diverse collection of fundamental computational problems.

Interestingly enough, approximation algorithms were designed even before the discovery of the theory of **NP**-completeness, by Vizing [263] for the minimum edge coloring problem, by Graham [119] for the minimum makespan problem (Problem 10.1), and by Erdös [79] for the MAX-CUT problem (Problem 2.14). However, the real significance of designing such algorithms emerged only after belief in the $\mathbf{P} \neq \mathbf{NP}$ conjecture grew. The notion of an approximation algorithm was formally introduced by Garey, Graham, and Ullman [97] and Johnson [157]. The first use of linear programming in approximation algorithms was due to Lovász [199], for analyzing the greedy set cover algorithm (see Chapter 13). An early work exploring the use of randomization in the design of algorithms was due to Rabin [232] – this notion is useful in the design of approximation algorithms as well. Theorem 1.7 is due to Edmonds [75] and Algorithm 1.2 is due independently to Gavril and Yannakakis (see [225]).

For basic books on algorithms, see Cormen, Leiserson, Rivest, and Stein [56], Papadimitriou and Steiglitz [225], and Tarjan [254]. For a good treatment of min–max relations, see Lovász and Plummer [202]. For books on approximation algorithms, see Hochbaum [133] and Ausiello, Crescenzi, Gambosi, Kann, Marchetti, and Protasi [18]. Books on linear programming, complexity theory, and randomized algorithms are listed in Sections 12.5, A.6, and B.4, respectively.

Part I

Combinatorial Algorithms

2 Set Cover

The set cover problem plays the same role in approximation algorithms that the maximum matching problem played in exact algorithms – as a problem whose study led to the development of fundamental techniques for the entire field. For our purpose this problem is particularly useful, since it offers a very simple setting in which many of the basic algorithm design techniques can be explained with great ease. In this chapter, we will cover two combinatorial techniques: the fundamental greedy technique and the technique of layering. In Part II we will explain both the basic LP-based techniques of rounding and the primal–dual schema using this problem. Because of its generality, the set cover problem has wide applicability, sometimes even in unexpected ways. In this chapter we will illustrate such an application – to the shortest superstring problem (see Chapter 7 for an improved algorithm for the latter problem).

Among the first strategies one tries when designing an algorithm for an optimization problem is some form of the greedy strategy. Even if this strategy does not work for a specific problem, proving this via a counterexample can provide crucial insights into the structure of the problem. Surprisingly enough, the straightforward, simple greedy algorithm for the set cover problem is essentially the best one can hope for for this problem (see Chapter 29 for a formal proof of this statement).

Problem 2.1 (Set cover) Given a universe U of n elements, a collection of subsets of U, $\mathcal{S} = \{S_1, \ldots, S_k\}$, and a cost function $c : \mathcal{S} \to \mathbf{Q}^+$, find a minimum cost subcollection of $\mathcal{S}$ that covers all elements of U.

Define the *frequency* of an element to be the number of sets it is in. A useful parameter is the frequency of the most frequent element. Let us denote this by f. The various approximation algorithms for set cover achieve one of two factors: $O(\log n)$ or f. Clearly, neither dominates the other in all instances. The special case of set cover with $f = 2$ is essentially the vertex cover problem (see Exercise 2.7), for which we gave a factor 2 approximation algorithm in Chapter 1.

2.1 The greedy algorithm

The greedy strategy applies naturally to the set cover problem: iteratively pick the most cost-effective set and remove the covered elements, until all elements are covered. Let C be the set of elements already covered at the beginning of an iteration. During this iteration, define the *cost-effectiveness* of a set S to be the average cost at which it covers new elements, i.e., $c(S)/|S-C|$. Define the *price* of an element to be the average cost at which it is covered. Equivalently, when a set S is picked, we can think of its cost being distributed equally among the new elements covered, to set their prices.

Algorithm 2.2 (Greedy set cover algorithm)

1. $C \leftarrow \emptyset$
2. While $C \neq U$ do
   Find the set whose cost-effectiveness is smallest, say S.
   Let $\alpha = \frac{c(S)}{|S-C|}$, i.e., the cost-effectiveness of S.
   Pick S, and for each $e \in S - C$, set price$(e) = \alpha$.
   $C \leftarrow C \cup S$.
3. Output the picked sets.

Number the elements of U in the order in which they were covered by the algorithm, resolving ties arbitrarily. Let $e_1, \ldots, e_n$ be this numbering.

Lemma 2.3 *For each* $k \in \{1, \ldots, n\}$, $\text{price}(e_k) \leq \text{OPT}/(n-k+1)$.

Proof: In any iteration, the leftover sets of the optimal solution can cover the remaining elements at a cost of at most OPT. Therefore, among these sets, there must be one having cost-effectiveness of at most $\text{OPT}/|\overline{C}|$, where $\overline{C} = U - C$. In the iteration in which element e_k was covered, $\overline{C}$ contained at least $n-k+1$ elements. Since e_k was covered by the most cost-effective set in this iteration, it follows that

$$\text{price}(e_k) \leq \frac{\text{OPT}}{|\overline{C}|} \leq \frac{\text{OPT}}{n-k+1}.$$

□

From Lemma 2.3 we immediately obtain:

Theorem 2.4 *The greedy algorithm is an H_n factor approximation algorithm for the minimum set cover problem, where* $H_n = 1 + \frac{1}{2} + \cdots + \frac{1}{n}$.

Proof: Since the cost of each set picked is distributed among the new elements covered, the total cost of the set cover picked is equal to $\sum_{k=1}^{n} \text{price}(e_k)$. By Lemma 2.3, this is at most $\left(1 + \frac{1}{2} + \cdots + \frac{1}{n}\right) \cdot \text{OPT}$. □

Example 2.5 The following is a tight example for Algorithm 2.2:

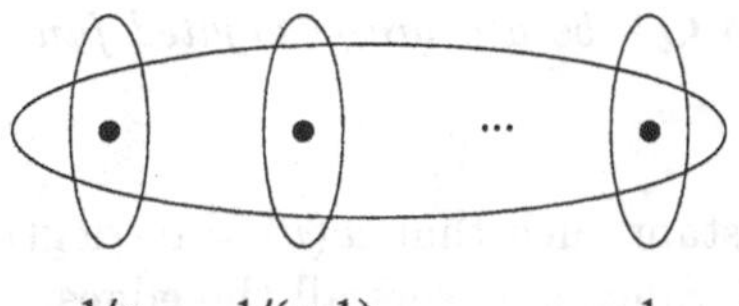

When run on this instance the greedy algorithm outputs the cover consisting of the n singleton sets, since in each iteration some singleton is the most cost-effective set. Thus, the algorithm outputs a cover of cost

$$\frac{1}{n} + \frac{1}{n-1} + \cdots + 1 = H_n.$$

On the other hand, the optimal cover has a cost of $1 + \varepsilon$. □

Surprisingly enough, for the minimum set cover problem the obvious algorithm given above is essentially the best one can hope for; see Sections 29.7 and 29.9.

In Chapter 1 we pointed out that finding a good lower bound on OPT is a basic starting point in the design of an approximation algorithm for a minimization problem. At this point the reader may be wondering whether there is any truth to this claim. We will show in Section 13.1 that the correct way to view the greedy set cover algorithm is in the setting of the LP-duality theory – this will not only provide the lower bound on which this algorithm is based, but will also help obtain algorithms for several generalizations of this problem.

2.2 Layering

The algorithm design technique of layering is also best introduced via set cover. We note, however, that this is not a very widely applicable technique. We will give a factor 2 approximation algorithm for vertex cover, assuming arbitrary weights, and leave the problem of generalizing this to a factor f approximation algorithm for set cover, where f is the frequency of the most frequent element (see Exercise 2.13).

The idea in layering is to decompose the given weight function on vertices into convenient functions, called degree-weighted, on a nested sequence of subgraphs of G. For degree-weighted functions, we will show that we will be within twice the optimal even if we pick all vertices in the cover.

Let us introduce some notation. Let $w : V \to \mathbf{Q}^+$ be the function assigning weights to the vertices of the given graph $G = (V, E)$. We will say that a function assigning vertex weights is *degree-weighted* if there is a constant

$c > 0$ such that the weight of each vertex $v \in V$ is $c \cdot \deg(v)$. The significance of such a weight function is captured in:

Lemma 2.6 *Let $w : V \to \mathbf{Q}^+$ be a degree-weighted function. Then $w(V) \leq 2 \cdot \text{OPT}$.*

Proof: Let c be the constant such that $w(v) = c \cdot \deg(v)$, and let U be an optimal vertex cover in G. Since U covers all the edges,

$$\sum_{v \in U} \deg(v) \geq |E|.$$

Therefore, $w(U) \geq c|E|$. Now, since $\sum_{v \in V} \deg(v) = 2|E|$, $w(V) = 2c|E|$. The lemma follows. □

Let us define the *largest degree-weighted function in w* as follows: remove all degree zero vertices from the graph, and over the remaining vertices, compute $c = \min\{w(v)/\deg(v)\}$. Then, $t(v) = c \cdot \deg(v)$ is the desired function. Define $w'(v) = w(v) - t(v)$ to be the *residual weight function.*

The algorithm for decomposing w into degree-weighted functions is as follows. Let $G_0 = G$. Remove degree zero vertices from G_0, say this set is D_0, and compute the largest degree-weighted function in w. Let W_0 be vertices of zero residual weight; these vertices are included in the vertex cover. Let G_1 be the graph induced on $V - (D_0 \cup W_0)$. Now, the entire process is repeated on G_1 w.r.t. the residual weight function. The process terminates when all vertices are of degree zero; let G_k denote this graph. The process is schematically shown in the following figure.

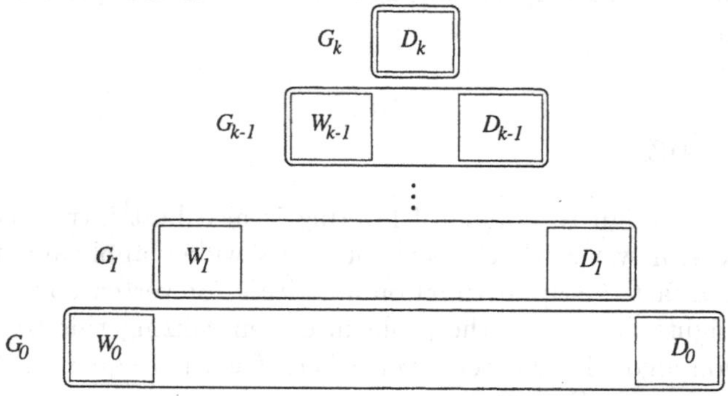

Let $t_0, ..., t_{k-1}$ be the degree-weighted functions defined on graphs $G_0, ..., G_{k-1}$. The vertex cover chosen is $C = W_0 \cup \ldots \cup W_{k-1}$. Clearly, $V - C = D_0 \cup \ldots \cup D_k$.

Theorem 2.7 *The layer algorithm achieves an approximation guarantee of factor 2 for the vertex cover problem, assuming arbitrary vertex weights.*

Proof: We need to show that set C is a vertex cover for G and $w(C) \leq 2 \cdot \text{OPT}$. Assume, for contradiction, that C is not a vertex cover for G. Then there must be an edge (u, v) with $u \in D_i$ and $v \in D_j$, for some i, j. Assume $i \leq j$. Therefore, (u, v) is present in G_i, contradicting the fact that u is a degree zero vertex.

Let C^* be an optimal vertex cover. For proving the second part, consider a vertex $v \in C$. If $v \in W_j$, its weight can be decomposed as

$$w(v) = \sum_{i \leq j} t_i(v).$$

Next, consider a vertex $v \in V - C$. If $v \in D_j$, a lower bound on its weight is given by

$$w(v) \geq \sum_{i<j} t_i(v).$$

The important observation is that in each layer i, $C^* \cap G_i$ is a vertex cover for G_i, since G_i is a vertex-induced graph. Therefore, by Lemma 2.6, $t_i(C \cap G_i) \leq 2 \cdot t_i(C^* \cap G_i)$. By the decomposition of weights given above, we get

$$w(C) = \sum_{i=0}^{k-1} t_i(C \cap G_i) \leq 2 \sum_{i=0}^{k-1} t_i(C^* \cap G_i) \leq 2 \cdot w(C^*).$$

□

Example 2.8 A tight example is provided by the family of complete bipartite graphs, $K_{n,n}$, with all vertices of unit weight. The layering algorithm will pick all $2n$ vertices of $K_{n,n}$ in the cover, whereas the optimal cover picks only one side of the bipartition. □

2.3 Application to shortest superstring

The following algorithm is given primarily to demonstrate the wide applicability of set cover. A constant factor approximation algorithm for shortest superstring will be given in Chapter 7.

Let us first provide motivation for this problem. The human DNA can be viewed as a very long string over a four-letter alphabet. Scientists are attempting to decipher this string. Since it is very long, several overlapping short segments of this string are first deciphered. Of course, the locations of these segments on the original DNA are not known. It is hypothesized that the shortest string which contains these segments as substrings is a good approximation to the original DNA string.

Problem 2.9 (Shortest superstring) Given a finite alphabet Σ, and a set of n strings, $S = \{s_1, \ldots, s_n\} \subseteq \Sigma^+$, find a shortest string s that contains each s_i as a substring. Without loss of generality, we may assume that no string s_i is a substring of another string s_j, $j \neq i$.

This problem is **NP**-hard. Perhaps the first algorithm that comes to mind for finding a short superstring is the following greedy algorithm. Define the *overlap* of two strings $s, t \in \Sigma^*$ as the maximum length of a suffix of s that is also a prefix of t. The algorithm maintains a set of strings T; initially $T = S$. At each step, the algorithm selects from T two strings that have maximum overlap and replaces them with the string obtained by overlapping them as much as possible. After $n - 1$ steps, T will contain a single string. Clearly, this string contains each s_i as a substring. This algorithm is conjectured to have an approximation factor of 2. To see that the approximation factor of this algorithm is no better than 2, consider an input consisting of 3 strings: ab^k, b^kc, and b^{k+1}. If the first two strings are selected in the first iteration, the greedy algorithm produces the string ab^kcb^{k+1}. This is almost twice as long as the shortest superstring, $ab^{k+1}c$.

We will obtain a $2H_n$ factor approximation algorithm, using the greedy set cover algorithm. The set cover instance, denoted by $\mathcal{S}$, is constructed as follows. For $s_i, s_j \in S$ and $k > 0$, if the last k symbols of s_i are the same as the first k symbols of s_j, let σ_{ijk} be the string obtained by overlapping these k positions of s_i and s_j:

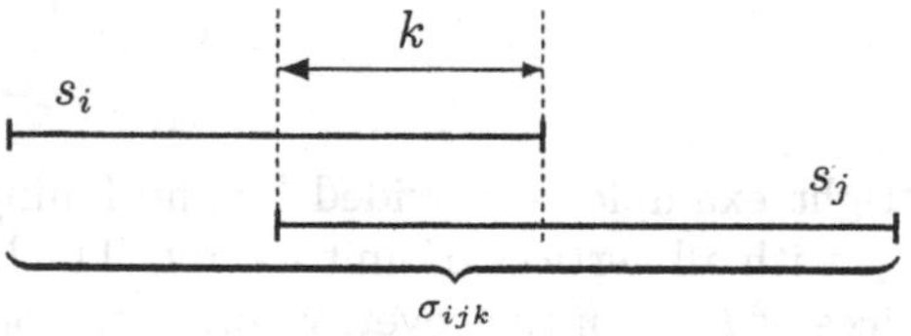

Let M be the set that consists of the strings σ_{ijk}, for all valid choices of i, j, k. For a string $\pi \in \Sigma^+$, define $\text{set}(\pi) = \{s \in S \mid s \text{ is a substring of } \pi\}$. The universal set of the set cover instance $\mathcal{S}$ is S, and the specified subsets of S are $\text{set}(\pi)$, for each string $\pi \in S \cup M$. The cost of $\text{set}(\pi)$ is $|\pi|$, i.e., the length of string π.

Let $\text{OPT}_{\mathcal{S}}$ and OPT denote the cost of an optimal solution to $\mathcal{S}$ and the length of the shortest superstring of S, respectively. As shown in Lemma 2.11, $\text{OPT}_{\mathcal{S}}$ and OPT are within a factor of 2 of each other, and so an approximation algorithm for set cover can be used to obtain an approximation algorithm for shortest superstring. The complete algorithm is:

Algorithm 2.10 (Shortest superstring via set cover)

1. Use the greedy set cover algorithm to find a cover for the instance $\mathcal{S}$. Let $\text{set}(\pi_1), \ldots, \text{set}(\pi_k)$ be the sets picked by this cover.
2. Concatenate the strings $\pi_1, \ldots, \pi_k$, in any order.
3. Output the resulting string, say s.

Lemma 2.11 $\text{OPT} \le \text{OPT}_{\mathcal{S}} \le 2 \cdot \text{OPT}$.

Proof: Consider an optimal set cover, say $\{\text{set}(\pi_{i_j}) | 1 \le j \le l\}$, and obtain a string, say s, by concatenating the strings π_{i_j}, $1 \le j \le l$, in any order. Clearly, $|s| = \text{OPT}_{\mathcal{S}}$. Since each string of S is a substring of some π_{i_j}, $1 \le j \le l$, it is also a substring of s. Hence $\text{OPT}_{\mathcal{S}} = |s| \ge \text{OPT}$.

To prove the second inequality, let s be a shortest superstring of $s_1, \ldots, s_n$, $|s| = \text{OPT}$. It suffices to produce *some* set cover of cost at most $2 \cdot \text{OPT}$.

Consider the leftmost occurrence of the strings $s_1, \ldots, s_n$ in string s. Since no string among $s_1, \ldots, s_n$ is a substring of another, these n leftmost occurrences start at distinct places in s. For the same reason, they also end at distinct places. Renumber the n strings in the order in which their leftmost occurrences start. Again, since no string is a substring of another, this is also the order in which they end.

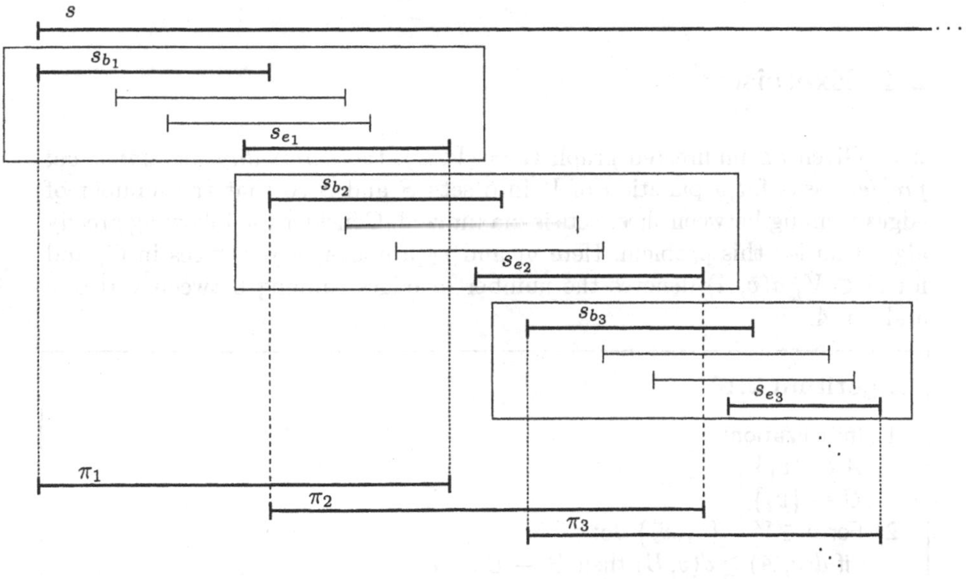

We will partition the ordered list of strings $s_1, \ldots, s_n$ in groups as described below. Each group will consist of a contiguous set of strings from this list. Let b_i and e_i denote the index of the first and last string in the ith group ($b_i = e_i$ is allowed). Thus, $b_1 = 1$. Let e_1 be the largest index of a string that overlaps with s_1 (there exists at least one such string, namely s_1 itself). In general, if $e_i < n$ we set $b_{i+1} = e_i + 1$ and denote by e_{i+1} the largest index of a string that overlaps with $s_{b_{i+1}}$. Eventually, we will get $e_t = n$ for some $t \leq n$.

For each pair of strings (s_{b_i}, s_{e_i}), let $k_i > 0$ be the length of the overlap between their leftmost occurrences in s (this may be different from their maximum overlap). Let $\pi_i = \sigma_{b_i e_i k_i}$. Clearly, $\{\text{set}(\pi_i) | 1 \leq i \leq t\}$ is a solution for $\mathcal{S}$, of cost $\sum_i |\pi_i|$.

The critical observation is that π_i does not overlap π_{i+2}. We will prove this claim for $i = 1$; the same argument applies to an arbitrary i. Assume, for contradiction, that π_1 overlaps π_3. Then the occurrence of s_{b_3} in s overlaps the occurrence of s_{e_1}. However, s_{b_3} does not overlap s_{b_2} (otherwise, s_{b_3} would have been put in the second group). This implies that s_{e_1} ends later than s_{b_2}, contradicting the property of endings of strings established earlier.

Because of this observation, each symbol of s is covered by at most two of the π_i's. Hence $\text{OPT}_{\mathcal{S}} \leq \sum_i |\pi_i| \leq 2 \cdot \text{OPT}$. □

The size of the universal set in the set cover instance $\mathcal{S}$ is n, the number of strings in the given shortest superstring instance. This fact, Lemma 2.11, and Theorem 2.4 immediately give the following theorem.

Theorem 2.12 *Algorithm 2.10 is a $2H_n$ factor algorithm for the shortest superstring problem, where n is the number of strings in the given instance.*

2.4 Exercises

2.1 Given an undirected graph $G = (V, E)$, the *cardinality maximum cut problem* asks for a partition of V into sets S and $\overline{S}$ so that the number of edges running between these sets is maximized. Consider the following greedy algorithm for this problem. Here v_1 and v_2 are arbitrary vertices in G, and for $A \subset V$, $d(v, A)$ denotes the number of edges running between vertex v and set A.

Algorithm 2.13

1. Initialization:
   $A \leftarrow \{v_1\}$
   $B \leftarrow \{v_2\}$
2. For $v \in V - \{v_1, v_2\}$ do:
   if $d(v, A) \geq d(v, B)$ then $B \leftarrow B \cup \{v\}$,
   else $A \leftarrow A \cup \{v\}$.
3. Output A and B.

Show that this is a factor 1/2 approximation algorithm and give a tight example. What is the upper bound on OPT that you are using? Give examples of graphs for which this upper bound is as bad as twice OPT. Generalize the problem and the algorithm to weighted graphs.

2.2 Consider the following algorithm for the maximum cut problem, based on the technique of *local search.* Given a partition of V into sets, the basic step of the algorithm, called *flip*, is that of moving a vertex from one side of the partition to the other. The following algorithm finds a *locally optimal solution* under the flip operation, i.e., a solution which cannot be improved by a single flip.

The algorithm starts with an arbitrary partition of V. While there is a vertex such that flipping it increases the size of the cut, the algorithm flips such a vertex. (Observe that a vertex qualifies for a flip if it has more neighbors in its own partition than in the other side.) The algorithm terminates when no vertex qualifies for a flip. Show that this algorithm terminates in polynomial time, and achieves an approximation guarantee of 1/2.

2.3 Consider the following generalization of the maximum cut problem.

Problem 2.14 (MAX k-CUT) Given an undirected graph $G = (V, E)$ with nonnegative edge costs, and an integer k, find a partition of V into sets $S_1, \ldots, S_k$ so that the total cost of edges running between these sets is maximized.

Give a greedy algorithm for this problem that achieves a factor of $(1 - \frac{1}{k})$. Is the analysis of your algorithm tight?

2.4 Give a greedy algorithm for the following problem achieving an approximation guarantee of factor 1/4.

Problem 2.15 (Maximum directed cut) Given a directed graph $G = (V, E)$ with nonnegative edge costs, find a subset $S \subset V$ so as to maximize the total cost of edges out of S, i.e., $\text{cost}(\{(u \rightarrow v) \mid u \in S \text{ and } v \in \overline{S}\})$.

2.5 (N. Vishnoi) Use the algorithm in Exercise 2.2 and the fact that the vertex cover problem is polynomial time solvable for bipartite graphs to give a factor $\lceil \log_2 \Delta \rceil$ algorithm for vertex cover, where Δ is the degree of the vertex having highest degree.

Hint: Let H denote the subgraph consisting of edges in the maximum cut found by Algorithm 2.13. Clearly, H is bipartite, and for any vertex v, $\text{deg}_H(v) \geq (1/2)\text{deg}_G(v)$.

2.6 (Wigderson [265]) Consider the following problem.

Problem 2.16 (Vertex coloring) Given an undirected graph $G = (V, E)$, color its vertices with the minimum number of colors so that the two endpoints of each edge receive distinct colors.

1. Give a greedy algorithm for coloring G with $\Delta+1$ colors, where Δ is the maximum degree of a vertex in G.
2. Give an algorithm for coloring a 3-colorable graph with $O(\sqrt{n})$ colors.
Hint: For any vertex v, the induced subgraph on its neighbors, $N(v)$, is bipartite, and hence optimally colorable. If v has degree $> \sqrt{n}$, color $v \cup N(v)$ using 3 distinct colors. Continue until every vertex has degree $\leq \sqrt{n}$. Then use the algorithm in the first part.

2.7 Let 2SC denote the restriction of set cover to instances having $f = 2$. Show that 2SC is equivalent to the vertex cover problem, with arbitrary costs, under approximation factor preserving reductions.

2.8 Prove that Algorithm 2.2 achieves an approximation factor of H_k, where k is the cardinality of the largest specified subset of U.

2.9 Give a greedy algorithm that achieves an approximation guarantee of H_n for set multicover, which is a generalization of set cover in which an integral coverage requirement is also specified for each element and sets can be picked multiple numbers of times to satisfy all coverage requirements. Assume that the cost of picking α copies of set S_i is $\alpha \cdot \text{cost}(S_i)$.

2.10 By giving an appropriate tight example, show that the analysis of Algorithm 2.2 cannot be improved even for the cardinality set cover problem, i.e., if all specified sets have unit cost.
Hint: Consider running the greedy algorithm on a vertex cover instance.

2.11 Consider the following algorithm for the weighted vertex cover problem. For each vertex v, $t(v)$ is initialized to its weight, and when $t(v)$ drops to 0, v is picked in the cover. $c(e)$ is the amount charged to edge e.

Algorithm 2.17

1. Initialization:
   $C \leftarrow \emptyset$
   $\forall v \in V$, $t(v) \leftarrow w(v)$
   $\forall e \in E$, $c(e) \leftarrow 0$
2. While C is not a vertex cover do:
   Pick an uncovered edge, say (u, v). Let $m = \min(t(u), t(v))$.
   $t(u) \leftarrow t(u) - m$
   $t(v) \leftarrow t(v) - m$
   $c(u, v) \leftarrow m$
   Include in C all vertices having $t(v) = 0$.
3. Output C.

Show that this is a factor 2 approximation algorithm.
Hint: Show that the total amount charged to edges is a lower bound on OPT and that the weight of cover C is at most twice the total amount charged to edges.

2.12 Consider the layering algorithm for vertex cover. Another weight function for which we have a factor 2 approximation algorithm is the constant function – by simply using the factor 2 algorithm for the cardinality vertex cover problem. Can layering be made to work by using this function instead of the degree-weighted function?

2.13 Use layering to get a factor f approximation algorithm for set cover, where f is the frequency of the most frequent element. Provide a tight example for this algorithm.

2.14 A *tournament* is a directed graph $G = (V, E)$, such that for each pair of vertices, $u, v \in V$, exactly one of (u, v) and (v, u) is in E. A *feedback vertex set* for G is a subset of the vertices of G whose removal leaves an acyclic graph. Give a factor 3 algorithm for the problem of finding a minimum feedback vertex set in a directed graph.
Hint: Show that it is sufficient to "kill" all length 3 cycles. Use the factor f set cover algorithm.

2.15 (Hochbaum [132]) Consider the following problem.

Problem 2.18 (Maximum coverage) Given a universal set U of n elements, with nonnegative weights specified, a collection of subsets of U, $S_1, \ldots, S_l$, and an integer k, pick k sets so as to maximize the weight of elements covered.

Show that the obvious algorithm, of greedily picking the best set in each iteration until k sets are picked, achieves an approximation factor of

$$1 - \left(1 - \frac{1}{k}\right)^k > 1 - \frac{1}{e}.$$

2.16 Using set cover, obtain approximation algorithms for the following variants of the shortest superstring problem (here s^R is the reverse of string s):

1. Find the shortest string that contains, for each string $s_i \in S$, both s_i and s_i^R as substrings.
   Hint: The universal set for the set cover instance will contain $2n$ elements, s_i and s_i^R, for $1 \le i \le n$.
2. Find the shortest string that contains, for each string $s_i \in S$, either s_i or s_i^R as a substring.
   Hint: Define $\text{set}(\pi) = \{s \in S \mid s \text{ or } s^R \text{ is a substring of } \pi\}$. Choose the strings π appropriately.

2.5 Notes

Algorithm 2.2 is due to Johnson [157], Lovász [199], and Chvátal [50]. The hardness result for set cover, showing that this algorithm is essentially the best possible, is due to Feige [86], improving on the result of Lund and Yannakakis [206]. The application to shortest superstring is due to Li [194].

3 Steiner Tree and TSP

In this chapter, we will present constant factor algorithms for two fundamental problems, metric Steiner tree and metric TSP. The reasons for considering the metric case of these problems are quite different. For Steiner tree, this is the core of the problem – the rest of the problem reduces to this case. For TSP, without this restriction, the problem admits no approximation factor, assuming $\mathbf{P} \neq \mathbf{NP}$. The algorithms, and their analyses, are similar in spirit, which is the reason for presenting these problems together.

3.1 Metric Steiner tree

The Steiner tree problem was defined by Gauss in a letter he wrote to Schumacher (reproduced on the cover of this book). Today, this problem occupies a central place in the field of approximation algorithms. The problem has a wide range of applications, all the way from finding minimum length interconnection of terminals in VLSI design to constructing phylogeny trees in computational biology. This problem and its generalizations will be studied extensively in this book, see Chapters 22 and 23.

Problem 3.1 (Steiner tree) Given an undirected graph $G = (V, E)$ with nonnegative edge costs and whose vertices are partitioned into two sets, *required* and *Steiner*, find a minimum cost tree in G that contains all the required vertices and any subset of the Steiner vertices.

We will first show that the core of this problem lies in its restriction to instances in which the edge costs satisfy *the triangle inequality*, i.e., G is a complete undirected graph, and for any three vertices u, v, and w, $\text{cost}(u, v) \leq \text{cost}(u, w) + \text{cost}(v, w)$. Let us call this restriction the *metric Steiner tree problem.*

Theorem 3.2 *There is an approximation factor preserving reduction from the Steiner tree problem to the metric Steiner tree problem.*

Proof: We will transform, in polynomial time, an instance I of the Steiner tree problem, consisting of graph $G = (V, E)$, to an instance I' of the metric Steiner tree problem as follows. Let G' be the complete undirected graph on vertex set V. Define the cost of edge (u, v) in G' to be the cost of a shortest u–v path in G. G' is called the *metric closure* of G. The partition of V into required and Steiner vertices in I' is the same as in I.

For any edge $(u, v) \in E$, its cost in G' is no more than its cost in G. Therefore, the cost of an optimal solution in I' does not exceed the cost of an optimal solution in I.

Next, given a Steiner tree T' in I', we will show how to obtain, in polynomial time, a Steiner tree T in I of at most the same cost. The cost of an edge (u, v) in G' corresponds to the cost of a path in G. Replace each edge of T' by the corresponding path to obtain a subgraph of G. Clearly, in this subgraph, all the required vertices are connected. However, this subgraph may, in general, contain cycles. If so, remove edges to obtain tree T. This completes the approximation factor preserving reduction. □

As a consequence of Theorem 3.2, any approximation factor established for the metric Steiner tree problem carries over to the entire Steiner tree problem.

3.1.1 MST-based algorithm

Let R denote the set of required vertices. Clearly, a minimum spanning tree (MST) on R is a feasible solution for this problem. Since the problem of finding an MST is in **P** and the metric Steiner tree problem is **NP**-hard, we cannot expect the MST on R to always give an optimal Steiner tree; below is an example in which the MST is strictly costlier.

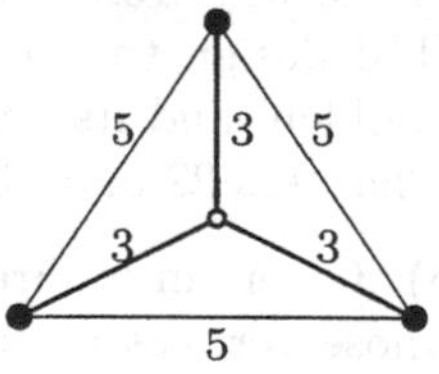

Even so, an MST on R is not much more costly than an optimal Steiner tree:

Theorem 3.3 *The cost of an MST on R is within* $2 \cdot \text{OPT}$.

Proof: Consider a Steiner tree of cost OPT. By doubling its edges we obtain an Eulerian graph connecting all vertices of R and, possibly, some Steiner vertices. Find an Euler tour of this graph, for example by traversing the edges in DFS (depth first search) order:

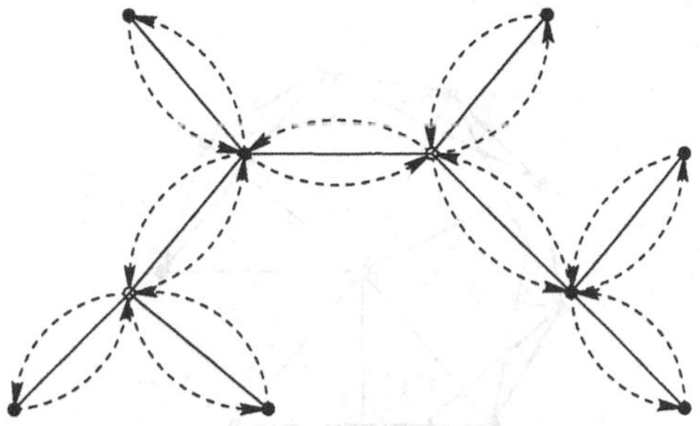

The cost of this Euler tour is $2 \cdot \text{OPT}$. Next obtain a Hamiltonian cycle on the vertices of R by traversing the Euler tour and "short-cutting" Steiner vertices and previously visited vertices of R:

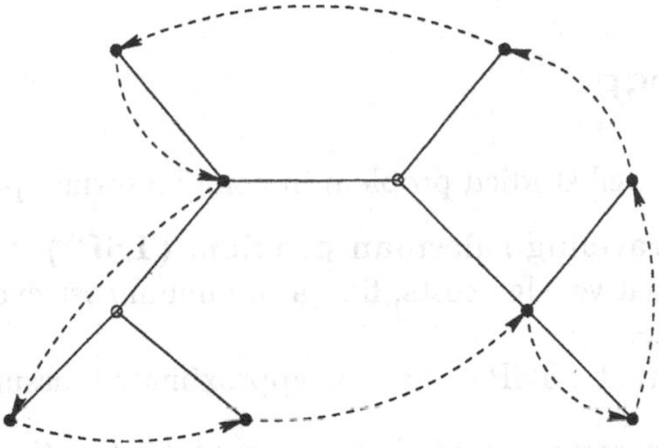

Because of triangle inequality, the shortcuts do not increase the cost of the tour. If we delete one edge of this Hamiltonian cycle, we obtain a path that spans R and has cost at most $2 \cdot \text{OPT}$. This path is also a spanning tree on R. Hence, the MST on R has cost at most $2 \cdot \text{OPT}$. □

Theorem 3.3 gives a straightforward factor 2 algorithm for the metric Steiner tree problem: simply find an MST on the set of required vertices. As in the case of set cover, the "correct" way of viewing this algorithm is in the setting of LP-duality theory. In Chapters 22 and 23 we will see that LP-duality provides the lower bound on which this algorithm is based and also helps solve generalizations of this problem.

Example 3.4 For a tight example, consider a graph with n required vertices and one Steiner vertex. An edge between the Steiner vertex and a required vertex has cost 1, and an edge between two required vertices has cost 2 (not all edges of cost 2 are shown below). In this graph, any MST on R has cost $2(n-1)$, while $\text{OPT} = n$.

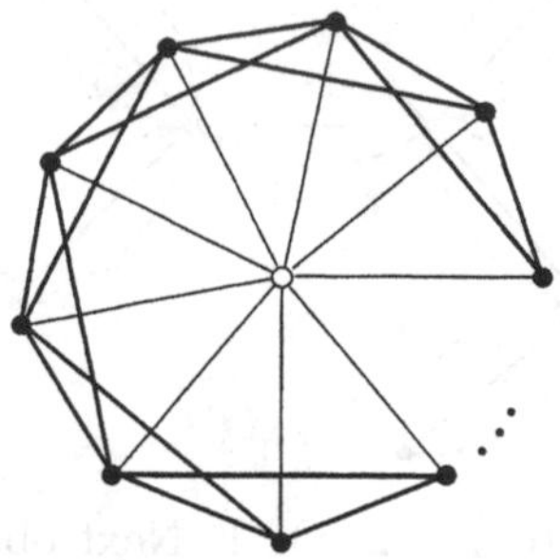

□

3.2 Metric TSP

The following is a well-studied problem in combinatorial optimization.

Problem 3.5 (Traveling salesman problem (TSP)) Given a complete graph with nonnegative edge costs, find a minimum cost cycle visiting every vertex exactly once.

In its full generality, TSP cannot be approximated, assuming $\mathbf{P} \neq \mathbf{NP}$.

Theorem 3.6 *For any polynomial time computable function $\alpha(n)$, TSP cannot be approximated within a factor of $\alpha(n)$, unless* $\mathbf{P} = \mathbf{NP}$.

Proof: Assume, for a contradiction, that there is a factor $\alpha(n)$ polynomial time approximation algorithm, $\mathcal{A}$, for the general TSP problem. We will show that $\mathcal{A}$ can be used for deciding the Hamiltonian cycle problem (which is **NP**-hard) in polynomial time, thus implying $\mathbf{P} = \mathbf{NP}$.

The central idea is a reduction from the Hamiltonian cycle problem to TSP, that transforms a graph G on n vertices to an edge-weighted complete graph G' on n vertices such that

- if G has a Hamiltonian cycle, then the cost of an optimal TSP tour in G' is n, and
- if G does not have a Hamiltonian cycle, then an optimal TSP tour in G' is of cost $> \alpha(n) \cdot n$.

Observe that when run on graph G', algorithm $\mathcal{A}$ must return a solution of cost $\leq \alpha(n) \cdot n$ in the first case, and a solution of cost $> \alpha(n) \cdot n$ in the second case. Thus, it can be used for deciding whether G contains a Hamiltonian cycle.

The reduction is simple. Assign a weight of 1 to edges of G, and a weight of $\alpha(n) \cdot n$ to nonedges, to obtain G'. Now, if G has a Hamiltonian cycle, then the corresponding tour in G' has cost n. On the other hand, if G has no Hamiltonian cycle, any tour in G' must use an edge of cost $\alpha(n) \cdot n$, and therefore has cost $> \alpha(n) \cdot n$. □

Notice that in order to obtain such a strong nonapproximability result, we had to assign edge costs that violate triangle inequality. If we restrict ourselves to graphs in which edge costs satisfy triangle inequality, i.e., consider *metric TSP*, the problem remains **NP**-complete, but it is no longer hard to approximate.

3.2.1 A simple factor 2 algorithm

We will first present a simple factor 2 algorithm. The lower bound we will use for obtaining this factor is the cost of an MST in G. This is a lower bound because deleting any edge from an optimal solution to TSP gives us a spanning tree of G.

Algorithm 3.7 (Metric TSP – factor 2)

1. Find an MST, T, of G.
2. Double every edge of the MST to obtain an Eulerian graph.
3. Find an Eulerian tour, $\mathcal{T}$, on this graph.
4. Output the tour that visits vertices of G in the order of their first appearance in $\mathcal{T}$. Let $\mathcal{C}$ be this tour.

Notice that Step 4 is similar to the "short-cutting" step in Theorem 3.3.

Theorem 3.8 *Algorithm 3.7 is a factor* 2 *approximation algorithm for metric TSP.*

Proof: As noted above, $\text{cost}(T) \leq \text{OPT}$. Since $\mathcal{T}$ contains each edge of T twice, $\text{cost}(\mathcal{T}) = 2 \cdot \text{cost}(T)$. Because of triangle inequality, after the "short-cutting" step, $\text{cost}(\mathcal{C}) \leq \text{cost}(\mathcal{T})$. Combining these inequalities we get that $\text{cost}(\mathcal{C}) \leq 2 \cdot \text{OPT}$. □

Example 3.9 A tight example for this algorithm is given by a complete graph on n vertices with edges of cost 1 and 2. We present the graph for $n = 6$ below, where thick edges have cost 1 and remaining edges have cost 2. For arbitrary n the graph has $2n - 2$ edges of cost 1, with these edges forming the union of a star and an $n - 1$ cycle; all remaining edges have cost 2. The optimal TSP tour has cost n, as shown below for $n = 6$:

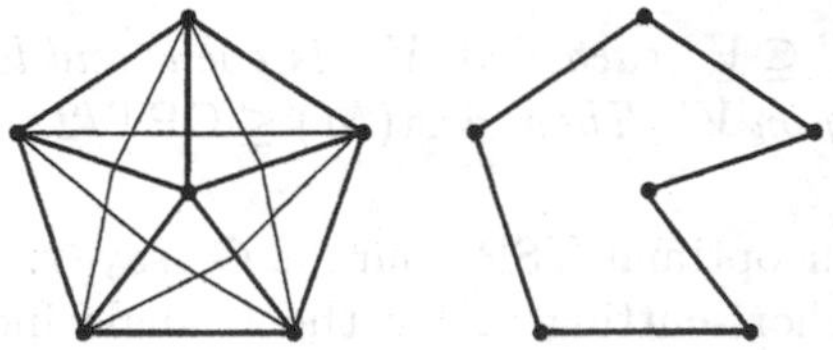

Suppose that the MST found by the algorithm is the spanning star created by edges of cost 1. Moreover, suppose that the Euler tour constructed in Step 3 visits vertices in order shown below for $n = 6$:

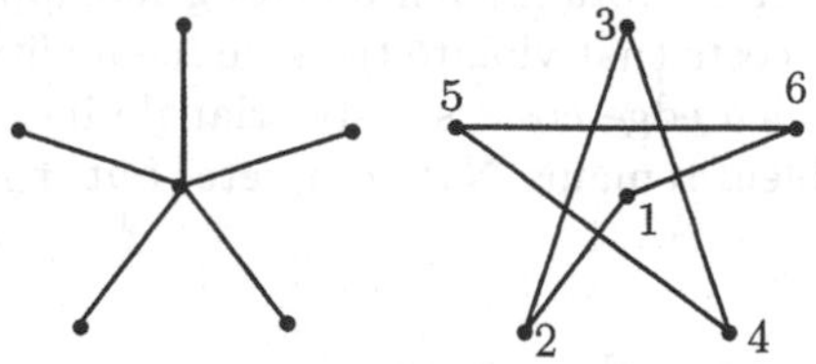

Then the tour obtained after short-cutting contains $n-2$ edges of cost 2 and has a total cost of $2n-2$. Asymptotically, this is twice the cost of the optimal TSP tour. □

3.2.2 Improving the factor to 3/2

Algorithm 3.7 first finds a low cost Euler tour spanning the vertices of G, and then short-cuts this tour to find a traveling salesman tour. Is there a cheaper Euler tour than that found by doubling an MST? Recall that a graph has an Euler tour iff all its vertices have even degrees. Thus, we only need to be concerned about the vertices of odd degree in the MST. Let V' denote this set of vertices. $|V'|$ must be even since the sum of degrees of all vertices in the MST is even. Now, if we add to the MST a minimum cost perfect matching on V', every vertex will have an even degree, and we get an Eulerian graph. With this modification, the algorithm achieves an approximation guarantee of 3/2.

Algorithm 3.10 (Metric TSP – factor 3/2)

1. Find an MST of G, say T.
2. Compute a minimum cost perfect matching, M, on the set of odd-degree vertices of T. Add M to T and obtain an Eulerian graph.
3. Find an Euler tour, $\mathcal{T}$, of this graph.
4. Output the tour that visits vertices of G in order of their first appearance in $\mathcal{T}$. Let $\mathcal{C}$ be this tour.

Interestingly, the proof of this algorithm is based on a second lower bound on OPT.

Lemma 3.11 *Let $V' \subseteq V$, such that $|V'|$ is even, and let M be a minimum cost perfect matching on V'. Then,* $\text{cost}(M) \leq \text{OPT}/2$.

Proof: Consider an optimal TSP tour of G, say τ. Let τ' be the tour on V' obtained by short-cutting τ. By the triangle inequality, $\text{cost}(\tau') \leq$

cost(τ). Now, τ' is the union of two perfect matchings on V', each consisting of alternate edges of τ'. Thus, the cheaper of these matchings has cost $\leq$ cost$(\tau')/2 \leq$ OPT/2. Hence the optimal matching also has cost at most OPT/2. □

Theorem 3.12 *Algorithm 3.10 achieves an approximation guarantee of* 3/2 *for metric TSP.*

Proof: The cost of the Euler tour,

$$\text{cost}(\mathcal{T}) \leq \text{cost}(T) + \text{cost}(M) \leq \text{OPT} + \frac{1}{2}\text{OPT} = \frac{3}{2}\text{OPT},$$

where the second inequality follows by using the two lower bounds on OPT. Using the triangle inequality, cost$(\mathcal{C}) \leq$ cost$(\mathcal{T})$, and the theorem follows. □

Example 3.13 A tight example for this algorithm is given by the following graph on n vertices, with n odd:

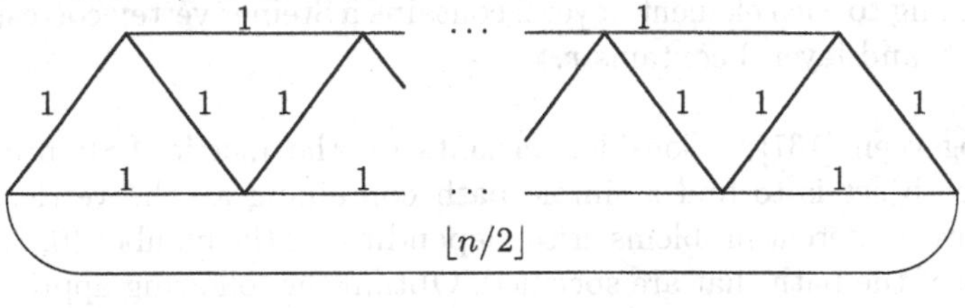

Thick edges represent the MST found in step 1. This MST has only two odd degree vertices, and by adding the edge joining them we obtain a traveling salesman tour of cost $(n-1) + \lfloor n/2 \rfloor$. In contrast, the optimal tour has cost n. □

Finding a better approximation algorithm for metric TSP is currently one of the outstanding open problems in this area. Many researchers have conjectured that an approximation factor of 4/3 may be achievable.

3.3 Exercises

3.1 The hardness of the Steiner tree problem lies in determining the optimal subset of Steiner vertices that need to be included in the tree. Show this by proving that if this set is provided, then the optimal Steiner tree can be computed in polynomial time.
Hint: Find an MST on the union of this set and the set of required vertices.

3.2 Let $G = (V, E)$ be a graph with nonnegative edge costs. S, the *senders* and R, the *receivers*, are disjoint subsets of V. The problem is to find a

minimum cost subgraph of G that has a path connecting each receiver to a sender (any sender suffices). Partition the instances into two cases: $S \cup R = V$ and $S \cup R \neq V$. Show that these two cases are in **P** and **NP**-hard, respectively. For the second case, give a factor 2 approximation algorithm.
Hint: Add a new vertex which is connected to each sender by a zero cost edge. Consider the new vertex and all receivers as required and the remaining vertices as Steiner, and find a minimum cost Steiner tree.

3.3 Give an approximation factor preserving reduction from the set cover problem to the following problem, thereby showing that it is unlikely to have a better approximation guarantee than $O(\log n)$.

Problem 3.14 (Directed Steiner tree) $G = (V, E)$ is a directed graph with nonnegative edge costs. The vertex set V is partitioned into two sets, *required* and *Steiner*. One of the required vertices, r, is special. The problem is to find a minimum cost tree in G rooted into r that contains all the required vertices and any subset of the Steiner vertices.
Hint: Construct a three layer graph: layer 1 contains a required vertex corresponding to each element, layer 2 contains a Steiner vertex corresponding to each set, and layer 3 contains r.

3.4 (Hoogeveen [137]) Consider variants on the metric TSP problem in which the object is to find a simple path containing all the vertices of the graph. Three different problems arise, depending on the number (0, 1, or 2) of endpoints of the path that are specified. Obtain the following approximation algorithms.

- If zero or one endpoints are specified, obtain a 3/2 factor algorithm.
- If both endpoints are specified, obtain a 5/3 factor algorithm.

Hint: Use the idea behind Algorithm 3.10.

3.5 (Papadimitriou and Yannakakis [227]) Let G be a complete undirected graph in which all edge lengths are either 1 or 2 (clearly, G satisfies the triangle inequality). Give a 4/3 factor algorithm for TSP in this special class of graphs.
Hint: Start by finding a minimum 2-matching in G. A 2-matching is a subset S of edges so that every vertex has exactly 2 edges of S incident at it.

3.6 (Frieze, Galbiati, and Maffioli [95]) Give an $O(\log n)$ factor approximation algorithm for the following problem.

Problem 3.15 (Asymmetric TSP) We are given a directed graph G on vertex set V, with a nonnegative cost specified for edge $(u \to v)$, for each pair $u, v \in V$. The edge costs satisfy *the directed triangle inequality*, i.e., for any three vertices u, v, and w, $\text{cost}(u \to v) \leq \text{cost}(u \to w) + \text{cost}(w \to v)$. The problem is to find a minimum cost cycle visiting every vertex exactly once.

Hint: Use the fact that a minimum cost cycle cover (i.e., disjoint cycles covering all the vertices) can be found in polynomial time. Shrink the cycles and recurse.

3.7 Let $G = (V, E)$ be a graph with edge costs satisfying the triangle inequality, and $V' \subseteq V$ be a set of even cardinality. Prove or disprove: The cost of a minimum cost perfect matching on V' is bounded above by the cost of a minimum cost perfect matching on V.

3.8 Given n points in $\mathbf{R}^2$, define the optimal Euclidean Steiner tree to be a minimum length tree containing all n points and any other subset of points from $\mathbf{R}^2$. Prove that each of the additional points must have degree three, with all three angles being 120°.

3.9 (Rao, Sadayappan, Hwang, and Shor [238]) This exercise develops a factor 2 approximation algorithm for the following problem.

Problem 3.16 (Rectilinear Steiner arborescence) Let $p_1, \ldots, p_n$ be points given in $\mathbf{R}^2$ in the positive quadrant. A path from the origin to point p_i is said to be *monotone* if it consists of segments traversing in the positive x direction or the positive y direction (informally, going right or up). The problem is to find a minimum length tree containing monotone paths from the origin to each of the n points; such a tree is called *rectilinear Steiner arborescence.*

For point p, define x_p and y_p to be its x and y coordinates, and $|p|_1 = |x_p| + |y_p|$. Say that point p *dominates* point q if $x_p \leq x_q$ and $y_p \leq y_q$. For sets of points A and B, we will say that A *dominates* B if for each point $b \in B$, there is a point $a \in A$ such that a dominates b. For points p and q, define $\text{dom}(p, q) = (x, y)$, where $x = \min(x_p, x_q)$ and $y = \min(y_p, y_q)$. If p dominates q, define segments(p, q) to be a monotone path from p to q. Consider the following algorithm.

Algorithm 3.17 (Rectilinear Steiner arborescence)

1. $T \leftarrow \emptyset$.
2. $P \leftarrow \{p_1, \ldots, p_n\} \cup \{(0,0)\}$.
3. while $|P| > 1$ do:
   Pick $p, q = \arg\max_{p,q \in P}(|\text{dom}(p,q)|_1)$.
   $P \leftarrow (P - \{p, q\}) \cup \{\text{dom}(p,q)\}$.
   $T \leftarrow T \cup \text{segments}(\text{dom}(p,q), p) \cup \text{segments}(\text{dom}(p,q), q)$.
4. Output T.

For $z \geq 0$, define ℓ_z to be the line $x + y = z$. For a rectilinear Steiner arborescence T, let $T(z) = |T \cap \ell_z|$. Prove that the length of T is

$$\int_{z=0}^{\infty} T(z)\,\mathrm{d}z.$$

Also, for every $z \geq 0$ define $P_z = \{p \in P \;:\; |p|_1 > z\}$, and

$$N(z) = \min\{|C| \;:\; C \subset \ell_z \text{ and } C \text{ dominates } P_z\}.$$

Prove that

$$\int_{z=0}^{\infty} N(z)\,\mathrm{d}z$$

is a lower bound on OPT.

Use these facts to show that Algorithm 3.17 achieves an approximation guarantee of 2.

3.10 (I. Măndoiu) This exercise develops a factor 9 approximation algorithm for the following problem, which finds applications in VLSI clock routing.

Problem 3.18 (Rectilinear zero-skew tree) Given a set S of points in the rectilinear plane, find a minimum length *zero-skew tree* (ZST) for S, i.e., a rooted tree T embedded in the rectilinear plane such that points in S are leaves of T and all root-to-leaf paths in T have equal length. By *length* of a path we mean the sum of the lengths of edges on it.

1. Let T be an arbitrary zero-skew tree, and let R' denote the common length of all root-to-leaf paths. For $r \geq 0$, let $T(r)$ denote the number of points of T that are at a length of $R' - r$ from the root. Prove that the length of T is

$$\int_0^{R'} T(r)\mathrm{d}r.$$

2. A closed ℓ_1 ball of radius r centered at point p is the set of all points whose ℓ_1-distance from p is $\leq r$. Let R denote the radius of the smallest ℓ_1-ball that contains all points of S. For $r \geq 0$, let $N(r)$ denote the minimum number of closed ℓ_1-balls of radius r needed to cover all points of S. Prove that

$$\int_0^{R} N(r)\mathrm{d}r$$

is a lower bound on the length of the optimum ZST.

3. Consider the following algorithm. First, compute R and find a radius R ℓ_1-ball enclosing all points of S. The center of this ball is chosen as the root of the resulting ZST. This ball can be partitioned into 4 balls, called its *quadrants*, of radius $R/2$ each. The root can be connected to the center of any of these balls by an edge of length $R/2$. These balls can be further partitioned into 4 balls each of radius $R/4$, and so on.
   The ZST is constructed recursively, starting with the ball of radius R. The center of the current ball is connected to the centers of each of its quadrants that has a point of S. The algorithm then recurses on each of these quadrants. If the current ball contains exactly one point of S, then this ball is not partitioned into quadrants. Let r' be the radius of this ball, c its center, and $p \in S$ the point in it. Clearly, the ℓ_1 distance between c and p is $\leq r'$. Connect c to p by a rectilinear path of length exactly r'.
   Show that for $0 \leq r \leq R$, $T(r) \leq 9N(r)$. Hence, show that this is a factor 9 approximation algorithm.

3.4 Notes

The Steiner tree problem has its origins in a problem posed by Fermat, and was defined by Gauss in a letter he wrote to his student Schumacher on March 21, 1836. Parts of the letter are reproduced on the cover of this book. Courant and Robbins [57] popularized this problem under the name of Steiner, a well known 19th century geometer. See Hwang, Richards, and Winter [140] and Schreiber [244] for the fascinating history of this problem.

The factor 2 Steiner tree algorithm was discovered independently by Choukhmane [46], Iwainsky, Canuto, Taraszow, and Villa [143], Kou, Markowsky, and Berman [184], and Plesník [229]. The factor 3/2 metric TSP algorithm is due to Christofides [47], and Theorem 3.6 is due to Sahni and Gonzalez [240]. The lower bound in Exercise 3.10 is from Charikar, Kleinberg, Kumar, Rajagopalan, Sahai, and Tomkins [42]. The best factor known for the rectilinear zero-skew tree problem, due to Zelikovsky and Măndoiu [272], is 3.

Given n points on the Euclidean plane, the minimum spanning tree on these points is within a factor of $2/\sqrt{3}$ of the minimum Steiner tree (which is allowed to use any set of points on the plane as Steiner points). This was shown by Du and Hwang [67], thereby settling the conjecture of Gilbert and Pollak [106].

4 Multiway Cut and k-Cut

The theory of cuts occupies a central place in the study of exact algorithms. In this chapter, we will present approximation algorithms for natural generalizations of the minimum cut problem. These generalizations are **NP**-hard.

Given a connected, undirected graph $G = (V, E)$ with an assignment of weights to edges, $w : E \rightarrow \mathbf{R}^+$, a *cut* is defined by a partition of V into two sets, say V' and $V - V'$, and consists of all edges that have one endpoint in each partition. Clearly, the removal of the cut from G disconnects G. Given *terminals* $s, t \in V$, consider a partition of V that separates s and t. The cut defined by such a partition will be called an *s–t cut.* The problems of finding a minimum weight cut and a minimum weight s–t cut can be efficiently solved using a maximum flow algorithm. Let us generalize these two notions:

Problem 4.1 (Multiway cut) Given a set of terminals $S = \{s_1, s_2, \ldots, s_k\} \subseteq V$, a *multiway cut* is a set of edges whose removal disconnects the terminals from each other. The multiway cut problem asks for the minimum weight such set.

Problem 4.2 (Minimum k-cut) A set of edges whose removal leaves k connected components is called a *k-cut.* The minimum k-cut problem asks for a minimum weight k-cut.

The problem of finding a minimum weight multiway cut is **NP**-hard for any fixed $k \geq 3$. Observe that the case $k = 2$ is precisely the minimum s–t cut problem. The minimum k-cut problem is polynomial time solvable for fixed k; however, it is **NP**-hard if k is specified as part of the input. In this chapter, we will obtain factor $2 - 2/k$ approximation algorithms for both problems. In Chapter 19 we will improve the guarantee for the multiway cut problem to 3/2.

4.1 The multiway cut problem

Define an *isolating cut* for s_i to be a set of edges whose removal disconnects s_i from the rest of the terminals.

Algorithm 4.3 (Multiway cut)

1. For each $i = 1, \ldots, k$, compute a minimum weight isolating cut for s_i, say C_i.
2. Discard the heaviest of these cuts, and output the union of the rest, say C.

Each computation in Step 1 can be accomplished by identifying the terminals in $S - \{s_i\}$ into a single node, and finding a minimum cut separating this node from s_i; this takes one max-flow computation. Clearly, removing C from the graph disconnects every pair of terminals, and so is a multiway cut.

Theorem 4.4 *Algorithm 4.3 achieves an approximation guarantee of* $2-2/k$.

Proof: Let A be an optimal multiway cut in G. We can view A as the union of k cuts as follows: The removal of A from G will create k connected components, each having one terminal (since A is a minimum weight multiway cut, no more than k components will be created). Let A_i be the cut separating the component containing s_i from the rest of the graph. Then $A = \bigcup_{i=1}^{k} A_i$.

Since each edge of A is incident at two of these components, each edge will be in two of the cuts A_i. Hence,

$$\sum_{i=1}^{k} w(A_i) = 2w(A).$$

Clearly, A_i is an isolating cut for s_i. Since C_i is a minimum weight isolating cut for s_i, $w(C_i) \leq w(A_i)$. Notice that this already gives a factor 2 algorithm, by taking the union of all k cuts C_i. Finally, since C is obtained by discarding the heaviest of the cuts C_i,

$$w(C) \leq \left(1 - \frac{1}{k}\right) \sum_{i=1}^{k} w(C_i) \leq \left(1 - \frac{1}{k}\right) \sum_{i=1}^{k} w(A_i) = 2\left(1 - \frac{1}{k}\right) w(A).$$

□

Once again, Algorithm 4.3 is not based on a lower bounding scheme. Exercise 19.2 gives an algorithm with the same guarantee using an LP-relaxation as the lower bound. The use of LP-relaxations is fruitful for this problem as well. Section 19.1 gives an algorithm with an improved guarantee, using another LP-relaxation.

Example 4.5 A tight example for this algorithm is given by a graph on $2k$ vertices consisting of a k-cycle and a distinct terminal attached to each vertex of the cycle. The edges of the cycle have weight 1 and edges attaching terminals to the cycle have weight $2 - \varepsilon$ for a small fraction $\varepsilon > 0$.

For example, the graph corresponding to $k = 4$ is:

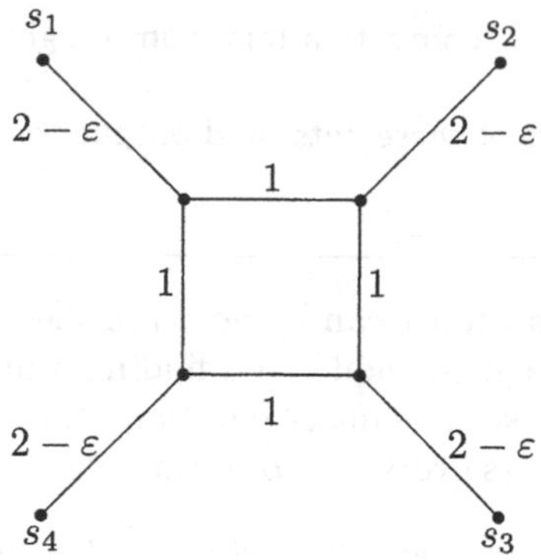

For each terminal s_i, the minimum weight isolating cut for s_i is given by the edge incident to s_i. So, the cut C returned by the algorithm has weight $(k-1)(2-\varepsilon)$. On the other hand, the optimal multiway cut is given by the cycle edges, and has weight k. □

4.2 The minimum k-cut problem

A natural algorithm for finding a k-cut is as follows. Starting with G, compute a minimum cut in each connected component and remove the lightest one; repeat until there are k connected components. This algorithm does achieve a guarantee of $2 - 2/k$, however, the proof is quite involved. Instead we will use the *Gomory–Hu tree representation of minimum cuts* to give a simpler algorithm achieving the same guarantee.

Minimum cuts, as well as sub-optimal cuts, in undirected graphs have several interesting structural properties, as opposed to cuts in directed graphs (the algorithm of Section 28.2 is based on exploiting some of these properties). The existence of Gomory–Hu trees is one of the remarkable consequences of these properties.

Let T be a tree on vertex set V; the edges of T need not be in E. Let e be an edge in T. Its removal from T creates two connected components. Let S and $\overline{S}$ be the vertex sets of these components. The cut defined in graph G by the partition $(S, \overline{S})$ is the *cut associated with e in G*. Define a weight function w' on the edges of T. Tree T will be said to be a Gomory–Hu tree for G if

1. for each pair of vertices $u, v \in V$, the weight of a minimum u–v cut in G is the same as that in T, and
2. for each edge $e \in T$, $w'(e)$ is the weight of the cut associated with e in G.

A Gomory–Hu tree encodes, in a succinct manner, a minimum u–v cut in G, for each pair of vertices $u, v \in V$ as follows. A minimum u–v cut in T is given by a minimum weight edge on the unique path from u to v in T, say e. By the properties stated above, the cut associated with e in G is a minimum u–v cut, and has weight $w'(e)$. So, for the $\binom{n}{2}$ pairs of vertices $u, v \in V$, we need only $n-1$ cuts, those encoded by the edges of a Gomory–Hu tree, to give minimum u–v cuts in G.

The following figure shows a weighted graph and its associated Gomory–Hu tree. Exercise 4.6 shows how to construct a Gomory–Hu tree for an undirected graph, using only $n-1$ max-flow computations.

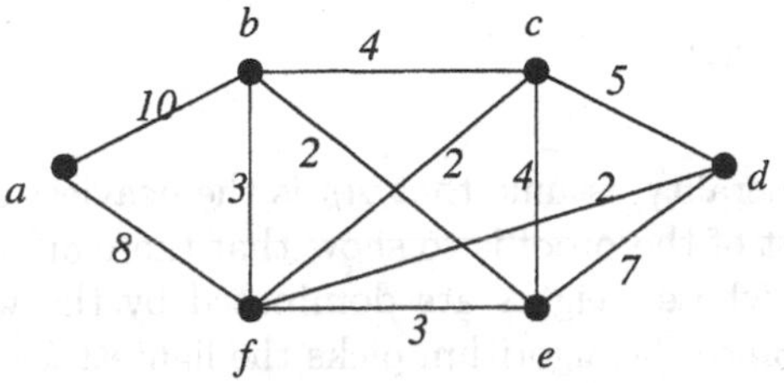

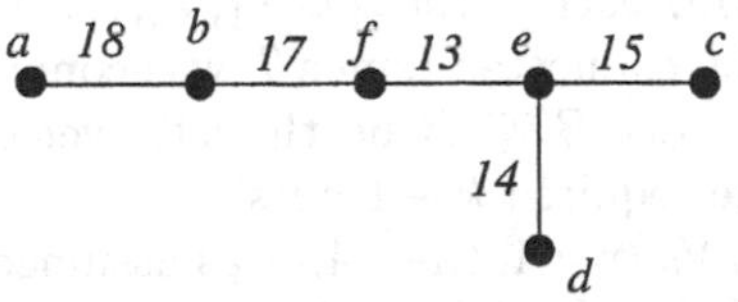

We will need the following lemma.

Lemma 4.6 *Let S be the union of cuts in G associated with l edges of T. Then, the removal of S from G leaves a graph with at least $l+1$ components.*

Proof: Removing the corresponding l edges from T leaves exactly $l+1$ connected components, say with vertex sets $V_1, V_2, \ldots, V_{l+1}$. Clearly, removing S from G will disconnect each pair V_i and V_j. Hence we must get at least $l+1$ connected components. □

As a consequence of Lemma 4.6, the union of $k-1$ cuts picked from T will form a k-cut in G. The complete algorithm is given below.

Algorithm 4.7 (Minimum k-cut)

1. Compute a Gomory–Hu tree T for G.
2. Output the union of the lightest $k-1$ cuts of the $n-1$ cuts associated with edges of T in G; let C be this union.

By Lemma 4.6, the removal of C from G will leave at least k components. If more than k components are created, throw back some of the removed edges until there are exactly k components.

Theorem 4.8 *Algorithm 4.7 achieves an approximation factor of* $2 - 2/k$.

Proof: Let A be an optimal k-cut in G. As in Theorem 4.4, we can view A as the union of k cuts: Let $V_1, V_2, \ldots, V_k$ be the k components formed by removing A from G, and let A_i denote the cut separating V_i from the rest of the graph. Then $A = A_1 \cup \cdots \cup A_k$, and, since each edge of A lies in two of these cuts,

$$\sum_{i=1}^{k} w(A_i) = 2w(A).$$

Without loss of generality assume that A_k is the heaviest of these cuts. The idea behind the rest of the proof is to show that there are $k-1$ cuts defined by the edges of T whose weights are dominated by the weight of the cuts $A_1, A_2, \ldots, A_{k-1}$. Since the algorithm picks the lightest $k-1$ cuts defined by T, the theorem follows.

The $k-1$ cuts are identified as follows. Let B be the set of edges of T that connect across two of the sets $V_1, V_2, \ldots, V_k$. Consider the graph on vertex set V and edge set B, and shrink each of the sets $V_1, V_2, \ldots, V_k$ to a single vertex. This shrunk graph must be connected (since T was connected). Throw edges away until a tree remains. Let $B' \subseteq B$ be the left over edges, $|B'| = k - 1$. The edges of B' define the required $k - 1$ cuts.

Next, root this tree at V_k (recall that A_k was assumed to be the heaviest cut among the cuts A_i). This helps in defining a correspondence between the edges in B' and the sets $V_1, V_2, \ldots, V_{k-1}$: each edge corresponds to the set it comes out of in the rooted tree.

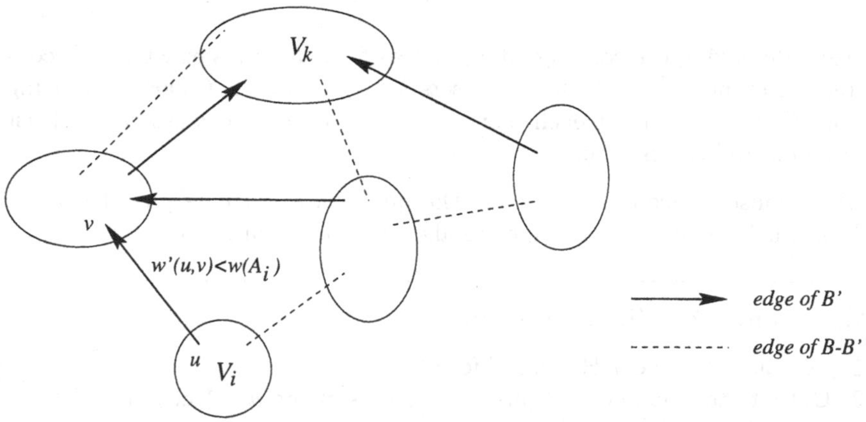

Suppose edge $(u, v) \in B'$ corresponds to set V_i in this manner. The weight of a minimum u–v cut in G is $w'(u, v)$. Since A_i is a u–v cut in G,

$$w(A_i) \geq w'(u, v).$$

Thus each cut among $A_1, A_2, \ldots, A_{k-1}$ is at least as heavy as the cut defined in G by the corresponding edge of B'. This, together with the fact that C is the union of the lightest $k-1$ cuts defined by T, gives:

$$w(C) \leq \sum_{e \in B'} w'(e) \leq \sum_{i=1}^{k-1} w(A_i) \leq \left(1 - \frac{1}{k}\right) \sum_{i=1}^{k} w(A_i) = 2\left(1 - \frac{1}{k}\right) w(A).$$

□

Example 4.9 The tight example given above for multiway cuts on $2k$ vertices also serves as a tight example for the k-cut algorithm (of course, there is no need to mark vertices as terminals). Below we give the example for $k = 4$, together with its Gomory–Hu tree.

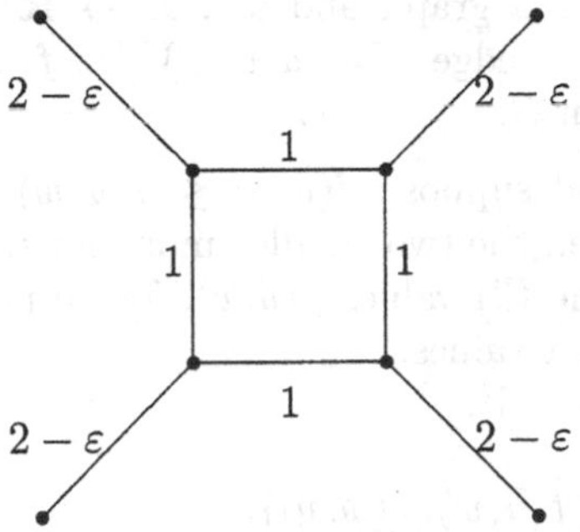

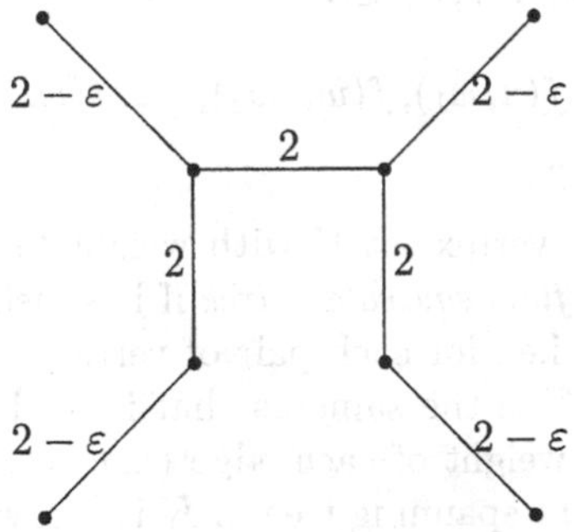

The lightest $k - 1$ cuts in the Gomory–Hu tree have weight $2 - \varepsilon$ each, corresponding to picking edges of weight $2 - \varepsilon$ of G. So, the k-cut returned by the algorithm has weight $(k-1)(2-\varepsilon)$. On the other hand, the optimal k-cut picks all edges of weight 1, and has weight k. □

4.3 Exercises

4.1 Show that Algorithm 4.3 can be used as a subroutine for finding a k-cut within a factor of $2-2/k$ of the minimum k-cut. How many subroutine calls are needed?

4.2 A natural greedy algorithm for computing a multiway cut is the following. Starting with G, compute minimum s_i–s_j cuts for all pairs s_i, s_j that are still connected and remove the lightest of these cuts; repeat this until all pairs s_i, s_j are disconnected. Prove that this algorithm also achieves a guarantee of $2-2/k$.

The next 4 exercises provide background and an algorithm for finding Gomory–Hu trees.

4.3 Let $G=(V,E)$ be a graph and $w: E \rightarrow \mathbf{R}^+$ be an assignment of nonnegative weights to its edges. For $u, v \in V$ let $f(u,v)$ denote the weight of a minimum u–v cut in G.

1. Let $u, v, w \in V$, and suppose $f(u,v) \leq f(u,w) \leq f(v,w)$. Show that $f(u,v) = f(u,w)$, i.e., the two smaller numbers are equal.
2. Show that among the $\binom{n}{2}$ values $f(u,v)$, for all pairs $u, v \in V$, there are at most $n-1$ distinct values.
3. Show that for $u, v, w \in V$,
$$f(u,v) \geq \min\{f(u,w), f(w,v)\}.$$
4. Show that for $u, v, w_1, \ldots, w_r \in V$
$$f(u,v) \geq \min\{f(u,w_1), f(w_1,w_2), \ldots, f(w_r,v).\} \tag{4.1}$$

4.4 Let T be a tree on vertex set V with weight function w' on its edges. We will say that T is a *flow equivalent tree* if it satisfies the first of the two Gomory–Hu conditions, i.e., for each pair of vertices $u, v \in V$, the weight of a minimum u–v cut in G is the same as that in T. Let K be the complete graph on V. Define the weight of each edge (u,v) in K to be $f(u,v)$. Show that any maximum weight spanning tree in K is a flow equivalent tree for G. **Hint:** For $u, v \in V$, let $u, w_1, \ldots, w_r, v$ be the unique path from u to v in T. Use (4.1) and the fact that since T is a maximum weight spanning tree, $f(u,v) \leq \min\{f(u,w_1), \ldots, f(w_r,v)\}$.

4.5 Let $(A, \bar{A})$ be a minimum s–t cut such that $s \in A$. Let x and y be any two vertices in A. Consider the graph G' obtained by collapsing all vertices of $\bar{A}$ to a single vertex $v_{\bar{A}}$. The weight of any edge $(a, v_{\bar{A}})$ in G' is defined to be the sum of the weights of edges (a, b) where $b \in \bar{A}$. Clearly, any cut in G' defines a cut in G. Show that a minimum x–y cut in G' defines a minimum x–y cut in G.

4.6 Now we are ready to state the Gomory–Hu algorithm. The algorithm maintains a partition of V, $(S_1, S_2, \ldots S_t)$, and a spanning tree T on the vertex set $\{S_1, \ldots, S_t\}$. Let w' be the function assigning weights to the edges of T. Tree T satisfies the following invariant.

Invariant: For any edge (S_i, S_j) in T there are vertices a and b in S_i and S_j respectively, such that $w'(S_i, S_j) = f(a, b)$, and the cut defined by edge (S_i, S_j) is a minimum a–b cut in G.

The algorithm starts with the trivial partition V, and proceeds in $n - 1$ iterations. In each iteration, it selects a set S_i in the partition such that $|S_i| \geq 2$ and refines the partition by splitting S_i, and finding a tree on the refined partition satisfying the invariant. This is accomplished as follows. Let x and y be two distinct vertices in S_i. Root the current tree T at S_i, and consider the subtrees rooted at the children of S_i. Each of these subtrees is collapsed into a single vertex, to obtain graph G' (besides these collapsed vertices, G' contains all vertices of S_i). A minimum x–y cut is found in G'. Let (A, B) be the partition of the vertices of G' defining this cut, with $x \in A$ and $y \in B$, and let w_{xy} be the weight of this cut. Compute $S_i^x = S_i \cap A$ and $S_i^y = S_i \cap B$, the two sets into which S_i splits.

The algorithm updates the partition and the tree as follows. It refines the partition by replacing S_i with two sets S_i^x and S_i^y. The new tree has the edge (S_i^x, S_i^y), with weight w_{xy}. Consider a subtree T' that was incident at S_i in T. Assume w.l.o.g. that the node corresponding to T' lies in A. Then, T' is connected by an edge to S_i^x. The weight of this connecting edge is the same as the weight of the edge connecting T' to S_i. All edges in T' retain their weights.

Show that the new tree satisfies the invariant. Hence show that the algorithm terminates (when the partition consists of singleton vertices) with a Gomory–Hu tree for G.

Consider the graph:

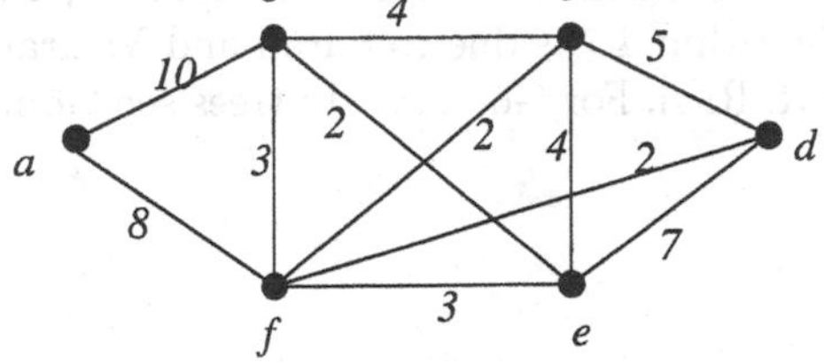

The execution of the Gomory–Hu algorithm is demonstrated below:

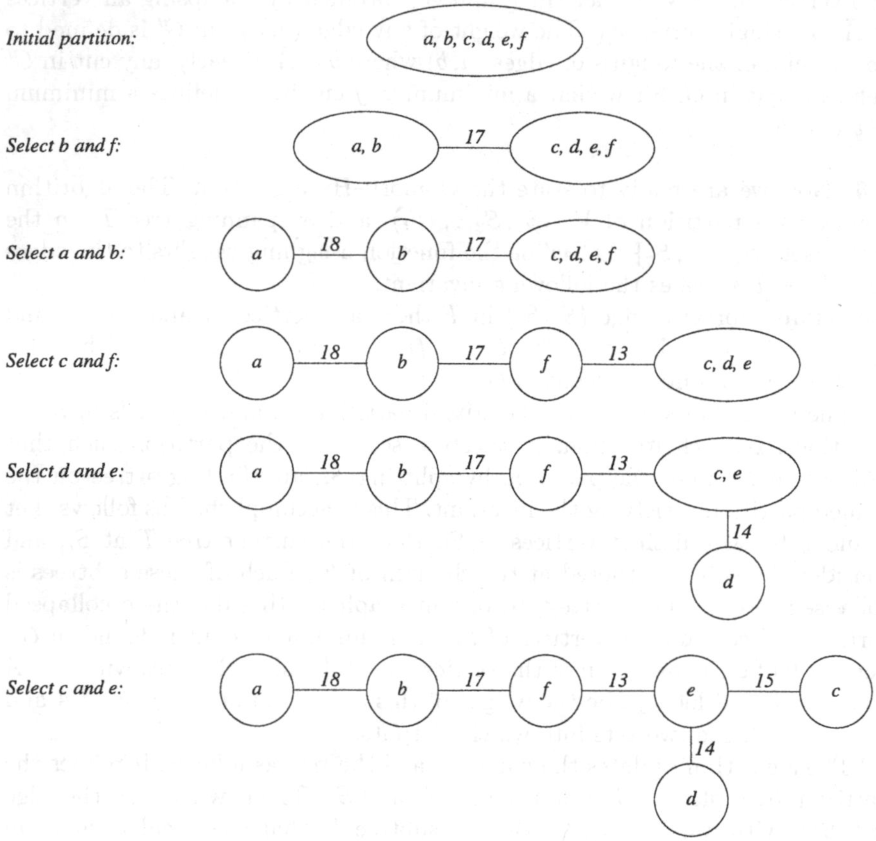

4.7 Prove that if the Gomory–Hu tree for an edge-weighted undirected graph G contains all $n-1$ distinct weights, then G can have only one minimum weight cut.

4.4 Notes

Algorithm 4.3 is due to Dahlhaus, Johnson, Seymour, Papadimitriou and Yannakakis [60]. Algorithm 4.7 is due to Saran and Vazirani [241]; the proof given here is due to R. Ravi. For Gomory–Hu trees see Gomory and Hu [116].

5 k-Center

Consider the following application. Given a set of cities, with intercity distances specified, pick k cities for locating warehouses in so as to minimize the maximum distance of a city from its closest warehouse. We will study this problem, called the k-center problem, and its weighted version, under the restriction that the edge costs satisfy the triangle inequality. Without this restriction, the k-center problem cannot be approximated within factor $\alpha(n)$, for any computable function $\alpha(n)$, assuming $\mathbf{P} \neq \mathbf{NP}$ (see Exercise 5.1).

We will introduce the algorithmic technique of *parametric pruning* for solving this problem. In Chapter 17 we will use this technique in a linear programming setting.

Problem 5.1 (Metric k-center) Let $G = (V, E)$ be a complete undirected graph with edge costs satisfying the triangle inequality, and k be a positive integer. For any set $S \subseteq V$ and vertex $v \in V$, define connect(v, S) to be the cost of the cheapest edge from v to a vertex in S. The problem is to find a set $S \subseteq V$, with $|S| = k$, so as to minimize $\max_v\{\text{connect}(v, S)\}$.

5.1 Parametric pruning applied to metric k-center

If we know the cost of an optimal solution, we may be able to prune away irrelevant parts of the input and thereby simplify the search for a good solution. However, as stated in Chapter 1, computing the cost of an optimal solution is precisely the difficult core of **NP**-hard **NP**-optimization problems. The technique of parametric pruning gets around this difficulty as follows. A parameter t is chosen, which can be viewed as a "guess" on the cost of an optimal solution. For each value of t, the given instance I is pruned by removing parts that will not be used in any solution of cost $\leq t$. Denote the pruned instance by $I(t)$. The algorithm consists of two steps. In the first step, the family of instances $I(t)$ is used for computing a lower bound on OPT, say t^*. In the second step, a solution is found in instance $I(\alpha \cdot t^*)$, for a suitable choice of α.

A restatement of the k-center problem shows how parametric pruning applies naturally to it. Sort the edges of G in nondecreasing order of cost, i.e., $\text{cost}(e_1) \leq \text{cost}(e_2) \leq \ldots \leq \text{cost}(e_m)$, and let $G_i = (V, E_i)$, where $E_i =$

$\{e_1, e_2, \ldots, e_i\}$. A *dominating set* in an undirected graph $H := (U, F)$ is a subset $S \subseteq U$ such that every vertex in $U - S$ is adjacent to a vertex in S. Let $\text{dom}(H)$ denote the size of a minimum cardinality dominating set in H. Computing $\text{dom}(H)$ is **NP**-hard. The k-center problem is equivalent to finding the smallest index i such that G_i has a dominating set of size at most k, i.e., G_i contains k stars spanning all vertices, where a *star* is the graph $K_{1,p}$, with $p \geq 1$. If i^* is the smallest such index, then $\text{cost}(e_{i^*})$ is the cost of an optimal k-center. We will denoted this by OPT. We will work with the family of graphs $G_1, \ldots, G_m$.

Define the *square of graph* H to be the graph containing an edge (u, v) whenever H has a path of length at most two between u and v, $u \neq v$. We will denote it by H^2. The following structural result gives a method for lower bounding OPT.

Lemma 5.2 *Given a graph H, let I be an independent set in H^2. Then,* $|I| \leq \text{dom}(H)$.

Proof: Let D be a minimum dominating set in H. Then, H contains $|D|$ stars spanning all vertices. Since each of these stars will be a clique in H^2, H^2 contains $|D|$ cliques spanning all vertices. Clearly, I can pick at most one vertex from each clique, and the lemma follows. □

The k-center algorithm is:

Algorithm 5.3 (Metric k-center)

1. Construct $G_1^2, G_2^2, \ldots, G_m^2$.
2. Compute a maximal independent set, M_i, in each graph G_i^2.
3. Find the smallest index i such that $|M_i| \leq k$, say j.
4. Return M_j.

The lower bound on which this algorithm is based is:

Lemma 5.4 *For j as defined in the algorithm,* $\text{cost}(e_j) \leq \text{OPT}$.

Proof: For every $i < j$ we have that $|M_i| > k$. Now, by Lemma 5.2, $\text{dom}(G_i) > k$, and so $i^* > i$. Hence, $j \leq i^*$. □

Theorem 5.5 *Algorithm 5.3 achieves an approximation factor of 2 for the metric k-center problem.*

Proof: The key observation is that a maximal independent set, I, in a graph is also a dominating set (for, if some vertex v is not dominated by I, then $I \cup \{v\}$ must also be an independent set, contradicting I's maximality). Thus, there exist stars in G_j^2, centered on the vertices of M_j, covering all vertices. By the triangle inequality, each edge used in constructing these stars has cost at most $2 \cdot \text{cost}(e_j)$. The theorem follows from Lemma 5.4. □

Example 5.6 A tight example for the previous algorithm is given by a wheel graph on $n + 1$ vertices, where all edges incident to the center vertex have cost 1, and the rest of the edges have cost 2:

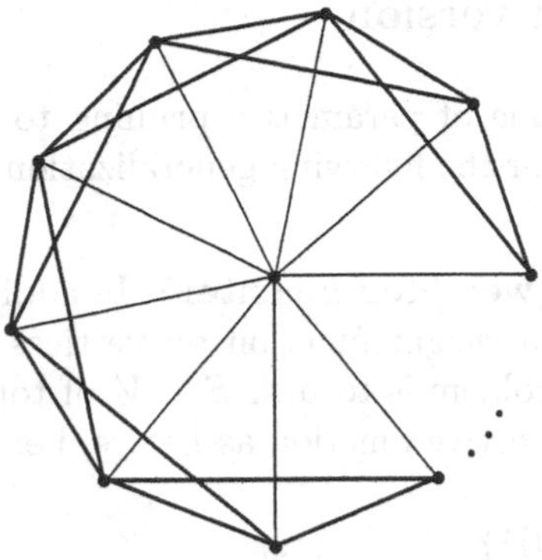

(Here, thin edges have cost 1 and thick edges have cost 2; not all edges of cost 2 are shown.)

For $k = 1$, the optimal solution is the center of the wheel, and OPT = 1. The algorithm will compute index $j = n$. Now, G_n^2 is a clique and, if a peripheral vertex is chosen as the maximal independent set, then the cost of the solution found is 2. □

Next, we will show that 2 is essentially the best approximation factor achievable for the metric k-center problem.

Theorem 5.7 *Assuming* **P** $\neq$ **NP**, *there is no polynomial time algorithm achieving a factor of* $2 - \varepsilon$, $\varepsilon > 0$, *for the metric k-center problem.*

Proof: We will show that such an algorithm can solve the dominating set problem in polynomial time. The idea is similar to that of Theorem 3.6 and involves giving a reduction from the dominating set problem to metric k-center. Let $G = (V, E)$, k be an instance of the dominating set problem. Construct a complete graph $G' = (V, E')$ with edge costs given by

$$\text{cost}(u, v) = \begin{cases} 1, \text{ if } (u, v) \in E, \\ 2, \text{ if } (u, v) \notin E. \end{cases}$$

Clearly, G' satisfies the triangle inequality. This reduction satisfies the conditions:

- if $\text{dom}(G) \leq k$, then G' has a k-center of cost 1, and

- if $\mathrm{dom}(G) > k$, then the optimum cost of a k-center in G' is 2.

In the first case, when run on G', the $(2-\varepsilon)$-approximation algorithm must give a solution of cost 1, since it cannot use an edge of cost 2. Hence, using this algorithm, we can distinguish between the two possibilities, thus solving the dominating set problem. □

5.2 The weighted version

We will use the technique of parametric pruning to obtain a factor 3 approximation algorithm for the following generalization of the metric k-center problem.

Problem 5.8 (Metric weighted k-center) In addition to a cost function on edges, we are given a weight function on vertices, $w : V \to R^+$, and a bound $W \in R^+$. The problem is to pick $S \subseteq V$ of total weight at most W, minimizing the same objective function as before, i.e.,

$$\max_{v \in V}\{\min_{u \in S}\{\mathrm{cost}(u,v)\}\}.$$

Let $\mathrm{wdom}(G)$ denote the weight of a minimum weight dominating set in G. Then, with respect to the graphs G_i defined above, we need to find the smallest index i such that $\mathrm{wdom}(G_i) \le W$. If i^* is this index, then the cost of the optimal solution is $\mathrm{OPT} = \mathrm{cost}(e_{i^*})$.

Given a vertex weighted graph H, let I be an independent set in H^2. For each $u \in I$, let $s(u)$ denote a lightest neighbor of u in H, where u is also considered a neighbor of itself. (Notice that the neighbor is picked in H and not in H^2.) Let $S = \{s(u) \mid u \in I\}$. The following fact, analogous to Lemma 5.2, will be used to derive a lower bound on OPT:

Lemma 5.9 $w(S) \le \mathrm{wdom}(H)$.

Proof: Let D be a minimum weight dominating set of H. Then there exists a set of disjoint stars in H, centered on the vertices of D and covering all the vertices. Since each of these stars becomes a clique in H^2, I can pick at most one vertex from each of them. Thus, each vertex in I has the center of the corresponding star available as a neighbor in H. Hence, $w(S) \le w(D)$. □

The algorithm is given below. In it, $s_i(u)$ will denote a lightest neighbor of u in G_i; for this definition, u will also be considered a neighbor of itself.

Algorithm 5.10 (Metric weighted k-center)

1. Construct $G_1^2, G_2^2, \ldots, G_m^2$.
2. Compute a maximal independent set, M_i, in each graph G_i^2.
3. Compute $S_i = \{s_i(u) |\ u \in M_i\}$.
4. Find the minimum index i such that $w(S_i) \leq W$, say j.
5. Return S_j.

Theorem 5.11 *Algorithm 5.10 achieves an approximation factor of 3 for the weighted k-center problem.*

Proof: By Lemma 5.9, cost(e_j) is a lower bound on OPT; the argument is identical to that in Lemma 5.4 and is omitted here. Since M_j is a dominating set in G_j^2, we can cover V with stars of G_j^2 centered in vertices of M_j. By the triangle inequality these stars use edges of cost at most $2 \cdot \text{cost}(e_j)$.

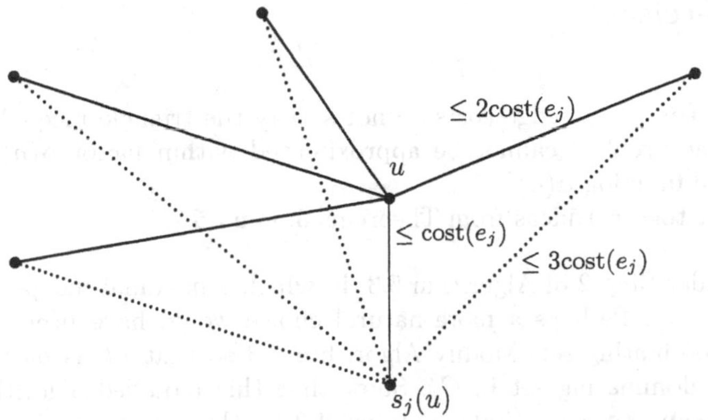

Each star center is adjacent to a vertex in S_j, using an edge of cost at most cost(e_j). Move each of the centers to the adjacent vertex in S_j and redefine the stars. Again, by the triangle inequality, the largest edge cost used in constructing the final stars is at most $3 \cdot \text{cost}(e_j)$. □

Example 5.12 A tight example is provided by the following graph on $n+4$ vertices. Vertex weights and edge costs are as marked; all missing edges have a cost given by the shortest path.

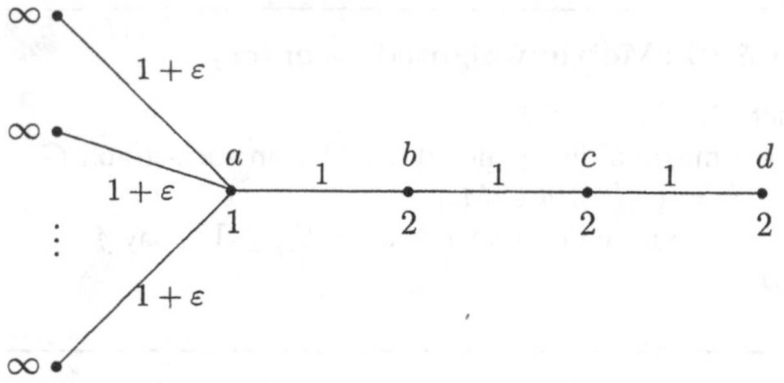

It is not difficult to see that for $W = 3$ the optimum cost of a k-center is $1 + \varepsilon$: a k-center achieving this cost is $\{a, c\}$. For any $i < n + 3$, the set S_i computed by the algorithm will contain a vertex of infinite weight. Suppose that, for $i = n + 3$, the algorithm chooses $M_{n+3} = \{b\}$ as a maximal independent set. Then $S_{n+3} = \{a\}$, and this is the output of the algorithm. The cost of this solution is 3. □

5.3 Exercises

5.1 Show that if the edge costs do not satisfy the triangle inequality, then the k-center problem cannot be approximated within factor $\alpha(n)$ for any computable function $\alpha(n)$.
Hint: Put together ideas from Theorems 3.6 and 5.7.

5.2 Consider Step 2 of Algorithm 5.3, in which a maximal independent set is found in G_i^2. Perhaps a more natural choice would have been to find a minimal dominating set. Modify Algorithm 5.3 so that M_i is picked to be a minimal dominating set in G_i^2. Show that this modified algorithm does not achieve an approximation guarantee of 2 for the k-center problem. What approximation factor can you establish for this algorithm?
Hint: With this modification, the lower bounding method does not work, since Lemma 5.2 does not hold if I is picked to be a minimal dominating set in H^2.

5.3 (Gonzalez [117]) Consider the following problem.

Problem 5.13 (Metric k-cluster) Let $G = (V, E)$ be a complete undirected graph with edge costs satisfying the triangle inequality, and let k be a positive integer. The problem is to partition V into sets $V_1, \ldots, V_k$ so as to minimize the costliest edge between two vertices in the same set, i.e., minimize

$$\max_{1 \le i \le k,\ u, v \in V_i} \mathrm{cost}(u, v).$$

1. Give a factor 2 approximation algorithm for this problem, together with a tight example.
2. Show that this problem cannot be approximated within a factor of $2-\varepsilon$, for any $\varepsilon > 0$, unless $\mathbf{P} = \mathbf{NP}$.

5.4 (Khuller, Pless, and Sussmann [176]) The *fault-tolerant* version of the metric k-center problem has an additional input, $\alpha \leq k$, which specifies the number of centers that each city should be connected to. The problem again is to pick k centers so that the length of the longest edge used is minimized.

A set $S \subseteq V$ in an undirected graph $H = (V, E)$ is an *α-dominating set* if each vertex $v \in V$ is adjacent to at least α vertices in S (assuming that a vertex is adjacent to itself). Let $\text{dom}_\alpha(H)$ denote the size of a minimum cardinality α-dominating set in H.

1. Let I be an independent set in H^2. Show that $\alpha|I| \leq \text{dom}_\alpha(H)$.
2. Give a factor 3 approximation algorithm for the fault-tolerant k-center problem.
   Hint: Compute a maximal independent set M_i in G_i^2, for $1 \leq i \leq m$. Find the smallest index i such that $|M_i| \leq \lfloor \frac{k}{\alpha} \rfloor$, and moreover, the degree of each vertex of M_i in G_i is $\geq \alpha - 1$.

5.5 (Khuller, Pless, and Sussmann [176]) Consider a modification of the problem of Exercise 5.4 in which vertices of S have no connectivity requirements and only vertices of $V - S$ have connectivity requirements. Each vertex of $V - S$ needs to be connected to α vertices in S. The object again is to pick S, $|S| = k$, so that the length of the longest edge used is minimized.

The algorithm for this problem works on each graph G_i. It starts with $S_i = \emptyset$. Vertex $v \in V - S_i$ is said to be *j-connected* if it is adjacent to j vertices in S_i, using edges of G_i^2. While there is a vertex $v \in V - S_i$ that is not α-connected, pick the vertex with minimum connectivity, and include it in S_i. Finally, find the minimum index i such that $|S_i| \leq k$, say l. Output S_l. Prove that this is a factor 2 approximation algorithm.

5.4 Notes

Both k-center algorithms presented in this chapter are due to Hochbaum and Shmoys [134], and Theorem 5.7 is due to Hsu and Nemhauser [139].

6 Feedback Vertex Set

In this chapter we will use the technique of *layering*, introduced in Chapter 2, to obtain a factor 2 approximation algorithm for the following problem. Recall that the idea behind layering was to decompose the given weight function into convenient functions on a nested sequence of subgraphs of G.

Problem 6.1 (Feedback vertex set) Given an undirected graph $G = (V, E)$ and a function w assigning nonnegative weights to its vertices, find a minimum weight subset of V whose removal leaves an acyclic graph.

6.1 Cyclomatic weighted graphs

Order the edges of G in an arbitrary order. The *characteristic vector* of a simple cycle C in G is a vector in $\mathrm{GF}[2]^m$, $m = |E|$, which has 1's in components corresponding to edges of C and 0's in the remaining components. The *cycle space of* G is the subspace of $\mathrm{GF}[2]^m$ that is spanned by the characteristic vectors of all simple cycles of G, and the *cyclomatic number of* G, denoted $\mathrm{cyc}(G)$, is the dimension of this space. $\mathrm{comps}(G)$ will denote the number of connected components of G.

Theorem 6.2 $\mathrm{cyc}(G) = |E| - |V| + \mathrm{comps}(G)$.

Proof: The cycle space of a graph is the direct sum of the cycle spaces of its connected components, and so its cyclomatic number is the sum of the cyclomatic numbers of its connected components. Therefore, it is sufficient to prove the theorem for a connected graph G.

Let T be a spanning tree in G. For each nontree edge e, define its *fundamental cycle* to be the unique cycle formed in $T \cup \{e\}$. The set of characteristic vectors of all such cycles is linearly independent (each cycle includes an edge that is in no other fundamental cycle). Thus, $\mathrm{cyc}(G) \geq |E| - |V| + 1$.

Each edge e of T defines a *fundamental cut* $(S, \overline{S})$ in G, $S \subset V$ (S and $\overline{S}$ are the vertex sets of two connected components formed by removing e from T). Define the *characteristic vector* of a cut to be a vector in $\mathrm{GF}[2]^m$ that has 1's in components corresponding to the edges of G in the cut and 0's in the remaining components. Consider the $|V| - 1$ vectors defined by edges of T. Since each cycle must cross each cut an even number of times, these vectors are orthogonal to the cycle space of G. Furthermore, these $|V|-1$ vectors are linearly independent, since each cut has an edge (the tree edge defining this cut) that is not in any of the other $|V|-2$ cuts. Therefore the dimension of the orthogonal complement to the cycle space is at least $|V|-1$. Hence, $\text{cyc}(G) \leq |E|-|V|+1$. Combining with the previous inequality we get $\text{cyc}(G) = |E|-|V|+1$. □

Denote by $\delta_G(v)$ the decrease in the cyclomatic number of the graph on removing vertex v. Since the removal of a feedback vertex set $F = \{v_1, \dots, v_f\}$ decreases the cyclomatic number of G down to 0,

$$\text{cyc}(G) = \sum_{i=1}^{f} \delta_{G_{i-1}}(v_i),$$

where $G_0 = G$ and, for $i > 0$, $G_i = G - \{v_1, \dots, v_i\}$. By Lemma 6.4 below, we get:

$$\text{cyc}(G) \leq \sum_{v \in F} \delta_G(v). \tag{6.1}$$

Let us say that a function assigning vertex weights is *cyclomatic* if there is a constant $c > 0$ such that the weight of each vertex v is $c \cdot \delta_G(v)$. Suppose the given weight function, w, is cyclomatic and F is an optimal feedback vertex set. Then, by inequality (6.1),

$$c \cdot \text{cyc}(G) \leq c \sum_{v \in F} \delta_G(v) = w(F) = \text{OPT}.$$

Hence, $c \cdot \text{cyc}(G)$ is a lower bound on OPT. The importance of cyclomatic weight functions is established in Lemma 6.5 below, which shows that for such a weight function, any minimal feedback vertex set has a weight within twice the optimal.

Let $\deg_G(v)$ denote the degree of v in G, and $\text{comps}(G-v)$ denote the number of connected components formed by removing v from G. The claim below follows in a straightforward way by applying Theorem 6.2 to G and $G - v$.

Claim 6.3 *For a connected graph G, $\delta_G(v) = \deg_G(v) - \text{comps}(G-v)$.*

Lemma 6.4 *Let H be a subgraph of G (not necessarily vertex induced). Then, $\delta_H(v) \leq \delta_G(v)$.*

Proof: It is sufficient to prove the lemma for the connected components of G and H containing v. We may thus assume w.l.o.g. that G and H are connected (H may be on a smaller set of vertices). By Claim 6.3, proving the following inequality is sufficient:

$$\deg_H(v) - \text{comps}(H - v) \le \deg_G(v) - \text{comps}(G - v).$$

We will show that edges in $G - H$ can only help this inequality. Let $c_1, c_2, \ldots, c_k$ be components formed by removing v from H. Edges of $G - H$ not incident at v can only help merge some of these components (and of course, they don't change the degree of v). An edge of $G - H$ that is incident at v can lead to an additional component, but this is compensated by the contribution the edge has to the degree of v. □

Lemma 6.5 *If F is a minimal feedback vertex set of G, then*

$$\sum_{v \in F} \delta_G(v) \le 2 \cdot \text{cyc}(G).$$

Proof: Since the cycle space of G is the direct sum of the cycle spaces of its connected components, it suffices to prove the lemma for a connected graph G.

Let $F = \{v_1, \ldots, v_f\}$, and let k be the number of connected components obtained by deleting F from G. Partition these components into two types: those that have edges incident to only one of the vertices of F, and those that have edges incident to two or more vertices of F. Let t and $k - t$ be the number of components of the first and second type, respectively. We will prove that

$$\sum_{i=1}^{f} \delta_G(v_i) = \sum_{i=1}^{f} (\deg_G(v_i) - \text{comps}(G - v_i)) \le 2(|E| - |V|),$$

thereby proving the lemma. Clearly, $\sum_{i=1}^{f} \text{comps}(G - v_i) = f + t$. Therefore, we are left to prove

$$\sum_{i=1}^{f} \deg_G(v_i) \le 2(|E| - |V|) + f + t.$$

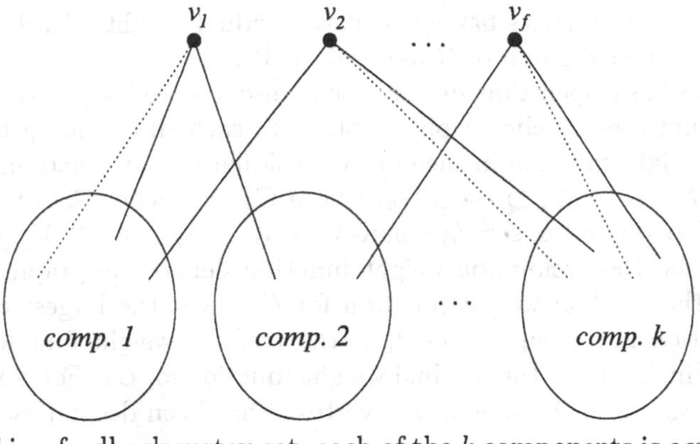

Since F is a feedback vertex set, each of the k components is acyclic and is therefore a tree. Thus, the number of edges in these components is $|V| - f - k$. Next, we put a lower bound on the number of edges in the cut $(F, V - F)$. Since F is minimal, each $v_i \in F$ must be in a cycle that contains no other vertices of F. Therefore, each v_i must have at least two edges incident at one of the components. For each v_i, arbitrarily remove one of these edges from G, thus removing a total of f edges. Now, each of the t components must still have at least one edge and each of the $k - t$ components must still have at least two edges incident at F. Therefore, the number of edges in the cut $(F, V - F)$ is at least $f + t + 2(k - t) = f + 2k - t$.

These two facts imply that

$$\sum_{i=1}^{f} \deg_G(v_i) \leq 2|E| - 2(|V| - f - k) - (f + 2k - t).$$

The lemma follows. □

Corollary 6.6 *Let w be a cyclomatic weight function on the vertices of G, and let F be a minimal feedback vertex set in it. Then $w(F) \leq 2 \cdot \mathrm{OPT}$.*

6.2 Layering applied to feedback vertex set

Let us now deal with arbitrary weighted graphs. Consider the following basic operation: Given graph $G = (V, E)$ and a weight function w, let

$$c = \min_{v \in V} \left\{ \frac{w(v)}{\delta_G(v)} \right\}.$$

The weight function $t(v) = c\delta_G(v)$ is the *largest cyclomatic weight function in w*. Define $w'(v) = w(v) - t(v)$ to be the *residual weight function*. Finally, let V' be the set of vertices having positive residual weight (clearly, $V' \subset V$), and let G' be the subgraph of G induced on V'.

Using this basic operation, decompose G into a nested sequence of induced subgraphs, until an acyclic graph is obtained, each time finding the largest cyclomatic weight function in the current residual weight function. Let these graphs be $G = G_0 \supset G_1 \supset \cdots \supset G_k$, where G_k is acyclic; G_i is the induced subgraph of G on vertex set V_i, where $V = V_0 \supset V_1 \supset \cdots \supset V_k$. Let $t_i, i = 0, \ldots, k-1$ be the cyclomatic weight function defined on graph G_i. Thus, $w_0 = w$ is the residual weight function for G_0, t_0 is the largest cyclomatic weight function in w_0, $w_1 = w_0 - t_0$ is the residual weight function for G_1, and so on. Finally, w_k is the residual weight function for G_k. For convenience, define $t_k = w_k$. Since the weight of a vertex v has been decomposed into the weights $t_0, t_1, \ldots, t_k$, we have

$$\sum_{i:\, v \in V_i} t_i(v) = w(v).$$

The next fact suggests an algorithm for constructing a feedback vertex set on which Lemma 6.5 can be applied.

Lemma 6.7 *Let H be a subgraph of $G = (V, E)$, induced on vertex set $V' \subset V$. Let F be a minimal feedback vertex set in H, and let $F' \subseteq V - V'$ be a minimal set such that $F \cup F'$ is a feedback vertex set for G. Then $F \cup F'$ is a minimal feedback vertex set for G.*

Proof: Since F is minimal for H, for each $v \in F$, there is a cycle, say C, in H that does not use any other vertex of F. Since $F' \cap V' = \emptyset$, C uses only one vertex, v, from $F \cup F'$ as well, and so v is not redundant. □

After the entire decomposition, $F_k = \emptyset$ is a minimal feedback vertex set of G_k. For $i = k, k-1, \ldots, 1$, the minimal feedback vertex set F_i found in G_i is extended in a minimal way using vertices of $V_{i-1} - V_i$ to yield a minimal feedback vertex set, say F_{i-1}, for G_{i-1}. The last set, F_0, is a feedback vertex set for G.

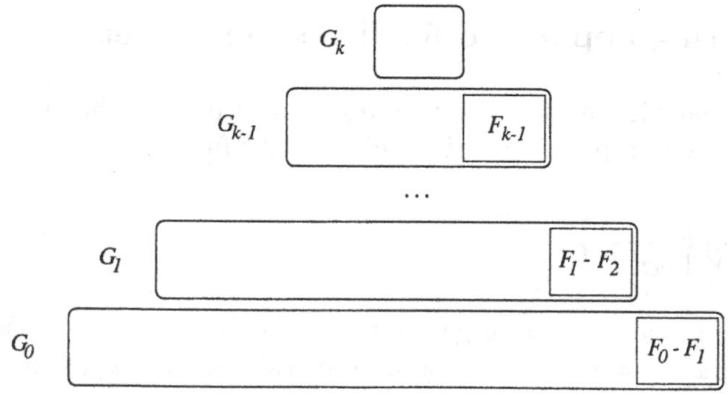

Algorithm 6.8 (Feedback vertex set)

1. Decomposition phase
   $H \leftarrow G$, $w' \leftarrow w$, $i \leftarrow 0$
   While H is not acyclic,
   $c \leftarrow \min_{u \in H} \left\{ \frac{w'(u)}{\delta_H(u)} \right\}$
   $G_i \leftarrow H$, $t_i \leftarrow c \cdot \delta_{G_i}$, $w' \leftarrow w' - t_i$
   $H \leftarrow$ the subgraph of G_i induced by vertices u with $w'(u) > 0$
   $i \leftarrow i + 1$,
   $k \leftarrow i$, $G_k \leftarrow H$
2. Extension phase
   $F_k \leftarrow \emptyset$
   For $i = k, \ldots, 1$, extend F_i to a feedback vertex set F_{i-1} of G_{i-1} by adding a minimal set of vertices from $V_{i-1} - V_i$.
   Output F_0.

Theorem 6.9 *Algorithm 6.8 achieves an approximation guarantee of factor 2 for the feedback vertex set problem.*

Proof: Let F^* be an optimal feedback vertex set for G. Since G_i is an induced subgraph of G, $F^* \cap V_i$ must be a feedback vertex set for G_i (not necessarily optimal). Since the weights of vertices have been decomposed into the functions t_i, we have

$$\mathrm{OPT} = w(F^*) = \sum_{i=0}^{k} t_i(F^* \cap V_i) \geq \sum_{i=0}^{k} \mathrm{OPT}_i,$$

where OPT_i is the weight of an optimal feedback vertex set of G_i with weight function t_i.

By decomposing the weight of F_0, we get

$$w(F_0) = \sum_{i=0}^{k} t_i(F_0 \cap V_i) = \sum_{i=0}^{k} t_i(F_i).$$

By Lemma 6.7, F_i is a minimal feedback vertex set in G_i. Since for $0 \leq i \leq k-1$, t_i is a cyclomatic weight function, by Lemma 6.5, $t_i(F_i) \leq 2\mathrm{OPT}_i$; recall that $F_k = \emptyset$. Therefore,

$$w(F_0) \leq 2 \sum_{i=0}^{k} \mathrm{OPT}_i \leq 2 \cdot \mathrm{OPT}.$$

□

Example 6.10 A tight example for the algorithm is given by the graph obtained by removing a perfect matching from a complete bipartite graph and duplicating every edge. (Note that the algorithm works for parallel edges as well. If a tight example without parallel edges is desired, then a vertex with very high weight can be placed on every edge.)

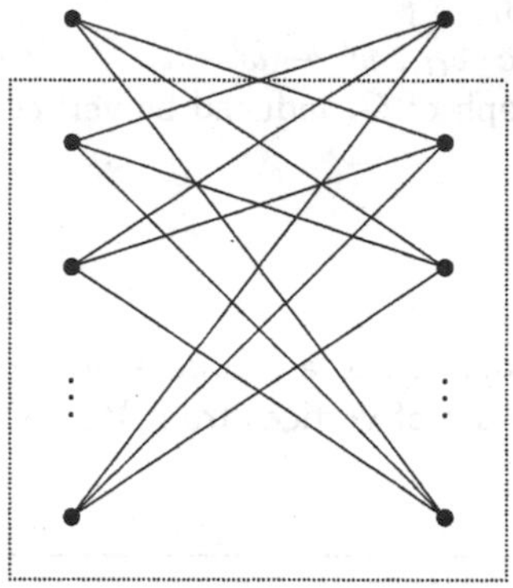

Assuming that the graph is cyclomatic weighted, each vertex receives the same weight. The decomposition obtained by the algorithm consists of only one nontrivial graph, G itself, on which the algorithm computes a minimal feedback vertex set. A possible output of the algorithm is the set shown above; this set contains $2n - 2$ vertices as compared with the optimum of n given by one side of the bipartition. □

6.3 Exercises

6.1 A natural greedy algorithm for finding a minimum feedback vertex set is to repeatedly pick and remove the most cost-effective vertex, i.e., a vertex minimizing $w(v)/\delta_H(v)$, where H is the current graph, until there are no more cycles left. Give examples to show that this is not a constant factor algorithm. What is the approximation guarantee of this algorithm?

6.2 Give an approximation factor preserving reduction from the vertex cover problem to the feedback vertex set problem (thereby showing that improving the factor for the latter problem will also improve it for the former; also see Section 30.1).

6.4 Notes

Algorithm 6.8 is due to Bafna, Berman, and Fujito [20] (see also Becker and Geiger [24] and Chudak, Goemans, Hochbaum, and Williamson [48] for other factor 2 algorithms for the feedback vertex set problem).

7 Shortest Superstring

In Chapter 2 we defined the shortest superstring problem (Problem 2.9) and gave a preliminary approximation algorithm using set cover. In this chapter, we will first give a factor 4 algorithm, and then we will improve this to factor 3.

7.1 A factor 4 algorithm

We begin by developing a good lower bound on OPT. Let us assume that $s_1, s_2, \ldots, s_n$ are numbered in order of leftmost occurrence in the shortest superstring, s.

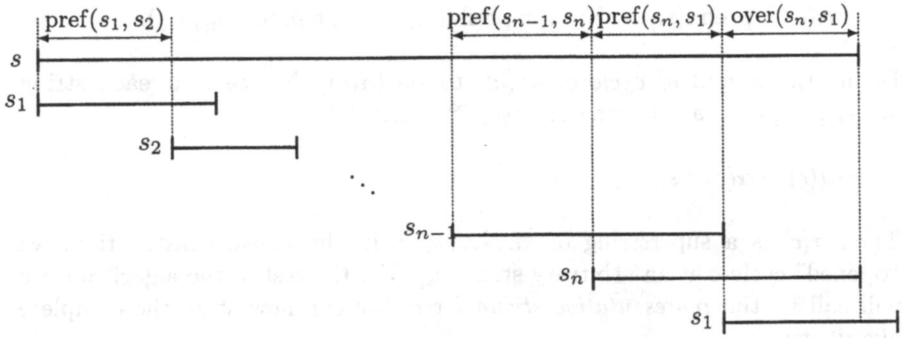

Let overlap(s_i, s_j) denote the maximum overlap between s_i and s_j, i.e., the longest suffix of s_i that is a prefix of s_j. Also, let prefix(s_i, s_j) be the prefix of s_i obtained by removing its overlap with s_j. The overlap in s between two consecutive s_i's is maximum possible, because otherwise a shorter superstring can be obtained. Hence, assuming that no s_i is a substring of another, we get

$$\begin{aligned} \mathrm{OPT} = &|\mathrm{prefix}(s_1, s_2)| + |\mathrm{prefix}(s_2, s_3)| + \ldots + |\mathrm{prefix}(s_n, s_1)| \\ &+ |\mathrm{overlap}(s_n, s_1)|. \end{aligned} \tag{7.1}$$

Notice that we have repeated s_1 at the end in order to obtain the last two terms of (7.1). This equality shows the close relation between the shortest superstring of S and the minimum traveling salesman tour on the *prefix graph of* S, defined as the directed graph on vertex set $\{1, \ldots, n\}$ that contains an edge $i \to j$ of weight $|\text{prefix}(s_i, s_j)|$ for each i, j, $i \neq j$ (i.e., self-loops are not included). Clearly, $|\text{prefix}(s_1, s_2)| + |\text{prefix}(s_2, s_3)| + \ldots + |\text{prefix}(s_n, s_1)|$ represents the weight of the tour $1 \to 2 \to \ldots \to n \to 1$. Hence, by (7.1), the minimum weight of a traveling salesman tour of the prefix graph gives a lower bound on OPT. As such, this lower bound is not very useful, since we cannot efficiently compute a minimum traveling salesman tour.

The key idea is to lower-bound OPT using the minimum weight of a *cycle cover* of the prefix graph (a cycle cover is a collection of disjoint cycles covering all vertices). Since the tour $1 \to 2 \to \ldots \to n \to 1$ is a cycle cover, from (7.1) we get that the minimum weight of a cycle cover lower-bounds OPT.

Unlike minimum TSP, a minimum weight cycle cover can be computed in polynomial time. Corresponding to the prefix graph, construct the following bipartite graph, H. $U = \{u_1, \ldots, u_n\}$ and $V = \{v_1, \ldots, v_n\}$ are the vertex sets of the two sides of the bipartition. For each $i, j \in \{1, \ldots, n\}$ add edge (u_i, v_j) of weight $|\text{prefix}(s_i, s_j)|$. It is easy to see that each cycle cover of the prefix graph corresponds to a perfect matching of the same weight in H and vice versa. Hence, finding a minimum weight cycle cover reduces to finding a minimum weight perfect matching in H.

If $c = (i_1 \to i_2 \to \ldots i_l \to i_1)$ is a cycle in the prefix graph, let

$$\alpha(c) = \text{prefix}(s_{i_1}, s_{i_2}) \circ \ldots \circ \text{prefix}(s_{i_{l-1}}, s_{i_l}) \circ \text{prefix}(s_{i_l}, s_{i_1}).$$

Define the *weight of cycle* c, $\text{wt}(c)$, to be $|\alpha(c)|$. Notice that each string $s_{i_1}, s_{i_2}, \ldots, s_{i_l}$ is a substring of $(\alpha(c))^\infty$. Next, let

$$\sigma(c) = \alpha(c) \circ s_{i_1}.$$

Then $\sigma(c)$ is a superstring of $s_{i_1}, \ldots, s_{i_l}$.[1] In the above construction, we "opened" cycle c at an arbitrary string s_{i_1}. For the rest of the algorithm, we will call s_{i_1} the *representative string* for c. We can now state the complete algorithm:

Algorithm 7.1 (Shortest superstring – factor 4)

1. Construct the prefix graph corresponding to strings in S.
2. Find a minimum weight cycle cover of the prefix graph, $\mathcal{C} = \{c_1, \ldots, c_k\}$.
3. Output $\sigma(c_1) \circ \ldots \circ \sigma(c_k)$.

[1] This remains true even for the shorter string $\alpha(c) \circ \text{overlap}(s_l, s_1)$. We will work with $\sigma(c)$, since it will be needed for the factor 3 algorithm presented in the next section, where we use the property that $\sigma(c)$ begins and ends with a copy of s_{i_1}.

Clearly, the output is a superstring of the strings in S. Notice that if in each of the cycles we can find a representative string of length at most the weight of the cycle, then the string output is within $2 \cdot \mathrm{OPT}$. Thus, the hard case is when all strings of some cycle c are long. But since they must all be substrings of $(\alpha(c))^\infty$, they must be periodic. This will be used to prove Lemma 7.3, which establishes another lower bound on OPT.

Lemma 7.2 *If each string in $S' \subseteq S$ is a substring of t^∞ for a string t, then there is a cycle of weight at most $|t|$ in the prefix graph covering all the vertices corresponding to strings in S'.*

Proof: For each string in S', locate the starting point of its first occurrence in t^∞. Clearly, all these starting points will be distinct (since no string in S is a substring of another) and will lie in the first copy of t. Consider the cycle in the prefix graph visiting the corresponding vertices in this order. Clearly, the weight of this cycle is at most $|t|$. □

Lemma 7.3 *Let c and c' be two cycles in $\mathcal{C}$, and let r, r' be representative strings from these cycles. Then*

$$|\mathrm{overlap}(r, r')| < \mathrm{wt}(c) + \mathrm{wt}(c').$$

Proof: Suppose, for contradiction, that $|\mathrm{overlap}(r, r')| \geq \mathrm{wt}(c) + \mathrm{wt}(c')$. Denote by α (α') the prefix of length $\mathrm{wt}(c)$ ($\mathrm{wt}(c')$, respectively) of $\mathrm{overlap}(r, r')$.

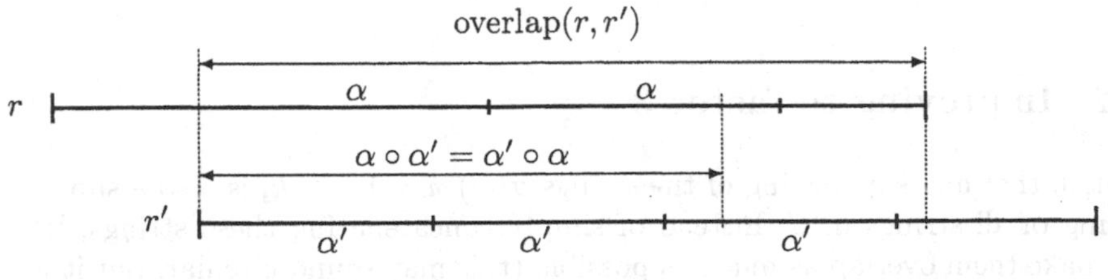

Clearly, $\mathrm{overlap}(r, r')$ is a prefix of both α^∞ and $(\alpha')^\infty$. In addition, α is a prefix of $(\alpha')^\infty$ and α' is a prefix of α^∞. Since $\mathrm{overlap}(r, r') \geq |\alpha| + |\alpha'|$, it follows that α and α' commute, i.e., $\alpha \circ \alpha' = \alpha' \circ \alpha$. But then, $\alpha^\infty = (\alpha')^\infty$. This is so because for any $k > 0$,

$$\alpha^k \circ (\alpha')^k = (\alpha')^k \circ \alpha^k.$$

Hence, for any $N > 0$, the prefix of length N of α^∞ is the same as that of $(\alpha')^\infty$.

Now, by Lemma 7.2, there is a cycle of weight at most $\mathrm{wt}(c)$ in the prefix graph covering all strings in c and c', contradicting the fact that $\mathcal{C}$ is a minimum weight cycle cover. □

Theorem 7.4 *Algorithm 7.1 achieves an approximation factor of 4 for the shortest superstring problem.*

Proof: Let $\text{wt}(\mathcal{C}) = \sum_{i=1}^{k} \text{wt}(c_i)$. The output of the algorithm has length

$$\sum_{i=1}^{k} |\sigma(c_i)| = \text{wt}(\mathcal{C}) + \sum_{i=1}^{k} |r_i|,$$

where r_i denotes the representative string from cycle c_i. We have shown that $\text{wt}(\mathcal{C}) \leq \text{OPT}$. Next, we show that the sum of the lengths of representative strings is at most $3 \cdot \text{OPT}$.

Assume that $r_1, \ldots, r_k$ are numbered in order of their leftmost occurrence in the shortest superstring of S. Using Lemma 7.3, we get the following lower bound on OPT:

$$\text{OPT} \geq \sum_{i=1}^{k} |r_i| - \sum_{i=1}^{k-1} |\text{overlap}(r_i, r_{i+1})| \geq \sum_{i=1}^{k} |r_i| - 2\sum_{i=1}^{k} \text{wt}(c_i).$$

Hence,

$$\sum_{i=1}^{k} |r_i| \leq \text{OPT} + 2\sum_{i=1}^{k} \text{wt}(c_i) \leq 3 \cdot \text{OPT}.$$

□

7.2 Improving to factor 3

Notice that any superstring of the strings $\sigma(c_i)$, $i = 1, \ldots, k$, is also a superstring of all strings in S. Instead of simply concatenating these strings, let us make them overlap as much as possible (this may sound circular, but it is not!).

Let X be a set of strings. We will denote by $||X||$ the sum of the lengths of the strings in X. Let us define the *compression* achieved by a superstring s as the difference between the sum of the lengths of the input strings and $|s|$, i.e., $||S|| - |s|$. Clearly, maximum compression is achieved by the shortest superstring. Several algorithms are known to achieve at least half the optimal compression. For instance, the greedy superstring algorithm, described in Section 2.3, does so; however, its proof is based on a complicated case analysis. For a less efficient algorithm, see Section 7.2.1. Either of these algorithms can be used in Step 3 of Algorithm 7.5.

Algorithm 7.5 (Shortest superstring – factor 3)

1. Construct the prefix graph corresponding to strings in S.
2. Find a minimum cycle cover of the prefix graph, $\mathcal{C} = \{c_1, \ldots, c_k\}$.
3. Run the greedy superstring algorithm on $\{\sigma(c_1), \ldots, \sigma(c_k)\}$ and output the resulting string, say τ.

Let OPT_σ denote the length of the shortest superstring of the strings in $S_\sigma = \{\sigma(c_1), \ldots, \sigma(c_k)\}$, and let r_i be the representative string of c_i.

Lemma 7.6 $\qquad |\tau| \leq \mathrm{OPT}_\sigma + \mathrm{wt}(\mathcal{C})$.

Proof: Assume w.l.o.g. that $\sigma(c_1), \ldots, \sigma(c_k)$ appear in this order in a shortest superstring of S_σ. The maximum compression that can be achieved on S_σ is given by

$$\sum_{i=1}^{k-1} |\mathrm{overlap}(\sigma(c_i), \sigma(c_{i+1}))|.$$

Since each string $\sigma(c_i)$ has r_i as a prefix as well as suffix, by Lemma 7.3,

$$|\mathrm{overlap}(\sigma(c_i), \sigma(c_{i+1}))| \leq \mathrm{wt}(c_i) + \mathrm{wt}(c_{i+1}).$$

Hence, the maximum compression achievable on S_σ is at most $2 \cdot \mathrm{wt}(\mathcal{C})$, i.e., $||S_\sigma|| - \mathrm{OPT}_\sigma \leq 2 \cdot \mathrm{wt}(\mathcal{C})$.

The compression achieved by the greedy superstring algorithm on S_σ is at least half the maximum compression. Therefore,

$$||S_\sigma|| - |\tau| \geq \frac{1}{2}(||S_\sigma|| - \mathrm{OPT}_\sigma).$$

Therefore,

$$2(|\tau| - \mathrm{OPT}_\sigma) \leq ||S_\sigma|| - \mathrm{OPT}_\sigma \leq 2 \cdot \mathrm{wt}(\mathcal{C}).$$

The lemma follows. □

Finally, we relate OPT_σ to OPT.

Lemma 7.7 $\qquad \mathrm{OPT}_\sigma \leq \mathrm{OPT} + \mathrm{wt}(\mathcal{C})$.

Proof: Let OPT_r denote the length of the shortest superstring of the strings in $S_r = \{r_1, \ldots, r_k\}$. The key observation is that each $\sigma(c_i)$ begins and ends with r_i. Therefore, the maximum compression achievable on S_σ is at least as large as that achievable on S_r, i.e.,

$$||S_\sigma|| - \mathrm{OPT}_\sigma \geq ||S_r|| - \mathrm{OPT}_r.$$

Clearly, $||S_\sigma|| = ||S_r|| + \mathrm{wt}(\mathcal{C})$. This gives

$$\mathrm{OPT}_\sigma \leq \mathrm{OPT}_r + \mathrm{wt}(\mathcal{C}).$$

The lemma follows by noticing that $\mathrm{OPT}_r \leq \mathrm{OPT}$. □

Combining the previous two lemmas we get:

Theorem 7.8 *Algorithm 7.5 achieves an approximation factor of 3 for the shortest superstring problem.*

7.2.1 Achieving half the optimal compression

We give a superstring algorithm that achieves at least half the optimal compression. Suppose that the strings to be compressed, $s_1, \cdots, s_n$, are numbered in the order in which they appear in a shortest superstring. Then, the optimal compression is given by

$$\sum_{i=1}^{n-1} |\mathrm{overlap}(s_i, s_{i+1})|.$$

This is the weight of the traveling salesman path $1 \to 2 \to \ldots \to n$ in the *overlap graph*, H, of the strings $s_1, \cdots, s_n$. H is a directed graph that has a vertex v_i corresponding to each string s_i, and contains an edge $(v_i \to v_j)$ of weight $|\mathrm{overlap}(s_i, s_j)|$ for each $i \neq j$, $1 \leq i, j \leq n$ (H has no self loops).

The optimal compression is upper bounded by the cost of a maximum traveling salesman tour in H, which in turn is upper bounded by the cost of a maximum cycle cover. The latter can be computed in polynomial time using matching, similar to the way we computed a minimum weight cycle cover. Since H has no self loops, each cycle has at least two edges. Remove the lightest edge from each cycle of the maximum cycle cover to obtain a set of disjoint paths. The sum of weights of edges on these paths is at least half the optimal compression. Overlap strings $s_1, \cdots, s_n$ according to the edges of these paths and concatenate the resulting strings. This gives a superstring achieving at least half the optimal compression.

7.3 Exercises

7.1 Show that Lemma 7.3 cannot be strengthened to

$$|\mathrm{overlap}(r, r')| < \max\{\mathrm{wt}(c), \mathrm{wt}(c')\}.$$

7.2 (Jiang, Li, and Du [155]) Obtain constant factor approximation algorithms for the variants of the shortest superstring problem given in Exercise 2.16.

7.4 Notes

The algorithms given in this chapter are due to Blum, Jiang, Li, Tromp, and Yannakakis [28].

8 Knapsack

In Chapter 1 we mentioned that some **NP**-hard optimization problems allow approximability to any required degree. In this chapter, we will formalize this notion and will show that the knapsack problem admits such an approximability.

Let Π be an **NP**-hard optimization problem with objective function f_Π. We will say that algorithm $\mathcal{A}$ is an *approximation scheme* for Π if on input (I, ε), where I is an instance of Π and $\varepsilon > 0$ is an error parameter, it outputs a solution s such that:

- $f_\Pi(I, s) \leq (1 + \varepsilon) \cdot \mathrm{OPT}$ if Π is a minimization problem.
- $f_\Pi(I, s) \geq (1 - \varepsilon) \cdot \mathrm{OPT}$ if Π is a maximization problem.

$\mathcal{A}$ will be said to be a *polynomial time approximation scheme*, abbreviated PTAS, if for each *fixed* $\varepsilon > 0$, its running time is bounded by a polynomial in the size of instance I.

The definition given above allows the running time of $\mathcal{A}$ to depend arbitrarily on ε. This is rectified in the following more stringent notion of approximability. If the previous definition is modified to require that the running time of $\mathcal{A}$ be bounded by a polynomial in the size of instance I and $1/\varepsilon$, then $\mathcal{A}$ will be said to be a *fully polynomial time approximation scheme*, abbreviated FPTAS.

In a very technical sense, an FPTAS is the best one can hope for an **NP**-hard optimization problem, assuming $\mathbf{P} \neq \mathbf{NP}$; see Section 8.3.1 for a short discussion on this issue. The knapsack problem admits an FPTAS.

Problem 8.1 (Knapsack) Given a set $S = \{a_1, \ldots, a_n\}$ of objects, with specified sizes and profits, $\mathrm{size}(a_i) \in \mathbf{Z}^+$ and $\mathrm{profit}(a_i) \in \mathbf{Z}^+$, and a "knapsack capacity" $B \in \mathbf{Z}^+$, find a subset of objects whose total size is bounded by B and total profit is maximized.

An obvious algorithm for this problem is to sort the objects by decreasing ratio of profit to size, and then greedily pick objects in this order. It is easy to see that as such this algorithm can be made to perform arbitrarily badly (Exercise 8.1).

8.1 A pseudo-polynomial time algorithm for knapsack

Before presenting an FPTAS for knapsack, we need one more concept. For any optimization problem Π, an instance consists of *objects*, such as sets or graphs, and *numbers*, such as cost, profit, size, etc. So far, we have assumed that all numbers occurring in a problem instance I are written in binary. The *size* of the instance, denoted $|I|$, was defined as the number of bits needed to write I under this assumption. Let us say that I_u will denote instance I with all numbers occurring in it written in unary. The *unary size* of instance I, denoted $|I_u|$, is defined as the number of bits needed to write I_u.

An algorithm for problem Π is said to be efficient if its running time on instance I is bounded by a polynomial in $|I|$. Let us consider the following weaker definition. An algorithm for problem Π whose running time on instance I is bounded by a polynomial in $|I_u|$ will be called a *pseudo-polynomial time algorithm.*

The knapsack problem, being **NP**-hard, does not admit a polynomial time algorithm; however, it does admit a pseudo-polynomial time algorithm. This fact is used critically in obtaining an FPTAS for it. All known pseudo-polynomial time algorithms for **NP**-hard problems are based on dynamic programming.

Let P be the profit of the most profitable object, i.e., $P = \max_{a \in S} \text{profit}(a)$. Then nP is a trivial upperbound on the profit that can be achieved by any solution. For each $i \in \{1, \dots, n\}$ and $p \in \{1, \dots, nP\}$, let $S_{i,p}$ denote a subset of $\{a_1, \dots, a_i\}$ whose total profit is exactly p and whose total size is minimized. Let $A(i,p)$ denote the size of the set $S_{i,p}$ ($A(i,p) = \infty$ if no such set exists). Clearly $A(1,p)$ is known for every $p \in \{0, 1, \dots, nP\}$. The following recurrence helps compute all values $A(i,p)$ in $O(n^2P)$ time:

$$A(i+1,p) = \\ \begin{cases} \min\{A(i,p),\ \text{size}(a_{i+1}) + A(i, p - \text{profit}(a_{i+1}))\} & \text{if } \text{profit}(a_{i+1}) \le p \\ A(i,p) & \text{otherwise} \end{cases}$$

The maximum profit achievable by objects of total size bounded by B is $\max\{p \mid A(n,p) \le B\}$. We thus get a pseudo-polynomial algorithm for knapsack.

8.2 An FPTAS for knapsack

Notice that if the profits of objects were small numbers, i.e., they were bounded by a polynomial in n, then this would be a regular polynomial time algorithm, since its running time would be bounded by a polynomial in $|I|$. The key idea behind obtaining an FPTAS is to exploit precisely this fact: we will ignore a certain number of least significant bits of profits of objects (depending on the error parameter ε), so that the modified profits can be viewed as numbers bounded by a polynomial in n and $1/\varepsilon$. This will enable us to find a solution whose profit is at least $(1-\varepsilon) \cdot \mathrm{OPT}$ in time bounded by a polynomial in n and $1/\varepsilon$.

Algorithm 8.2 (FPTAS for knapsack)

1. Given $\varepsilon > 0$, let $K = \frac{\varepsilon P}{n}$.
2. For each object a_i, define $\text{profit}'(a_i) = \left\lfloor \frac{\text{profit}(a_i)}{K} \right\rfloor$.
3. With these as profits of objects, using the dynamic programming algorithm, find the most profitable set, say S'.
4. Output S'.

Lemma 8.3 *The set, S', output by the algorithm satisfies:*

$$\text{profit}(S') \geq (1-\varepsilon) \cdot \mathrm{OPT}.$$

Proof: Let O denote the optimal set. For any object a, because of rounding down, $K \cdot \text{profit}'(a)$ can be smaller than $\text{profit}(a)$, but by not more than K. Therefore,

$$\text{profit}(O) - K \cdot \text{profit}'(O) \leq nK.$$

The dynamic programming step must return a set at least as good as O under the new profits. Therefore,

$$\begin{aligned} \text{profit}(S') &\geq K \cdot \text{profit}'(O) \geq \text{profit}(O) - nK = \mathrm{OPT} - \varepsilon P \\ &\geq (1-\varepsilon) \cdot \mathrm{OPT}\ , \end{aligned}$$

where the last inequality follows from the observation that $\mathrm{OPT} \geq P$. □

Theorem 8.4 *Algorithm 8.2 is a fully polynomial approximation scheme for knapsack.*

Proof: By Lemma 8.3, the solution found is within $(1-\varepsilon)$ factor of OPT. Since the running time of the algorithm is $O\left(n^2 \left\lfloor \frac{P}{K} \right\rfloor\right) = O\left(n^2 \left\lfloor \frac{n}{\varepsilon} \right\rfloor\right)$, which is polynomial in n and $1/\varepsilon$, the theorem follows. □

8.3 Strong NP-hardness and the existence of FPTAS's

In this section, we will prove in a formal sense that very few of the known **NP**-hard problems admit an FPTAS. First, here is a strengthening of the notion of **NP**-hardness in a similar sense in which a pseudo-polynomial algorithm is a weakening of the notion of an efficient algorithm. A problem Π is *strongly* **NP**-*hard* if every problem in **NP** can be polynomially reduced to Π in such a way that numbers in the reduced instance are always written in unary.

The restriction automatically forces the transducer to use polynomially bounded numbers only. Most known **NP**-hard problems are in fact strongly **NP**-hard; this includes all the problems in previous chapters for which approximation algorithms were obtained. A strongly **NP**-hard problem cannot have a pseudo-polynomial time algorithm, assuming $\mathbf{P} \neq \mathbf{NP}$ (Exercise 8.4). Thus, knapsack is not strongly **NP**-hard, assuming $\mathbf{P} \neq \mathbf{NP}$.

We will show below that under some very weak restrictions, any **NP**-hard problem admitting an FPTAS must admit a pseudo-polynomial time algorithm. Theorem 8.5 is proven for a minimization problem; a similar proof holds for a maximization problem.

Theorem 8.5 *Let p be a polynomial and Π be an* **NP**-*hard minimization problem such that the objective function f_Π is integer valued and on any instance I, $\mathrm{OPT}(I) < p(|I_u|)$. If Π admits an FPTAS, then it also admits a pseudo-polynomial time algorithm.*

Proof: Suppose there is an FPTAS for Π whose running time on instance I and error parameter ε is $q(|I|, 1/\varepsilon)$, where q is a polynomial.

On instance I, set the error parameter to $\varepsilon = 1/p(|I_u|)$, and run the FPTAS. Now, the solution produced will have objective function value less than or equal to:

$$(1+\varepsilon)\mathrm{OPT}(I) < \mathrm{OPT}(I) + \varepsilon p(|I_u|) = \mathrm{OPT}(I) + 1.$$

In fact, with this error parameter, the FPTAS will be forced to produce an optimal solution. The running time will be $q(|I|, p(|I_u|))$, i.e., polynomial in $|I_u|$. Therefore, we have obtained a pseudo-polynomial time algorithm for Π. □

The following corollary applies to most known **NP**-hard problems.

Corollary 8.6 *Let Π be an* **NP**-*hard optimization problem satisfying the restrictions of Theorem 8.5. If Π is strongly* **NP**-*hard, then Π does not admit an FPTAS, assuming $\mathbf{P} \neq \mathbf{NP}$.*

Proof: If Π admits an FPTAS, then it admits a pseudo-polynomial time algorithm by Theorem 8.5. But then it is not strongly **NP**-hard, assuming $\mathbf{P} \neq \mathbf{NP}$, leading to a contradiction. □

The stronger assumption that $\mathrm{OPT}(I) < p(|I|)$ in Theorem 8.5 would have enabled us to prove that there is a polynomial time algorithm for Π. However, this stronger assumption is less widely applicable. For instance, it is not satisfied by the minimum makespan problem, which we will study in Chapter 10.

8.3.1 Is an FPTAS the most desirable approximation algorithm?

The design of almost all known FPTAS's and PTAS's is based on the idea of trading accuracy for running time – the given problem instance is mapped to a coarser instance, depending on the error parameter ε, which is solved optimally by a dynamic programming approach. The latter ends up being an exhaustive search of polynomially many different possibilities (for instance, for knapsack, this involves computing $A(i,p)$ for all i and p). In most such algorithms, the running time is prohibitive even for reasonable n and ε. Further, if the algorithm had to resort to exhaustive search, does the problem really offer "footholds" to home in on a solution efficiently? Is an FPTAS or PTAS the best one can hope for for an **NP**-hard problem? Clearly, the issue is complex and there is no straightforward answer.

8.4 Exercises

8.1 Consider the greedy algorithm for the knapsack problem. Sort the objects by decreasing ratio of profit to size, and then greedily pick objects in this order. Show that this algorithm can be made to perform arbitrarily badly.

8.2 Consider the following modification to the algorithm given in Exercise 8.1. Let the sorted order of objects be $a_1, \ldots, a_n$. Find the lowest k such that the size of the first k objects exceeds B. Now, pick the more profitable of $\{a_1, \ldots, a_{k-1}\}$ and $\{a_k\}$ (we have assumed that the size of each object is at most B). Show that this algorithm achieves an approximation factor of 2.

8.3 (Bazgan, Santha, and Tuza [23]) Obtain an FPTAS for the following problem.

Problem 8.7 (Subset-sum ratio problem) Given n positive integers, $a_1 < \ldots < a_n$, find two disjoint nonempty subsets $S_1, S_2 \subseteq \{1, \ldots, n\}$ with $\sum_{i \in S_1} a_i \geq \sum_{i \in S_2} a_i$, such that the ratio

$$\frac{\sum_{i \in S_1} a_i}{\sum_{i \in S_2} a_i}$$

is minimized.

Hint: First, obtain a pseudo-polynomial time algorithm for this problem. Then, scale and round appropriately.

8.4 Show that a strongly **NP**-hard problem cannot have a pseudo-polynomial time algorithm, assuming $\mathbf{P} \neq \mathbf{NP}$.

8.5 Notes

Algorithm 8.2 is due to Ibarra and Kim [141]. Theorem 8.5 is due to Garey and Johnson [98].

9 Bin Packing

Consider the following problem.

Problem 9.1 (Bin packing) Given n items with sizes $a_1, \ldots, a_n \in (0, 1]$, find a packing in unit-sized bins that minimizes the number of bins used.

This problem finds many industrial applications. For instance, in the stock-cutting problem, bins correspond to a standard length of paper and items correspond to specified lengths that need to be cut.

It is easy to obtain a factor 2 approximation algorithm for this problem. For instance, let us consider the algorithm called First-Fit. This algorithm considers items in an arbitrary order. In the ith step, it has a list of partially packed bins, say $B_1, \ldots, B_k$. It attempts to put the next item, a_i, in one of these bins, in this order. If a_i does not fit into any of these bins, it opens a new bin B_{k+1}, and puts a_i in it. If the algorithm uses m bins, then at least $m - 1$ bins are more than half full. Therefore,

$$\sum_{i=1}^{n} a_i > \frac{m-1}{2}.$$

Since the sum of the item sizes is a lower bound on OPT, $m - 1 < 2\text{OPT}$, i.e., $m \leq 2\text{OPT}$ (see Notes for a better analysis). On the negative side:

Theorem 9.2 *For any $\varepsilon > 0$, there is no approximation algorithm having a guarantee of $3/2 - \varepsilon$ for the bin packing problem, assuming* $\mathbf{P} \neq \mathbf{NP}$.

Proof: If there were such an algorithm, then we show how to solve the **NP**-hard problem of deciding if there is a way to partition n nonnegative numbers $a_1, \ldots, a_n$ into two sets, each adding up to $\frac{1}{2}\sum_i a_i$. Clearly, the answer to this question is 'yes' iff the n items can be packed in 2 bins of size $\frac{1}{2}\sum_i a_i$. If the answer is 'yes' the $3/2 - \varepsilon$ factor algorithm will have to give an optimal packing, and thereby solve the partitioning problem. □

9.1 An asymptotic PTAS

Notice that the argument in Theorem 9.2 uses very special instances: those for which OPT is a small number, such as 2 or 3, even though the number of items is unbounded. What can we say about "typical" instances, those for which OPT increases with n?

Theorem 9.3 *For any ε, $0 < \varepsilon \le 1/2$, there is an algorithm $\mathcal{A}_\varepsilon$ that runs in time polynomial in n and finds a packing using at most $(1 + 2\varepsilon)\mathrm{OPT} + 1$ bins.*

The sequence of algorithms, $\mathcal{A}_\varepsilon$, form an *asymptotic polynomial time approximation scheme* for bin packing, since for each $\varepsilon > 0$ $\exists N > 0$, and a polynomial time algorithm in this sequence, say $\mathcal{B}$, such that $\mathcal{B}$ has an approximation guarantee of $1 + \varepsilon$ for all instances having $\mathrm{OPT} \ge N$. However, Theorem 9.3 should not be considered a practical solution to the bin packing problem, since the running times of the algorithms $\mathcal{A}_\varepsilon$ are very high.

We will prove Theorem 9.3 in three steps.

Lemma 9.4 *Let $\varepsilon > 0$ be fixed, and let K be a fixed nonnegative integer. Consider the restriction of the bin packing problem to instances in which each item is of size at least ε and the number of distinct item sizes is K. There is a polynomial time algorithm that optimally solves this restricted problem.*

Proof: The number of items in a bin is bounded by $\lfloor 1/\varepsilon \rfloor$. Denote this by M. Therefore, the number of different bin types is bounded by $R = \binom{M+K}{M}$ (see Exercise 9.4), which is a (large!) constant. Clearly, the total number of bins used is at most n. Therefore, the number of possible feasible packings is bounded by $P = \binom{n+R}{R}$, which is polynomial in n (see Exercise 9.4). Enumerating them and picking the best packing gives the optimal answer. □

Lemma 9.5 *Let $\varepsilon > 0$ be fixed. Consider the restriction of the bin packing problem to instances in which each item is of size at least ε. There is a polynomial time approximation algorithm that solves this restricted problem within a factor of $(1 + \varepsilon)$.*

Proof: Let I denote the given instance. Sort the n items by increasing size, and partition them into $K = \lceil 1/\varepsilon^2 \rceil$ groups each having at most $Q = \lfloor n\varepsilon^2 \rfloor$ items. Notice that two groups may contain items of the same size.

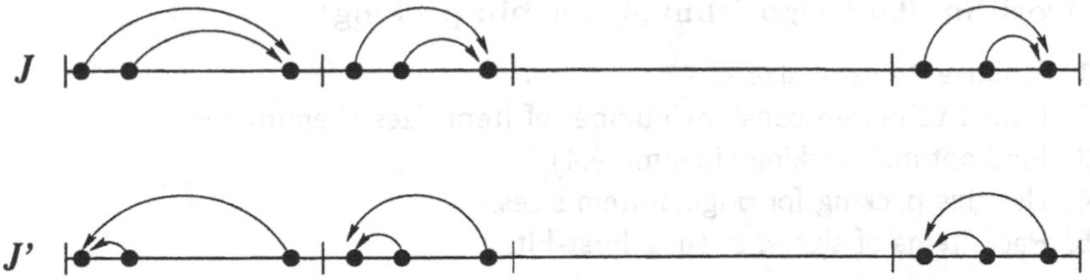

Construct instance J by rounding up the size of each item to the size of the largest item in its group. Instance J has at most K different item sizes.

Therefore, by Lemma 9.4, we can find an optimal packing for J. Clearly, this will also be a valid packing for the original item sizes. We show below that $\text{OPT}(J) \leq (1+\varepsilon)\text{OPT}(I)$, thereby proving the lemma.

The following clever argument accomplishes this. Let us construct another instance, say J', by rounding down the size of each item to that of the smallest item in its group. Clearly $\text{OPT}(J') \leq \text{OPT}(I)$. The crucial observation is that a packing for instance J' yields a packing for all but the largest Q items of instance J (Exercise 9.6 asks for a formal proof). Therefore,

$$\text{OPT}(J) \leq \text{OPT}(J') + Q \leq \text{OPT}(I) + Q.$$

Since each item in I has size at least ε, $\text{OPT}(I) \geq n\varepsilon$. Therefore, $Q = \lfloor n\varepsilon^2 \rfloor \leq \varepsilon\text{OPT}$. Hence, $\text{OPT}(J) \leq (1+\varepsilon)\text{OPT}(I)$. □

Proof of Theorem 9.3: Let I denote the given instance, and I' denote the instance obtained by discarding items of size $< \varepsilon$ from I. By Lemma 9.5, we can find a packing for I' using at most $(1+\varepsilon)\text{OPT}(I')$ bins. Next, we start packing the small items (of size $< \varepsilon$) in a First-Fit manner in the bins opened for packing I'. Additional bins are opened if an item does not fit into any of the already open bins.

If no additional bins are needed, then we have a packing in $(1+\varepsilon)\text{OPT}(I') \leq (1+\varepsilon)\text{OPT}(I)$ bins. In the second case, let M be the total number of bins used. Clearly, all but the last bin must be full to the extent of at least $1-\varepsilon$. Therefore, the sum of the item sizes in I is at least $(M-1)(1-\varepsilon)$. Since this is a lower bound on OPT, we get

$$M \leq \frac{\text{OPT}}{(1-\varepsilon)} + 1 \leq (1+2\varepsilon)\text{OPT} + 1,$$

where we have used the assumption that $\varepsilon \leq 1/2$. Hence, for each value of ε, $0 < \varepsilon \leq 1/2$, we have a polynomial time algorithm achieving a guarantee of $(1+2\varepsilon)\text{OPT} + 1$. □

Algorithm $\mathcal{A}_\varepsilon$ is summarized below.

Algorithm 9.6 (Algorithm $\mathcal{A}_\varepsilon$ for bin packing)

1. Remove items of size $< \varepsilon$.
2. Round to obtain constant number of item sizes (Lemma 9.5).
3. Find optimal packing (Lemma 9.4).
4. Use this packing for original item sizes.
5. Pack items of size $< \varepsilon$ using First-Fit.

9.2 Exercises

9.1 Give an example on which First-Fit does at least as bad as $5/3 \cdot \text{OPT}$.

9.2 (Johnson [156]) Consider a more restricted algorithm than First-Fit, called Next-Fit, which tries to pack the next item only in the most recently started bin. If it does not fit, it is packed in a new bin. Show that this algorithm also achieves factor 2. Give a factor 2 tight example.

9.3 (C. Kenyon) Say that a bin packing algorithm is *monotonic* if the number of bins it uses for packing a subset of the items is at most the number of bins it uses for packing all n items. Show that whereas Next-Fit is monotonic, First-Fit is not.

9.4 Prove the bounds on R and P stated in Lemma 9.4.
Hint: Use the fact that the number of ways of throwing n identical balls into k distinct bins is $\binom{n+k-1}{n}$.

9.5 Consider an alternative way of establishing Lemma 9.5. All items having sizes in the interval $(\varepsilon(1+\varepsilon)^r, \varepsilon(1+\varepsilon)^{r+1}]$ are rounded up to $\min(\varepsilon(1+\varepsilon)^{r+1}, 1)$, for $r \geq 0$. Clearly, this yields a constant number of item sizes. Does the rest of the proof go through?
Hint: Consider the situation that there are lots of items of size $1/2$, and $1/2 \neq \varepsilon(1+\varepsilon)^r$ for any $r \geq 0$.

9.6 Prove the following statement made in Lemma 9.5, "A packing for instance J' yields a packing for all but the largest Q items of instance J."
Hint: Throw away the Q largest items of J and the Q smallest items of J', and establish a domination.

9.7 Use the fact that integer programming with a fixed number of variables is in **P** to give an alternative proof of Lemma 9.4. (Because of the exorbitant running time of the integer programming algorithm, this variant is also impractical.)

9.8 Show that if there is an algorithm for bin packing having a guarantee of $\text{OPT}(I) + \log^2(\text{OPT}(I))$, then there is a fully polynomial approximation scheme for this problem.

9.9 (C. Kenyon) Consider the following problem.

Problem 9.7 (Bin covering) Given n items with sizes $a_1, \ldots, a_n \in (0, 1]$, maximize the number of bins opened so that each bin has items summing to at least 1.

Give an asymptotic PTAS for this problem when restricted to instances in which item sizes are bounded below by c, for a fixed constant $c > 0$.
Hint: The main idea of Algorithm 9.6 applies to this problem as well.

9.3 Notes

The first nontrivial bin packing result, showing that First-Fit requires at most (17/10)OPT + 3 bins, was due to Ullman [256]. The asymptotic PTAS is due to Fernandez de la Vega and Lueker [92]. An improved algorithm, having a guarantee of $\mathrm{OPT}(I) + \log^2(\mathrm{OPT}(I))$ was given by Karmarkar and Karp [170]. For further results, see the survey of Coffman, Garey, and Johnson [52]. The result cited in Exercise 9.7, showing that integer programming with a fixed number of variables is in **P**, is due to Lenstra [192]. Bin packing has also been extensively studied in the on-line model. For these and other on-line algorithms see Borodin and El-Yaniv [32].

10 Minimum Makespan Scheduling

A central problem in scheduling theory is the following.

Problem 10.1 (Minimum makespan scheduling) Given processing times for n jobs, $p_1, p_2, \ldots, p_n$, and an integer m, find an assignment of the jobs to m identical machines so that the completion time, also called the *makespan*, is minimized.

We will give a simple factor 2 algorithm for this problem before presenting a PTAS for it.

10.1 Factor 2 algorithm

The algorithm is very simple: schedule the jobs one by one, in an arbitrary order, each job being assigned to a machine with least amount of work so far. This algorithm is based on the following two lower bounds on the optimal makespan, OPT:

1. The average time for which a machine has to run, $\left(\sum_i p_i\right)/m$; and
2. The largest processing time.

Let LB denote the combined lower bound, i.e.,

$$\text{LB} = \max\left\{\frac{1}{m}\sum_i p_i,\ \max_i\{p_i\}\right\}.$$

Algorithm 10.2 (Minimum makespan scheduling)

1. Order the jobs arbitrarily.
2. Schedule jobs on machines in this order, scheduling the next job on the machine that has been assigned the least amount of work so far.

Theorem 10.3 *Algorithm 10.2 achieves an approximation guarantee of 2 for the minimum makespan problem.*

Proof: Let M_i be the machine that completes its jobs last in the schedule produced by the algorithm, and let j be the index of the last job scheduled on this machine.

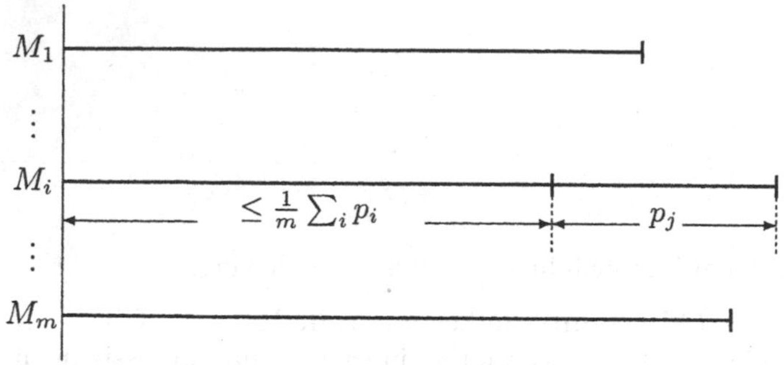

Let $start_j$ be the time at which job j starts execution on M_i. Since the algorithm assigns a job to the least loaded machine, it follows that all machines are busy until $start_j$. This implies that

$$start_j \leq \frac{1}{m}\sum_i p_i \leq \mathrm{OPT}.$$

Further, $p_j \leq \mathrm{OPT}$. Thus, the makespan of the schedule is $start_j + p_j \leq 2 \cdot \mathrm{OPT}$. □

Example 10.4 A tight example for this algorithm is provided by a sequence of m^2 jobs with unit processing time, followed by a single job of length m. The schedule obtained by the algorithm has a makespan of $2m$, while OPT $= m + 1$. □

10.2 A PTAS for minimum makespan

The minimum makespan problem is strongly **NP**-hard; thus, by Corollary 8.6, it does not admit an FPTAS, assuming $\mathbf{P} \neq \mathbf{NP}$. We will obtain a PTAS for it. The minimum makespan problem is closely related to the bin packing problem by the following observation. There exists a schedule with makespan t iff n objects of sizes $p_1, p_2, \ldots, p_n$ can be packed into m bins of capacity t each. This suggests a reduction from minimum makespan to bin packing as follows. Denoting the sizes of the n objects, $p_1, \ldots, p_n$, by I, let $\mathrm{bins}(I, t)$ represent the minimum number of bins of size t required to pack these n objects. Then, the minimum makespan is given by

$$\min\{t : \ \mathrm{bins}(I, t) \leq m\}.$$

As shown above, LB and $2 \cdot \text{LB}$ are lower and upper bounds on the minimum makespan. Thus, we can determine the minimum makespan by a binary search in this interval. At first sight, this reduction may not seem very useful since the bin packing problem is also **NP**-hard. However, it turns out that this problem is polynomial time solvable if the object sizes are drawn from a set of fixed cardinality. We will use this fact critically for solving the minimum makespan problem.

10.2.1 Bin packing with fixed number of object sizes

We first present a dynamic programming algorithm for the restricted bin packing problem, thereby improving on the result of Lemma 9.4 in two ways. We will not require a lower bound on item sizes and will improve on the running time. Let k be the fixed number of object sizes, and assume that bins have capacity 1. Fix an ordering on the object sizes. Now, an instance of the bin packing problem can be described by a k-tuple, $(i_1, i_2, \ldots, i_k)$, specifying the number of objects of each size. Let $\text{BINS}(i_1, i_2, \ldots, i_k)$ denote the minimum number of bins needed to pack these objects.

For a given instance, $(n_1, n_2, \ldots, n_k)$, $\sum_{i=1}^{k} n_i = n$, we first compute $\mathcal{Q}$, the set of all k-tuples $(q_1, q_2, \ldots, q_k)$ such that $\text{BINS}(q_1, q_2, \ldots, q_k) = 1$ and $0 \leq q_i \leq n_i, 1 \leq i \leq k$. Clearly, $\mathcal{Q}$ contains at most $O(n^k)$ elements. Next, we compute all entries of the k-dimensional table $\text{BINS}(i_1, i_2, \ldots, i_k)$ for every $(i_1, i_2, \ldots, i_k) \in \{0, \ldots, n_1\} \times \{0, \ldots, n_2\} \times \ldots \times \{0, \ldots, n_k\}$. The table is initialized by setting $\text{BINS}(q) = 1$ for every $q \in \mathcal{Q}$. Then, we use the following recurrence to compute the remaining entries:

$$\text{BINS}(i_1, i_2, \ldots, i_k) = 1 + \min_{q \in \mathcal{Q}} \text{BINS}(i_1 - q_1, \ldots, i_k - q_k). \tag{10.1}$$

Computing each entry takes $O(n^k)$ time. Thus, the entire table can be computed in $O(n^{2k})$ time, thereby determining $\text{BINS}(n_1, n_2, \ldots, n_k)$.

10.2.2 Reducing makespan to restricted bin packing

The basic idea is that if we can tolerate some error in computing the minimum makespan, then we can reduce this problem to the restricted version of bin packing in polynomial time. There will be two sources of error:

- rounding object sizes so that there are a bounded number of different sizes, and
- terminating the binary search to ensure polynomial running time.

Each error can be made as small as needed, at the expense of running time. Moreover, for any fixed error bound, the running time is polynomial in n, and thus we obtain a polynomial time approximation scheme.

Let ε be an error parameter and t be in the interval $[\text{LB}, 2 \cdot \text{LB}]$. We say that an object is *small* if its size is less than $t\varepsilon$; small objects are discarded for now. The rest of the objects are rounded down as follows: each p_j in the interval $\left[t\varepsilon(1+\varepsilon)^i,\ t\varepsilon(1+\varepsilon)^{i+1}\right)$ is replaced by $p'_j = t\varepsilon(1+\varepsilon)^i$, for $i \geq 0$. The resulting p'_j's can assume at most $k = \lceil \log_{1+\varepsilon} \frac{1}{\varepsilon} \rceil$ distinct values. Determine an optimal packing for the rounded objects in bins of size t using the dynamic programming algorithm. Since rounding reduces the size of each object by a factor of at most $1 + \varepsilon$, if we consider the original sizes of the objects, then the packing determined is valid for a bin size of $t(1+\varepsilon)$. Keeping this as the bin size, pack the small objects greedily in leftover spaces in the bins; open new bins only if needed. Clearly, any time a new bin is opened, all previous bins must be full to the extent of at least t. Denote with $\alpha(I, t, \varepsilon)$ the number of bins used by this algorithm; recall that these bins are of size $t(1+\varepsilon)$.

Let us call the algorithm presented above the *core algorithm* since it will form the core of the PTAS for computing makespan. As shown in Lemma 10.5 and its corollary, the core algorithm also helps establish a lower bound on the optimal makespan.

Lemma 10.5 $\quad \alpha(I, t, \varepsilon) \leq \text{bins}(I, t)$.

Proof: If the algorithm does not open any new bins for the small objects, then the assertion clearly holds since the rounded down pieces have been packed optimally in bins of size t. In the other case, all but the last bin are packed at least to the extent of t. Hence, the optimal packing of I in bins of size t must also use at least $\alpha(I, t, \varepsilon)$ bins. □

Since $\text{OPT} = \min\{t : \text{bins}(I, t) \leq m\}$, Lemma 10.5 gives:

Corollary 10.6 $\quad \min\{t : \alpha(I, t, \varepsilon) \leq m\} \leq \text{OPT}$.

If $\min\{t : \alpha(I, t, \varepsilon) \leq m\}$ could be determined with no additional error during the binary search, then clearly we could use the core algorithm to obtain a schedule with a makespan of $(1+\varepsilon)\text{OPT}$. Next, we will specify the details of the binary search and show how to control the error it introduces. The binary search is performed on the interval $[\text{LB}, 2 \cdot \text{LB}]$. Thus, the length of the available interval is LB at the start of the search, and it reduces by a factor of 2 in each iteration. We continue the search until the available interval drops to a length of $\varepsilon \cdot \text{LB}$. This will require $\lceil \log_2 \frac{1}{\varepsilon} \rceil$ iterations. Let T be the right endpoint of the interval we terminate with.

Lemma 10.7 $\quad T \leq (1+\varepsilon) \cdot \text{OPT}$.

Proof: Clearly, $\min\{t : \alpha(I, t, \varepsilon) \leq m\}$ must be in the interval $[T - \varepsilon \cdot LB, T]$. Hence,

$$T \leq \min\{t : \alpha(I, t, \varepsilon) \leq m\} + \varepsilon \cdot \text{LB}.$$

Now, using Corollary 10.6 and the fact that LB $\leq$ OPT, the lemma follows. □

Finally, the output of the core algorithm for $t = T$ gives a schedule whose makespan is at most $T \cdot (1 + \varepsilon)$. We get:

Theorem 10.8 *The algorithm produces a valid schedule having makespan at most*

$$(1 + \varepsilon)^2 \cdot \text{OPT} \leq (1 + 3\varepsilon) \cdot \text{OPT}.$$

The running time of the entire algorithm is $O\left(n^{2k} \lceil \log_2 \frac{1}{\varepsilon} \rceil\right)$, where $k = \lceil \log_{1+\varepsilon} \frac{1}{\varepsilon} \rceil$.

10.3 Exercises

10.1 (Graham [120]) The tight example for the factor 2 algorithm, Example 10.4, involves scheduling a very long job last. This suggests sorting the jobs by decreasing processing times before scheduling them. Show that this leads to a 4/3 factor algorithm. Provide a tight example for this algorithm.

10.2 (Horowitz and Sahni [138]) Give an FPTAS for the variant of the minimum makespan scheduling problem in which the number of machines, m, is a fixed constant.

10.4 Notes

Algorithm 10.2 is due to Graham [119]. The PTAS is due to Hochbaum and Shmoys [135].

11 Euclidean TSP

In this chapter, we will give a PTAS for the special case of the traveling salesman problem in which the points are given in a d-dimensional Euclidean space. As before, the central idea of the PTAS is to define a "coarse solution", depending on the error parameter ε, and to find it using dynamic programming. A feature this time is that we do not know a deterministic way of specifying the coarse solution – it is specified probabilistically.

Problem 11.1 (Euclidean TSP) For fixed d, given n points in $\mathbf{R}^d$, the problem is to find the minimum length tour of the n points. The distance between any two points x and y is defined to be the Euclidean distance between them, i.e., $\left(\sum_{i=1}^{d}(x_i - y_i)^2\right)^{1/2}$.

11.1 The algorithm

We will give the algorithm for points on the plane, i.e., $d = 2$. The extension to arbitrary d is straightforward. The algorithm involves numerous details. In the interest of highlighting the main ideas, some of these details will be left as exercises.

Define the *bounding box* of the instance to be the smallest axis-parallel square that contains all n points. Via a simple perturbation of the instance, we may assume that the length of this square, L, is $4n^2$ and that there is a unit grid defined on the square such that each point lies on a gridpoint (see Exercise 11.1). Further, assume w.l.o.g. that n is a power of 2, and let $L = 2^k, k = 2 + 2\log_2 n$.

The *basic dissection* of the bounding box is a recursive partitioning into smaller squares. Thus, the $L \times L$ square is divided into four $L/2 \times L/2$ squares, and so on. It will be convenient to view this dissection as a 4-ary tree, T, whose root is the bounding box. The four children of the root are the four $L/2 \times L/2$ squares, and so on. The nodes of T are assigned *levels*. The root is at level 0, its children at level 1, and so on. The squares represented by nodes get levels accordingly. Thus, squares at level i have dimensions $L/2^i \times L/2^i$. The dissection is continued until we obtain unit squares. Clearly, T has depth $k = O(\log n)$. By a *useful square* we mean a square represented by a node in T.

Next, let us define *levels for the horizontal and vertical lines* that accomplish the basic dissection (these are all the lines of the grid defined on the bounding box). The two lines that divide the bounding box into four squares have level 1. In general, the 2^i lines that divide the level $i-1$ squares into level i squares each have level i. Therefore, a line of level i forms the edge of useful squares at levels $i, i+1, \ldots$, i.e., the largest useful square on it has dimensions $L/2^i \times L/2^i$:

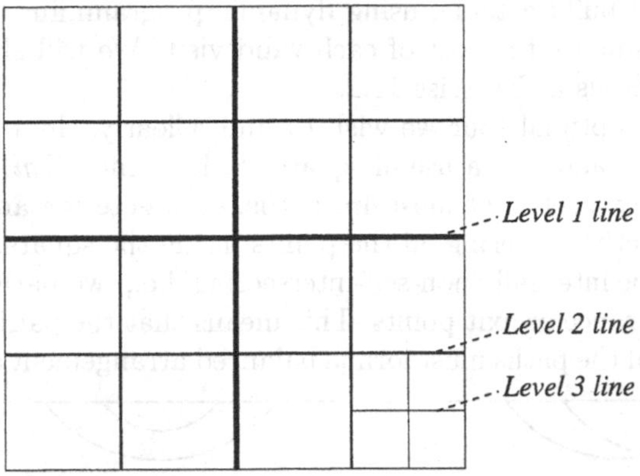

Each line will have a special set of points called *portals*. The coarse solution we will be seeking is allowed to cross a line only at a portal. The portals on each line are equidistant points. On a line of level i, these points are $L/(2^i m)$ apart, where the parameter m is fixed to be a power of 2 in the range $[k/\varepsilon, 2k/\varepsilon]$. Clearly, $m = O(\log n/\varepsilon)$. Since the largest useful square on a level i line has dimensions $L/2^i \times L/2^i$, each useful square has a total of at most $4m$ portals on its four sides and corners. We have chosen m to be a power of 2 so that a portal in a lower level square is a portal for all higher level squares it lies in.

We will say that a tour τ is *well behaved w.r.t. the basic dissection* if it is a tour on the n points and any subset of the portals. In addition, this tour is allowed to visit portals multiple times, but other than that it must be *non-self-intersecting*. The key structural fact to be established is that there is such a tour of length at most $(1+\varepsilon) \cdot \mathrm{OPT}$. This requires a probabilistic argument, and we will return to it. First let us show why a PTAS follows from this fact.

We will say that tour τ is *well behaved w.r.t. the basic dissection and has limited crossings* if it is well behaved w.r.t. the basic dissection, and furthermore, it visits each portal at most twice.

Lemma 11.2 *Let tour τ be well behaved w.r.t. the basic dissection. Then there must be a tour that is well behaved with limited crossings, whose length is at most that of τ.*

Proof: The basic reason is that removing self-intersections by "short-cutting" can only result in a shorter tour, since Euclidean distance satisfies the triangle inequality. If τ uses a portal on line l more than twice, we can keep "short-cutting" on the two sides of l until the portal is used at most twice. If this introduces additional self-intersections, they can also be removed. □

Lemma 11.3 *The optimal well behaved tour w.r.t. the basic dissection, having limited crossings, can be computed in time* $2^{O(m)} = n^{O(1/\varepsilon)}$.

Proof: We will build a table, using dynamic programming, that contains, for each useful square, the cost of each valid visit. We will sketch the main ideas, leaving details as Exercise 11.2.

Let τ be the optimal tour we wish to find. Clearly, the total number of times τ can enter and exit a useful square, S, is at most $8m$. The part of τ inside S is simply a set of at most $4m$ paths, each entering and exiting S at portals, and together covering all the points inside the square. Furthermore, the paths must be internally non-self-intersecting, i.e., two paths can intersect only at their entrance or exit points. This means that the pairing of entrance and exit points of the paths must form a balanced arrangement of parentheses.

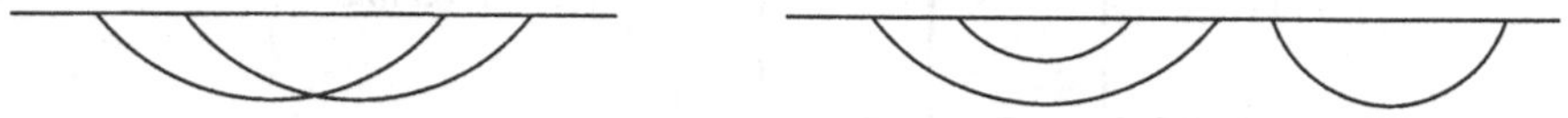

Invalid pairing *Valid pairing*

Let us call such a listing of portals, together with their pairing as entrance and exit points, a *valid visit.*

The number of useful squares is clearly poly(n). Let us first show that the number of valid visits in a useful square is at most $n^{O(1/\varepsilon)}$, thereby showing that the number of entries in the table is bounded by $n^{O(1/\varepsilon)}$.

Consider a useful square S. Each of its portals is used 0, 1, or 2 times, a total of $3^{4m} = n^{O(1/\varepsilon)}$ possibilities. Of these, retain only those possibilities that involve an even number of portal usages. Consider one such possibility, and suppose that it uses $2r$ portals. Next, we need to consider all possible pairings of these portals that form a balanced arrangement of parentheses. The number of such arrangements is the rth Catalan number, and is bounded by $2^{2r} = n^{O(1/\varepsilon)}$. Hence, the total number of valid visits in S is bounded by $n^{O(1/\varepsilon)}$.

For each entry in the table, we need to compute the optimal length of this valid visit. The table is built up the decomposition tree, starting at its leaves. Consider a valid visit V in a square S. Let S be a level i square. We have already fixed the entrances and exits on the boundary of S. Square S has four children at level $i+1$, which have four sides internal to S, with a total of at most $4m$ more portals. Each of these portals is used 0, 1, or 2 times, giving rise again to $n^{O(1/\varepsilon)}$ possibilities. Consider one such possibility, and consider all its portal usages together with portal usages of a valid visit V. Obtain all possible valid pairings of these portals that are consistent with those of visit V. Again, using Catalan numbers, their number is bounded by $n^{O(1/\varepsilon)}$. Each such pairing will give rise to valid visits in the four squares.

The cost of the optimal way of executing these valid visits in the four squares has already been computed. Compute their sum. The smallest of these sums is the optimal way of executing visit V in square S. □

11.2 Proof of correctness

For the proof of correctness, it suffices to show that there is a well behaved tour w.r.t. the basic dissection whose length is bounded by $(1+\varepsilon)\mathrm{OPT}$. It turns out that this is not always the case (see Exercise 11.3). Instead, we will construct a larger family of dissections and will show that, for any placement of the n points, at least half these dissections have short well behaved tours with limited crossings. So, picking a random dissection from this set suffices.

Let us define L^2 different dissections of the bounding box, which are shifts of the basic dissection. Given integers a, b with $0 \leq a, b < L$, the (a, b)-*shifted* dissection is obtained by moving each vertical line from its original location x to $(a+x) \bmod L$, and moving each horizontal line from its original location y to $(b+y) \bmod L$. Thus, the middle lines of the shifted dissection are located at $(a + L/2) \bmod L$ and $(b + L/2) \bmod L$, respectively.

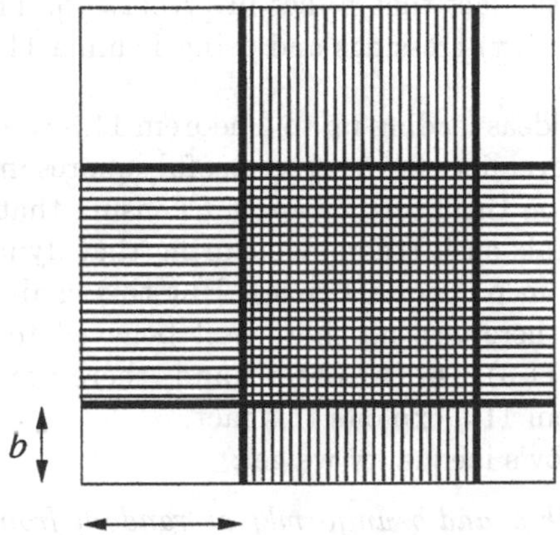

The entire bounding box is thought of as being "wrapped around". Useful squares that extend beyond L in their x or y coordinates will thus be thought of as "wrapped around", and will still be thought of as a single square. Of course, the positions of the given n points remains unchanged; only the dissection is shifted.

Let π be the optimal tour, and $N(\pi)$ be the total number of times π crosses horizontal and vertical grid lines. If π uses a point at the intersection of two grid lines, then we will count it as two crossings. The following fact is left as Exercise 11.4.

Lemma 11.4 $N(\pi) \leq 2 \cdot \mathrm{OPT}$.

Following is the central fact leading to the PTAS.

Theorem 11.5 *Pick a and b uniformly at random from $[0, L)$. Then, the expected increase in cost in making π well behaved w.r.t. the (a, b)-shifted dissection is bounded by $2\varepsilon \cdot \mathrm{OPT}$.*

Proof: Given any dissection, consider the process of making π well behaved w.r.t. it. This involves replacing a segment of π that does not cross a line l at a portal by two segments so that the crossing is at the closest portal on l. The corresponding increase in the length of the tour is bounded by the interportal distance on line l.

Consider the expected increase in length due to one of the crossings of tour π with a line. Let l be this line. l will be a level i line in the randomly picked dissection with probability $2^i/L$. If l is a level i line, then the interportal distance on it is $L/(2^i m)$. Thus, the expected increase in the length of the tour due to this crossing is at most

$$\sum_i \frac{L}{2^i m} \frac{2^i}{L} = \frac{k}{m} \leq \varepsilon,$$

where we have used the fact that m lies in $[k/\varepsilon, 2k/\varepsilon]$. The theorem follows by summing over all $N(\pi)$ crossings and using Lemma 11.4. □

Remark 11.6 The ideas leading up to Theorem 11.5 can be summarized as follows. Since lower level lines have bigger useful squares incident at them, we had to place portals on them further apart to ensure that any useful square had at most $4m$ portals on it (thereby ensuring that dynamic programming could be carried out in polynomial time). But this enabled us to construct instances for which there was no short well behaved tour w.r.t. the basic dissection (Exercise 11.3). On the other hand, there are fewer lines having lower levels – Theorem 11.5 exploits this fact.

Now, using Markov's inequality we get:

Corollary 11.7 *Pick a and b uniformly at random from $[0, L)$. Then, the probability that there is a well behaved tour of length at most $4\varepsilon \cdot \mathrm{OPT}$ w.r.t. the (a, b)-shifted dissection is greater or equal to $1/2$.*

Notice that Lemma 11.2 holds in the setting of an (a, b)-shifted dissection as well. The PTAS is now straightforward. Simply pick a random dissection, and find an optimal well behaved tour with limited crossings w.r.t. this dissection using the dynamic programming procedure of Lemma 11.3. Notice that the same procedure holds even for a shifted dissection. The algorithm can be derandomized by trying all possible shifts and outputting the shortest tour obtained. Thus, we get:

Theorem 11.8 *There is a PTAS for the Euclidean TSP problem in $\mathbf{R}^2$.*

11.3 Exercises

11.1 Show that we may assume that the length of the bounding square can be taken to be $L = 4n^2$ and that there is a unit grid defined on the square such that each point lies on a gridpoint.
Hint: Since we started with the smallest axis-parallel bounding square, its length is a lower bound on OPT. Therefore, moving each point to a grid point can increase the length of the tour by at most OPT/n^2.

11.2 Provide the missing details in the proof of Lemma 11.3.

11.3 Give an instance of the Euclidean TSP problem for which, w.r.t. the basic dissection, the process of making the optimal tour well behaved increases its length by a fixed constant factor.
Hint: Make the optimal tour cross the middle line of the dissection that has the largest interportal distance numerous times.

11.4 Prove Lemma 11.4.
Hint: Notice that the left-hand side simply measures the ℓ_1 length of tour π. The bound of $2\sqrt{2} \cdot \text{OPT}$ is easier to prove, since this applies to single edges as well. This bound suffices for the PTAS.

11.5 Extend the arguments given to obtain a PTAS for the Euclidean TSP problem in $\mathbf{R}^d$.

11.6 Generalize the algorithm to norms other than the Euclidean norm.

11.7 (Arora [11]) Obtain a PTAS for the Euclidean Steiner tree problem. Given n points in $\mathbf{R}^d$, find the minimum length tree containing all n points and any other subset of points. The latter points are called Steiner. The distance between two points is assumed to be their Euclidean distance.

11.8 Consider the Euclidean Steiner tree problem in $\mathbf{R}^2$. Show that in any optimal Steiner tree each Steiner point has degree 3 and the three angles so formed are of 120° each. (See Gauss' figures on cover for an illustration of this fact.)

11.4 Notes

The first PTAS for Euclidean TSP was given by Arora [10], following a PTAS for the planar graph TSP problem due to Grigni, Koutsoupias, and Papadimitriou [121]. Subsequently, Mitchell [215] independently obtained the same result. Later, Arora [11] went on to give an $n(\log n)^{O(1/\varepsilon)}$ algorithm for the problem for any fixed d. For a PTAS with an improved running time see Rao and Smith [237]. This chapter is based on Arora [11] and Arora, Raghavan, and Rao [14].

Part II

LP-Based Algorithms

12 Introduction to LP-Duality

A large fraction of the theory of approximation algorithms, as we know it today, is built around linear programming (LP). In Section 12.1 we will review some key concepts from this theory. In Section 12.2 we will show how the LP-duality theorem gives rise to min-max relations which have far-reaching algorithmic significance. Finally, in Section 12.3 we introduce the two fundamental algorithm design techniques of rounding and the primal–dual schema, as well as the method of dual fitting, which yield all the algorithms of Part II of this book.

12.1 The LP-duality theorem

Linear programming is the problem of optimizing (i.e., minimizing or maximizing) a linear function subject to linear inequality constraints. The function being optimized is called the *objective function*. Perhaps the most interesting fact about this problem from our perspective is that it is well-characterized (see definition in Section 1.2). Let us illustrate this through a simple example.

$$
\begin{array}{ll}
\text{minimize} & 7x_1 + x_2 + 5x_3 \\[4pt]
\text{subject to} & \begin{array}{rcrcrcl} x_1 & - & x_2 & + & 3x_3 & \geq & 10 \\ 5x_1 & + & 2x_2 & - & x_3 & \geq & 6 \end{array} \\[4pt]
 & x_1, x_2, x_3 \geq 0
\end{array}
$$

Notice that in this example all constraints are of the kind "$\geq$" and all variables are constrained to be nonnegative. This is the standard form of a minimization linear program; a simple transformation enables one to write any minimization linear program in this manner. The reason for choosing this form will become clear shortly.

Any solution, i.e., a setting for the variables in this linear program, that satisfies all the constraints is said to be a *feasible solution*. Let z^* denote the optimum value of this linear program. Let us consider the question, "Is z^* at most α?" where α is a given rational number. For instance, let us ask whether $z^* \leq 30$. A Yes certificate for this question is simply a feasible solution whose objective function value is at most 30. For example, $\boldsymbol{x} = (2, 1, 3)$ constitutes such a certificate since it satisfies the two constraints of the problem, and the objective function value for this solution is $7 \cdot 2 + 1 + 5 \cdot 3 = 30$. Thus, any Yes certificate to this question provides an upper bound on z^*.

How do we provide a No certificate for such a question? In other words, how do we place a good lower bound on z^*? In our example, one such bound is given by the first constraint: since the x_i's are restricted to be nonnegative, term-by-term comparison of coefficients shows that $7x_1 + x_2 + 5x_3 \geq x_1 - x_2 + 3x_3$. Since the right-hand side of the first constraint is 10, the objective function is at least 10 for any feasible solution. A better lower bound can be obtained by taking the sum of the two constraints: for any feasible solution $\boldsymbol{x}$,

$$7x_1 + x_2 + 5x_3 \geq (x_1 - x_2 + 3x_3) + (5x_1 + 2x_2 - x_3) \geq 16.$$

The idea behind this process of placing a lower bound is that we are finding suitable nonnegative multipliers for the constraints so that when we take their sum, the coefficient of each x_i in the sum is dominated by the coefficient in the objective function. Now, the right-hand side of this sum is a lower bound on z^* since any feasible solution has a nonnegative setting for each x_i. Notice the importance of ensuring that the multipliers are nonnegative: they do not reverse the direction of the constraint inequality.

Clearly, the rest of the game lies in choosing the multipliers in such a way that the right-hand side of the sum is as large as possible. Interestingly enough, the problem of finding the best such lower bound can be formulated as a linear program:

$$\begin{array}{ll} \text{maximize} & 10y_1 + 6y_2 \\ \text{subject to} & \begin{array}{rcrcl} y_1 & + & 5y_2 & \leq & 7 \\ -y_1 & + & 2y_2 & \leq & 1 \\ 3y_1 & - & y_2 & \leq & 5 \end{array} \\ & y_1, y_2 \geq 0 \end{array}$$

Here y_1 and y_2 were chosen to be the nonnegative multipliers for the first and the second constraint, respectively. Let us call the first linear program the *primal program* and the second the *dual program*. There is a systematic way of obtaining the dual of any linear program; one is a minimization problem and the other is a maximization problem. Further, the dual of the dual is the primal program itself (Exercise 12.1). By construction, every feasible solution to the dual program gives a lower bound on the optimum value of the primal. Observe that the reverse also holds. Every feasible solution to the primal program gives an upper bound on the optimal value of the dual. Therefore, if we can find feasible solutions for the dual and the primal with matching objective function values, then both solutions must be optimal. In our example, $\boldsymbol{x} = (7/4, 0, 11/4)$ and $\boldsymbol{y} = (2, 1)$ both achieve objective function values of 26, and thus both are optimal solutions (see figure below). The reader may wonder whether our example was ingeniously constructed to make this happen. Surprisingly enough, this is not an exception, but the rule! This is the central theorem of linear programming: the *LP-duality theorem.*

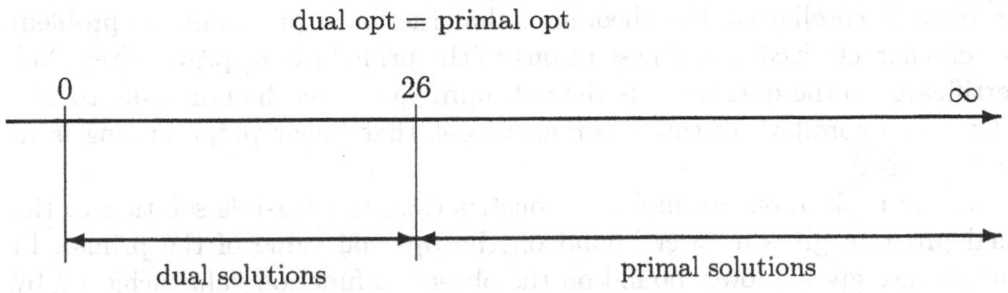

In order to state this theorem formally, let us consider the following minimization problem, written in standard form, as the primal program; equivalently, we could have started with a maximization problem as the primal program.

$$\begin{aligned} \text{minimize} \quad & \sum_{j=1}^{n} c_j x_j && && (12.1) \\ \text{subject to} \quad & \sum_{j=1}^{n} a_{ij} x_j \geq b_i, && i = 1, \ldots, m \\ & x_j \geq 0, && j = 1, \ldots, n \end{aligned}$$

where a_{ij}, b_i, and c_j are given rational numbers.

Introducing variables y_i for the ith inequality, we get the dual program:

$$\begin{aligned} \text{maximize} \quad & \sum_{i=1}^{m} b_i y_i && && (12.2) \\ \text{subject to} \quad & \sum_{i=1}^{m} a_{ij} y_i \leq c_j, && j = 1, \ldots, n \\ & y_i \geq 0, && i = 1, \ldots, m \end{aligned}$$

Theorem 12.1 (LP-duality theorem) *The primal program has finite optimum iff its dual has finite optimum. Moreover, if* $\boldsymbol{x}^* = (x_1^*, \ldots, x_n^*)$ *and*

$\boldsymbol{y}^* = (y_1^*, \ldots, y_m^*)$ *are optimal solutions for the primal and dual programs, respectively, then*

$$\sum_{j=1}^{n} c_j x_j^* = \sum_{i=1}^{m} b_i y_i^*.$$

Notice that the LP-duality theorem is really a min–max relation, since one program is a minimization problem and the other is a maximization problem. A corollary of this theorem is that the linear programming problem is well-characterized. Feasible solutions to the primal (dual) provide Yes (No) certificates to the question, "Is the optimum value less than or equal to α?" Thus, as a corollary of this theorem we get that linear programming is in **NP** ∩ co-**NP**.

Going back to our example, by construction, any feasible solution to the dual program gives a lower bound on the optimal value of the primal. In fact, it also gives a lower bound on the objective function value achieved by any feasible solution to the primal. This is the easy half of the LP-duality theorem, sometimes called the *weak duality theorem.* We give a formal proof of this theorem, since some steps in the proof will lead to the next important fact. The design of several exact algorithms have their basis in the LP-duality theorem. In contrast, in approximation algorithms, typically the weak duality theorem suffices.

Theorem 12.2 (Weak duality theorem) *If* $\boldsymbol{x} = (x_1, \ldots, x_n)$ *and* $\boldsymbol{y} = (y_1, \ldots, y_m)$ *are feasible solutions for the primal and dual program, respectively, then*

$$\sum_{j=1}^{n} c_j x_j \geq \sum_{i=1}^{m} b_i y_i. \tag{12.3}$$

Proof: Since $\boldsymbol{y}$ is dual feasible and x_j's are nonnegative,

$$\sum_{j=1}^{n} c_j x_j \geq \sum_{j=1}^{n} \left(\sum_{i=1}^{m} a_{ij} y_i \right) x_j. \tag{12.4}$$

Similarly, since $\boldsymbol{x}$ is primal feasible and y_i's are nonnegative,

$$\sum_{i=1}^{m} \left(\sum_{j=1}^{n} a_{ij} x_j \right) y_i \geq \sum_{i=1}^{m} b_i y_i. \tag{12.5}$$

The theorem follows by observing that

$$\sum_{j=1}^{n}\left(\sum_{i=1}^{m} a_{ij}y_i\right)x_j = \sum_{i=1}^{m}\left(\sum_{j=1}^{n} a_{ij}x_j\right)y_i.$$

□

By the LP-duality theorem, $\boldsymbol{x}$ and $\boldsymbol{y}$ are both optimal solutions iff (12.3) holds with equality. Clearly, this happens iff both (12.4) and (12.5) hold with equality. Hence, we get the following result about the structure of optimal solutions:

Theorem 12.3 (Complementary slackness conditions) *Let $\boldsymbol{x}$ and $\boldsymbol{y}$ be primal and dual feasible solutions, respectively. Then, $\boldsymbol{x}$ and $\boldsymbol{y}$ are both optimal iff all of the following conditions are satisfied:*

Primal complementary slackness conditions
For each $1 \le j \le n$: either $x_j = 0$ or $\sum_{i=1}^{m} a_{ij}y_i = c_j$; and
Dual complementary slackness conditions
For each $1 \le i \le m$: either $y_i = 0$ or $\sum_{j=1}^{n} a_{ij}x_j = b_i$.

The complementary slackness conditions play a vital role in the design of efficient algorithms, both exact and approximation; see Chapter 15 for details. (For a better appreciation of their importance, we recommend that the reader study algorithms for the weighted matching problem, see Section 12.5.)

12.2 Min–max relations and LP-duality

In order to appreciate the role of LP-duality theory in approximation algorithms, it is important to first understand its role in exact algorithms. To do so, we will review some of these ideas in the context of the max-flow min-cut theorem. In particular, we will show how this and other min–max relations follow from the LP-duality theorem. Some of the ideas on cuts and flows developed here will also be used in the study of multicommodity flow in Chapters 18, 20, and 21.

The problem of computing a maximum flow in a network is: given a directed[1] graph, $G = (V, E)$ with two distinguished nodes, *source* s and *sink* t, and positive arc capacities, $c : E \rightarrow \mathbf{R}^+$, find the maximum amount of flow that can be sent from s to t subject to

1. *capacity constraint:* for each arc e, the flow sent through e is bounded by its capacity, and

[1] The maximum flow problem in undirected graphs reduces to that in directed graphs: replace each edge (u, v) by two directed edges, $(u \rightarrow v)$ and $(v \rightarrow u)$, each of the same capacity as (u, v).

2. *flow conservation:* at each node v, other than s and t, the total flow into v should equal the total flow out of v.

An *s–t cut* is defined by a partition of the nodes into two sets X and $\overline{X}$ so that $s \in X$ and $t \in \overline{X}$, and consists of the set of arcs going from X to $\overline{X}$. The *capacity* of this cut, $c(X, \overline{X})$, is defined to be the sum of capacities of these arcs. Because of the capacity constraints on flow, the capacity of any s–t cut is an upper bound on any feasible flow. Thus, if the capacity of an s–t cut, say $(X, \overline{X})$, equals the value of a feasible flow, then $(X, \overline{X})$ must be a minimum s–t cut and the flow must be a maximum flow in G. The max-flow min-cut theorem proves that it is always possible to find a flow and an s–t cut so that equality holds.

Let us formulate the maximum flow problem as a linear program. First, introduce a fictitious arc of infinite capacity from t to s, thus converting the flow to a circulation; the objective now is to maximize the flow on this arc, denoted by f_{ts}. The advantage of making this modification is that we can now require flow conservation at s and t as well. If f_{ij} denotes the amount of flow sent through arc $(i, j) \in E$, we can formulate the maximum flow problem as follows:

$$
\begin{array}{lll}
\text{maximize} & f_{ts} & \\
\text{subject to} & f_{ij} \le c_{ij}, & (i,j) \in E \\
& \displaystyle\sum_{j:\ (j,i)\in E} f_{ji} - \sum_{j:\ (i,j)\in E} f_{ij} \le 0, & i \in V \\
& f_{ij} \ge 0, & (i,j) \in E
\end{array}
$$

The second set of inequalities say that for each node i, the total flow into i is at most the total flow out of i. Notice that if this inequality holds at each node, then in fact it must be satisfied with equality at each node, thereby implying flow conservation at each node (this is so because a deficit in flow balance at one node implies a surplus at some other node). With this trick, we get a linear program in standard form.

To obtain the dual program we introduce variables d_{ij} and p_i corresponding to the two types of inequalities in the primal. We will view these variables as distance labels on arcs and potentials on nodes, respectively. The dual program is:

$$
\begin{array}{lll}
\text{minimize} & \displaystyle\sum_{(i,j)\in E} c_{ij} d_{ij} & \\
\text{subject to} & d_{ij} - p_i + p_j \ge 0, & (i,j) \in E \\
& p_s - p_t \ge 1 & \\
& d_{ij} \ge 0, & (i,j) \in E
\end{array}
\tag{12.6}
$$

$$p_i \geq 0, \qquad i \in V \qquad (12.7)$$

For developing an intuitive understanding of the dual program, it will be best to first transform it into an integer program that seeks 0/1 solutions to the variables:

$$\begin{array}{lll} \text{minimize} & \displaystyle\sum_{(i,j)\in E} c_{ij} d_{ij} & \\ \text{subject to} & d_{ij} - p_i + p_j \geq 0, & (i,j) \in E \\ & p_s - p_t \geq 1 & \\ & d_{ij} \in \{0,1\}, & (i,j) \in E \\ & p_i \in \{0,1\}, & i \in V \end{array}$$

Let $(\boldsymbol{d}^*, \boldsymbol{p}^*)$ be an optimal solution to this integer program. The only way to satisfy the inequality $p_s^* - p_t^* \geq 1$ with a 0/1 substitution is to set $p_s^* = 1$ and $p_t^* = 0$. This solution naturally defines an s–t cut $(X, \overline{X})$, where X is the set of potential 1 nodes, and $\overline{X}$ the set of potential 0 nodes. Consider an arc (i,j) with $i \in X$ and $j \in \overline{X}$. Since $p_i^* = 1$ and $p_j^* = 0$, by the first constraint, $d_{ij}^* \geq 1$. But since we have a 0/1 solution, $d_{ij}^* = 1$. The distance label for each of the remaining arcs can be set to either 0 or 1 without violating the first constraint; however, in order to minimize the objective function value, it must be set to 0. The objective function value must thus be equal to the capacity of the cut $(X, \overline{X})$, and $(X, \overline{X})$ must be a minimum s–t cut.

Thus, the previous integer program is a formulation of the minimum s–t cut problem! What about the dual program? The dual program can be viewed as a relaxation of the integer program where the integrality constraint on the variables is dropped. This leads to the constraints $1 \geq d_{ij} \geq 0$ for $(i,j) \in E$ and $1 \geq p_i \geq 0$ for $i \in V$. Next, we notice that the upper bound constraints on the variables are redundant; their omission cannot give a better solution. Dropping these constraints gives the dual program in the form given above. We will say that this program is the *LP-relaxation* of the integer program.

Consider an s–t cut C. Set C has the property that any path from s to t in G contains at least one edge of C. Using this observation, we can interpret any feasible solution to the dual program as a *fractional s–t cut*: the distance labels it assigns to arcs satisfy the property that on any path from s to t the distance labels add up to at least 1. To see this, consider an s–t path $(s = v_0, v_1, \ldots, v_k = t)$. Now, the sum of the potential differences on the endpoints of arcs on this path is

$$\sum_{i=0}^{k-1} (p_i - p_{i+1}) = p_s - p_t.$$

By the first constraint, the sum of the distance labels on the arcs must add up to at least $p_s - p_t$, which is ≥ 1. Let us define the *capacity* of this fractional s–t cut to be the dual objective function value achieved by it.

In principle, the best fractional s–t cut could have lower capacity than the best integral cut. Surprisingly enough, this does not happen. Consider the polyhedron defining the set of feasible solutions to the dual program. Let us call a feasible solution an *extreme point solution* if it is a vertex of this polyhedron, i.e., it cannot be expressed as a convex combination of two feasible solutions. From linear programming theory we know that for any objective function, i.e., assignment of capacities to the arcs of G, there is an extreme point solution that is optimal (for this discussion let us assume that for the given objective function, an optimal solution exists). Now, it can be proven that each extreme point solution of the polyhedron is integral, with each coordinate being 0 or 1 (see Exercise 12.7). Thus, the dual program always has an integral optimal solution.

By the LP-duality theorem maximum flow in G must equal capacity of a minimum fractional s–t cut. But since the latter equals the capacity of a minimum s–t cut, we get the max-flow min-cut theorem.

The max-flow min-cut theorem is therefore a special case of the LP-duality theorem; it holds because the dual polyhedron has integral vertices. In fact, most min–max relations in combinatorial optimization hold for a similar reason.

Finally, let us illustrate the usefulness of complementary slackness conditions by utilizing them to derive additional properties of optimal solutions to the flow and cut programs. Let $\boldsymbol{f}^*$ be an optimum solution to the primal LP (i.e., a maximum s–t flow). Also, let $(\boldsymbol{d}^*, \boldsymbol{p}^*)$ be an integral optimum solution to the dual LP, and let $(X, \overline{X})$ be the cut defined by $(\boldsymbol{d}^*, \boldsymbol{p}^*)$. Consider an arc (i, j) such that $i \in X$ and $j \in \overline{X}$. We have proven above that $d^*_{ij} = 1$. Since $d^*_{ij} \neq 0$, by the dual complementary slackness condition, $f^*_{ij} = c_{ij}$. Next, consider an arc (k, l) such that $k \in \overline{X}$ and $l \in X$. Since $p^*_k - p^*_l = -1$, and $d^*_{kl} \in \{0, 1\}$, the constraint $d^*_{kl} - p^*_k + p^*_l \geq 0$ must be satisfied as a strict inequality. By the primal complementary slackness condition, $f^*_{kl} = 0$. Thus, we have proven that arcs going from X to $\overline{X}$ are saturated by $\boldsymbol{f}^*$ and the reverse arcs carry no flow. (Observe that it was not essential to invoke complementary slackness conditions to prove these facts; they also follow from the fact that flow across cut $(X, \overline{X})$ equals its capacity.)

12.3 Two fundamental algorithm design techniques

We can now explain why linear programming is so useful in approximation algorithms. Many combinatorial optimization problems can be stated as integer programs. Once this is done, the linear relaxation of this program provides a natural way of lower bounding the cost of the optimal solution. As stated in Chapter 1, this is typically a key step in the design of an approximation algorithm. As in the case of the minimum s–t cut problem, a feasible solution to the LP-relaxation can be thought of as a *fractional solution* to the original problem. However, in the case of an **NP**-hard problem, we cannot expect the polyhedron defining the set of feasible solutions to have integer vertices. Thus, our task is not to look for an optimal solution to the LP-relaxation, but rather a near-optimal integral solution.

There are two basic techniques for obtaining approximation algorithms using linear programming. The first, and more obvious, method is to solve the linear program and then convert the fractional solution obtained into an integral solution, trying to ensure that in the process the cost does not increase much. The approximation guarantee is established by comparing the cost of the integral and fractional solutions. This technique is called *LP-rounding* or simply *rounding*.

The second, less obvious and perhaps more sophisticated, method is to use the dual of the LP-relaxation in the design of the algorithm. This technique is called the *primal–dual schema*. Let us call the LP-relaxation the primal program. Under this schema, an integral solution to the primal program and a feasible solution to the dual program are constructed iteratively. Notice that any feasible solution to the dual also provides a lower bound on OPT. The approximation guarantee is established by comparing the two solutions.

Both these techniques have been used extensively to obtain algorithms for many fundamental problems. Fortunately, once again, these techniques can be illustrated in the simple setting of the set cover problem. This is done in Chapters 14 and 15. Later chapters will present ever more sophisticated use of these techniques for solving a variety of problems.

LP-duality theory has also been useful in analyzing combinatorially obtained approximation algorithms, using the *method of dual fitting*. In Chapter 13 we will give an alternative analysis of the greedy set cover algorithm, Algorithm 2.2, using this method. This method has also been used to analyze greedy algorithms for the metric uncapacitated facility location problem (see Exercise 24.12). The method seems quite basic and should find other applications as well.

12.3.1 A comparison of the techniques and the notion of integrality gap

The reader may suspect that from the viewpoint of approximation guarantee, the primal–dual schema is inferior to rounding, since an optimal solution to the primal gives a tighter lower bound than a feasible solution to the dual. It turns out that this is not so. In order to give a formal explanation, we need to introduce the crucial notion of *integrality gap of an LP-relaxation*.

Given an LP-relaxation for a minimization problem Π, let $\mathrm{OPT}_f(I)$ denote the cost of an optimal fractional solution to instance I, i.e., the objective function value of an optimal solution to the LP-relaxation. Define the *inte-*

grality gap, sometimes also called the *integrality ratio*, of the relaxation to be

$$\sup_I \frac{\text{OPT}(I)}{\text{OPT}_f(I)},$$

i.e., the supremum of the ratio of the optimal integral and fractional solutions. In the case of a maximization problem, the integrality gap will be defined to be the infimum of this ratio. As stated in Section 12.2, most min–max relations arise from LP-relaxations that always have integral optimal solutions. Clearly, the integrality gap of such an LP is 1. We will call such an LP-relaxation an *exact relaxation.*

If the cost of the solution found by the algorithm is compared directly with the cost of an optimal fractional solution (or a feasible dual solution), as is done in most algorithms, the best approximation factor we can hope to prove is the integrality gap of the relaxation (see Exercise 12.5). Interestingly enough, for many problems, both techniques have been successful in yielding algorithms having guarantees essentially equal to the integrality gap of the relaxation.

The main difference in performance between the two techniques lies in the running times of the algorithms produced. An LP-rounding algorithm needs to find an optimal solution to the linear programming relaxation. Since linear programming is in **P**, this can be done in polynomial time if the relaxation has polynomially many constraints. Even if the relaxation has exponentially many constraints, this may still be achievable, if a polynomial time *separation oracle* can be constructed, i.e., a polynomial time algorithm that given a point in $\mathbf{R}^n$, where n is the number of variables in the relaxation, either confirms that this point is a feasible solution (i.e., satisfies all constraints), or produces a violated constraint (see the notes in Section 12.5 for references). The running time for both possibilities is high; for the second it may be exorbitant. Let us remark that for certain problems, extreme point solutions have additional structural properties and some LP-rounding algorithms require such a solution to the linear programming relaxation. Such solutions can also be found in polynomial time.

On the other hand, the primal–dual schema leaves enough room to exploit the special combinatorial structure of individual problems and is thereby able to yield algorithms having good running times. It provides only a broad outline of the algorithm – the details have to be designed individually for specific problems. In fact, for many problems, once the algorithm has been designed using the primal–dual schema, the scaffolding of linear programming can be completely dispensed with to get a purely combinatorial algorithm.

This brings us to another advantage of the primal–dual schema – this time not objectively quantifiable. A combinatorial algorithm is more malleable than an algorithm that requires an LP-solver. Once a basic problem is solved using the primal–dual schema, one can also solve variants and generalizations of the basic problem. Exercises in Chapters 22 and 24 illustrate this point. From a practical standpoint, a combinatorial algorithm is more useful, since it is easier to adapt it to specific applications and fine tune its performance for specific types of inputs.

12.4 Exercises

12.1 Show that the dual of the dual of a linear program is the original program itself.

12.2 Show that any minimization program can be transformed into an equivalent program in standard form, i.e., the form of LP (12.1).

12.3 Change some of the constraints of the primal program (12.1) into equalities, i.e., so they are of the form

$$\sum_{j=1}^{n} a_{ij}x_j = b_i, i \in I.$$

Show that the dual of this program involves modifying program (12.2) so that the corresponding dual variables $y_i, i \in I$ are *unconstrained*, i.e., they are not constrained to be nonnegative. Additionally, if some of the variables $x_j, j \in J$ in program (12.1) are unconstrained, then the corresponding constraints in the dual become equalities.

12.4 Consider LP's (13.2) and (13.3), the LP-relaxation and dual LP for the set cover problem, Problem 2.1. Let $\boldsymbol{x}$ and $\boldsymbol{y}$ be primal and dual feasible solutions, respectively, and assume that they satisfy all complementary slackness conditions. Show that the dual pays for the primal exactly via a "local paying mechanism" as follows: if each element e pays $y_e x_S$ to each set S containing e, then the amount collected by each set S is precisely $c(S)x_S$. Hence show that $\boldsymbol{x}$ and $\boldsymbol{y}$ have the same objective function values.

12.5 Is the following a theorem: An approximation algorithm designed using an LP-relaxation cannot achieve a better approximation guarantee than the integrality gap of the relaxation.
Hint: In principle it may be possible to show, using additional structural properties, that whenever an instance has a bad gap, the cost of the solution found by the algorithm is much less that $\alpha\mathrm{OPT}$, where α is the integrality gap of the relaxation. (Observe that if the instance has a bad gap, the cost of the solution found cannot be much less than $\alpha\mathrm{OPT}_f$.)

12.6 Use the max-flow min-cut theorem to derive Menger's Theorem:

Theorem 12.4 *Let $G = (V, E)$ be a directed graph with $s, t \in V$. Then, the maximum number of edge-disjoint (vertex-disjoint) s–t paths is equal to the minimum number of edges (vertices) whose removal disconnects s from t.*

12.7 Show that each extreme point solution for LP (12.6) is 0/1, and hence represents a valid cut.
Hint: An $n \times m$ matrix A is said to be *totally unimodular* if the determinant of every square submatrix of A is $1, -1$, or 0. Show, by induction, that the constraint matrix of this LP is totally unimodular. Also, use the fact that a feasible solution for a set of linear inequalities in $\mathbf{R}^n$ is an extreme point solution iff it satisfies n linearly independent inequalities with equality.

12.8 This exercise develops a proof of the König-Egerváry Theorem (Theorem 1.6). Let $G = (V, E)$ be a bipartite graph.

1. Show that the following is an exact LP-relaxation (i.e., always has an integral optimal solution) for the maximum matching problem in G.

$$\begin{aligned} \text{maximize} \quad & \sum_e x_e && && (12.8) \\ \text{subject to} \quad & \sum_{e:\ e \text{ incident at } v} x_e \leq 1, && v \in V \\ & x_e \geq 0, && e \in E \end{aligned}$$

   Hint: Using the technique of Exercise 12.7 show that each extreme point solution for LP (12.8) is 0/1, and hence represents a valid matching.
2. Obtain the dual of this LP and show that it is an exact LP-relaxation for the problem of finding a minimum vertex cover in bipartite graph G.
3. Use the previous result to derive the König-Egerváry Theorem.

12.9 (Edmonds [74])

1. Let $G = (V, E)$ be an undirected graph, with weights w_e on edges. The following is an exact LP-relaxation for the problem of finding a maximum weight matching in G. (By $e:\ e \in S$ we mean edges e that have both endpoints in S.)

$$\begin{aligned} \text{maximize} \quad & \sum_e w_e x_e && && (12.9) \\ \text{subject to} \quad & \sum_{e:\ e \text{ incident at } v} x_e \leq 1, && v \in V \\ & \sum_{e:\ e \in S} x_e \leq \frac{|S| - 1}{2}, && S \subset V, |S| \text{ odd} \\ & x_e \geq 0, && e \in E \end{aligned}$$

Obtain the dual of this LP. If the weight function is integral, the dual is also exact. Observe that Theorem 1.7 follows from these facts.

2. Assume that $|V|$ is even. The following is an exact LP-relaxation for the minimum weight perfect matching problem in G (a matching is *perfect* if it matches all vertices). Obtain the dual of this LP. Use complementary slackness conditions to give conditions satisfied by a pair of optimal primal (integral) and dual solutions for both formulations.

$$\begin{aligned} \text{minimize} \quad & \sum_e w_e x_e && && (12.10) \\ \text{subject to} \quad & \sum_{e:\ e \text{ incident at } v} x_e = 1, && v \in V \\ & \sum_{e:\ e \in S} x_e \le \frac{|S|-1}{2}, && S \subset V, |S| \text{ odd} \\ & x_e \ge 0, && e \in E \end{aligned}$$

12.10 (Edmonds [77]) Show that the following is an integer programming formulation for the minimum spanning tree (MST) problem. Assume we are given graph $G = (V, E)$, $|V| = n$, with cost function $c : E \to \mathbf{Q}^+$. For $A \subseteq E$, we will denote by $\kappa(A)$ the number of connected components in graph $G_A = (V, A)$.

$$\begin{aligned} \text{minimize} \quad & \sum_e c_e x_e && && (12.11) \\ \text{subject to} \quad & \sum_{e \in A} x_e = n - \kappa(A), && A \subset E \\ & \sum_{e \in E} x_e = n - 1 \\ & x_e \in \{0, 1\}, && e \in E \end{aligned}$$

The rest of this exercise develops a proof that the LP-relaxation of this integer program is exact for the MST problem.

1. First, it will be convenient to change the objective function of IP (12.11) to $\max \sum_e -c_e x_e$. Obtain the LP-relaxation and dual of this modified formulation.
2. Consider the primal solution produced by Kruskal's algorithm. Let $e_1, \ldots, e_m$ be the edges sorted by increasing cost, $|E| = m$. This algorithm greedily picks a maximal acyclic subgraph from this sorted list. Obtain a suitable dual feasible solution so that all complementary slackness conditions are satisfied.

Hint: Let $A_t = \{e_1, \ldots, e_t\}$. Set $y_{A_t} = e_{t+1} - e_t$, for $1 \le t < m$, and $y_E = -c_m$, where y is the dual variable.

3. Show that $\boldsymbol{x}$ is a feasible solution to the above-stated primal program iff it is a feasible solution to the following LP. That is, prove that this is also an exact relaxation for the MST problem.

$$\begin{aligned}
\text{minimize} \quad & \sum_e c_e x_e && && (12.12)\\
\text{subject to} \quad & \sum_{e \in S} x_e \le |S| - 1, && S \subset V\\
& \sum_{e \in E} x_e = n - 1\\
& x_e \ge 0, && e \in E
\end{aligned}$$

12.11 In this exercise, you will derive von Neumann's minimax theorem in game theory from the LP-duality theorem. A finite two-person zero-sum game is specified by an $m \times n$ matrix $\boldsymbol{A}$ with real entries. In each round, the row player, R, selects a row, say i; simultaneously, the column player, C, selects a column, say j. The *payoff* to R at the end of this round is a_{ij}. Thus, $|a_{ij}|$ is the amount that C pays R (R pays C) if a_{ij} is positive (a_{ij} is negative); no money is exchanged if a_{ij} is zero. *Zero-sum game* refers to the fact that the total amount of money possessed by R and C together is conserved.

The *strategy* of each player is specified by a vector whose entries are non-negative and add up to one, giving the probabilities with which the player picks each row or column. Let R's strategy be given by m-dimensional vector $\boldsymbol{x}$, and C's strategy be given by n-dimensional vector $\boldsymbol{y}$. Then, the expected payoff to R in a round is $\boldsymbol{x}^T \boldsymbol{A} \boldsymbol{y}$. The job of each player is to pick a strategy that *guarantees* maximum possible expected winnings (equivalently, minimum possible expected losses), regardless of the strategy chosen by the other player. If R chooses strategy $\boldsymbol{x}$, he can be sure of winning only $\min_{\boldsymbol{y}} \boldsymbol{x}^T \boldsymbol{A} \boldsymbol{y}$, where the minimum is taken over all possible strategies of C. Thus, the optimal choice for R is given by $\max_{\boldsymbol{x}} \min_{\boldsymbol{y}} \boldsymbol{x}^T \boldsymbol{A} \boldsymbol{y}$. Similarly, C will minimize her losses by choosing the strategy given by $\min_{\boldsymbol{y}} \max_{\boldsymbol{x}} \boldsymbol{x}^T \boldsymbol{A} \boldsymbol{y}$. The minimax theorem states that for every matrix $\boldsymbol{A}$, $\max_{\boldsymbol{x}} \min_{\boldsymbol{y}} \boldsymbol{x}^T \boldsymbol{A} \boldsymbol{y} = \min_{\boldsymbol{y}} \max_{\boldsymbol{x}} \boldsymbol{x}^T \boldsymbol{A} \boldsymbol{y}$.

Let us say that a strategy is *pure* if it picks a single row or column, i.e., the vector corresponding to it consists of one 1 and the rest 0's. A key observation is that for any strategy $\boldsymbol{x}$ of R, $\min_{\boldsymbol{y}} \boldsymbol{x}^T \boldsymbol{A} \boldsymbol{y}$ is attained for a pure strategy of C: Suppose the minimum is attained for strategy $\boldsymbol{y}$. Consider the pure strategy corresponding to any nonzero component of $\boldsymbol{y}$. The fact that the components of $\boldsymbol{y}$ are nonnegative and add up to one leads to an easy proof that this pure strategy attains the same minimum. Thus, R's optimum strategy is given by $\max_{\boldsymbol{x}} \min_j \sum_{i=1}^{m} a_{ij}x_i$. The second critical observation is that the problem of computing R's optimal strategy can be expressed as a linear program:

$$\begin{aligned} \text{maximize} \quad & z \\ \text{subject to} \quad & z - \sum_{i=1}^{m} a_{ij}x_i \leq 0, \qquad j = 1, \ldots, n \\ & \sum_{i=1}^{m} x_i = 1 \\ & x_i \geq 0, \qquad\qquad i = 1, \ldots, m \end{aligned}$$

Find the dual of this LP and show that it computes the optimal strategy for C. (Use the fact that for any strategy $\boldsymbol{y}$ of C, $\max_{\boldsymbol{x}} \boldsymbol{x}^T A\boldsymbol{y}$ is attained for a pure strategy of R.) Hence, prove the minimax theorem using the LP-duality theorem.

12.5 Notes

For a good introduction to theory of linear programming, see Chvátal [51]. There are numerous other books on the topic, e.g., Dantzig [61], Karloff [167], Nemhauser and Wolsey [220], and Schrijver [245]. Linear programming has been extensively used in combinatorial optimization, see Ahuja, Magnanti, and Orlin [2], Cook, Cunningham, Pulleyblank, and Schrijver [54], Grötschel, Lovász, and Schrijver [123], Lovász [201], Lovász and Plummer [202], and Papadimitriou and Steiglitz [225]. For a good explanation of Edmonds' weighted matching algorithm, see Lovász and Plummer [202]. For algorithms for finding a solution to an LP, given a separation oracle, see Grötschel, Lovász, and Schrijver [122, 123] and Schrijver [245].

13 Set Cover via Dual Fitting

In this chapter we will introduce the method of dual fitting, which helps analyze combinatorial algorithms using LP-duality theory. Using this method, we will present an alternative analysis of the natural greedy algorithm (Algorithm 2.2) for the set cover problem (Problem 2.1). Recall that in Section 2.1 we deferred giving the lower bounding method on which this algorithm was based. We will provide the answer below. The power of this approach will become apparent when we show the ease with which it extends to solving several generalizations of the set cover problem (see Section 13.2).

The method of dual fitting can be described as follows, assuming a minimization problem: The basic algorithm is combinatorial – in the case of set cover it is in fact the simple greedy algorithm. Using the linear programming relaxation of the problem and its dual, one shows that the primal integral solution found by the algorithm is fully paid for by the dual computed; however, the dual is infeasible. By *fully paid for* we mean that the objective function value of the primal solution found is at most the objective function value of the dual computed. The main step in the analysis consists of dividing the dual by a suitable factor and showing that the shrunk dual is feasible, i.e., it fits into the given instance. The shrunk dual is then a lower bound on OPT, and the factor is the approximation guarantee of the algorithm.

13.1 Dual-fitting-based analysis for the greedy set cover algorithm

To formulate the set cover problem as an integer program, let us assign a variable x_S for each set $S \in \mathcal{S}$, which is allowed 0/1 values. This variable will be set to 1 iff set S is picked in the set cover. Clearly, the constraint is that for each element $e \in U$ we want that at least one of the sets containing it be picked.

$$\begin{aligned} \text{minimize} \quad & \sum_{S \in \mathcal{S}} c(S) x_S && (13.1) \\ \text{subject to} \quad & \sum_{S:\ e \in S} x_S \geq 1, \quad e \in U \end{aligned}$$

$$x_S \in \{0,1\}, \qquad S \in \mathcal{S}$$

The LP-relaxation of this integer program is obtained by letting the domain of variables x_S be $1 \geq x_S \geq 0$. Since the upper bound on x_S is redundant, we get the following LP. A solution to this LP can be viewed as a fractional set cover.

$$\begin{aligned} \text{minimize} \quad & \sum_{S \in \mathcal{S}} c(S) x_S \\ \text{subject to} \quad & \sum_{S:\ e \in S} x_S \geq 1, \qquad e \in U \\ & x_S \geq 0, \qquad S \in \mathcal{S} \end{aligned} \tag{13.2}$$

Example 13.1 Let us give a simple example to show that a fractional set cover may be cheaper than the optimal integral set cover. Let $U = \{e, f, g\}$ and the specified sets be $S_1 = \{e, f\}, S_2 = \{f, g\}, S_3 = \{e, g\}$, each of unit cost. An integral cover must pick two of the sets for a cost of 2. On the other hand, picking each set to the extent of 1/2 gives a fractional cover of cost 3/2. □

Introducing a variable y_e corresponding to each element $e \in U$, we obtain the dual program.

$$\begin{aligned} \text{maximize} \quad & \sum_{e \in U} y_e \\ \text{subject to} \quad & \sum_{e:\ e \in S} y_e \leq c(S), \qquad S \in \mathcal{S} \\ & y_e \geq 0, \qquad e \in U \end{aligned} \tag{13.3}$$

Intuitively, why is LP (13.3) the dual of LP (13.2)? In our experience, this is not the right question to be asked. As stated in Section 12.1, there is a purely mechanical procedure for obtaining the dual of a linear program. Once the dual is obtained, one can devise intuitive, and possibly physically meaningful, ways of thinking about it. Using this mechanical procedure, one can obtain the dual of a complex linear program in a fairly straightforward manner. Indeed, the LP-duality-based approach derives its wide applicability from this fact.

An intuitive way of thinking about LP (13.3) is that it is packing "stuff" into elements, trying to maximize the total amount packed, subject to the constraint that no set is overpacked. A set is said to be *overpacked* if the total amount packed into its elements exceeds the cost of the set. Whenever the coefficients in the constraint matrix, objective function, and right-hand side are all nonnegative, the minimization LP is called a *covering LP* and the maximization LP is called a *packing LP*. Thus, (13.2) and (13.3) are a covering-packing pair of linear programs. Such pairs of programs will arise frequently in subsequent chapters.

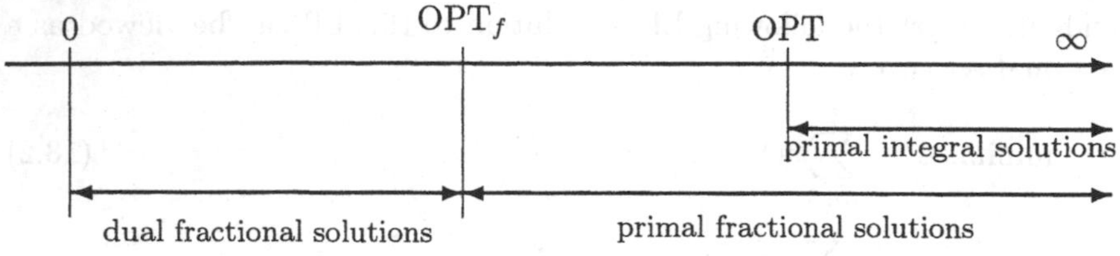

At this point, we can state the lower bounding scheme being used by Algorithm 2.2. Denote by OPT_f the cost of an optimal fractional set cover, i.e., an optimal solution to LP (13.2). Clearly $\mathrm{OPT}_f \le \mathrm{OPT}$, the cost of an optimal (integral) set cover. The cost of any feasible solution to the dual program, LP (13.3), is a lower bound on OPT_f, and hence also on OPT. Algorithm 2.2 uses this as the lower bound.

Algorithm 2.2 defines dual variables price(e), for each element, e. Observe that the cover picked by the algorithm is fully payed for by this dual solution. However, in general, this dual solution is not feasible (see Exercise 13.2). We will show below that if this dual is shrunk by a factor of H_n, it fits into the given set cover instance, i.e., no set is overpacked. For each element e define,

$$y_e = \frac{\mathrm{price}(e)}{H_n}.$$

Algorithm 2.2 uses the dual feasible solution, $\boldsymbol{y}$, as the lower bound on OPT.

Lemma 13.2 *The vector $\boldsymbol{y}$ defined above is a feasible solution for the dual program (13.3).*

Proof: We need to show that no set is overpacked by the solution $\boldsymbol{y}$. Consider a set $S \in \mathcal{S}$ consisting of k elements. Number the elements in the order in which they are covered by the algorithm, breaking ties arbitrarily, say $e_1, \ldots, e_k$.

Consider the iteration in which the algorithm covers element e_i. At this point, S contains at least $k-i+1$ uncovered elements. Thus, in this iteration, S itself can cover e_i at an average cost of at most $c(S)/(k-i+1)$. Since the algorithm chose the most cost-effective set in this iteration, $\mathrm{price}(e_i) \le c(S)/(k-i+1)$. Thus,

$$y_{e_i} \le \frac{1}{H_n} \cdot \frac{c(S)}{k-i+1}.$$

Summing over all elements in S,

$$\sum_{i=1}^{k} y_{e_i} \leq \frac{c(S)}{H_n} \cdot \left(\frac{1}{k} + \frac{1}{k-1} + \cdots + \frac{1}{1}\right) = \frac{H_k}{H_n} \cdot c(S) \leq c(S).$$

Therefore, S is not overpacked. □

Theorem 13.3 *The approximation guarantee of the greedy set cover algorithm is H_n.*

Proof: The cost of the set cover picked is

$$\sum_{e \in U} \text{price}(e) = H_n \left(\sum_{e \in U} y_e\right) \leq H_n \cdot \text{OPT}_f \leq H_n \cdot \text{OPT},$$

where the first inequality follows from the fact that $\boldsymbol{y}$ is dual feasible. □

13.1.1 Can the approximation guarantee be improved?

Consider the three questions raised in Section 1.1.2 regarding improving the approximation guarantee for vertex cover. Let us ask analogous questions for set cover. The first and third questions are already answered in Section 2.1.

As a corollary of Theorem 13.3 we get an upper bound of H_n on the integrality gap of relaxation (13.2). Example 13.4 shows that this bound is essentially tight. Since the integrality gap of the LP-relaxation used bounds the best approximation factor one can hope to achieve using this relaxation, the answer to the second question is also essentially "no".

Example 13.4 Consider the following set cover instance. Let $n = 2^k - 1$, where k is a positive integer, and let $U = \{e_1, e_2, \ldots, e_n\}$. For $1 \leq i \leq n$, consider i written as a k-bit number. We can view this as a k-dimensional vector over $GF[2]$. Let $\mathbf{i}$ denote this vector. For $1 \leq i \leq n$ define set $S_i = \{e_j | \ \mathbf{i} \cdot \mathbf{j} = 1\}$, where $\mathbf{i} \cdot \mathbf{j}$ denotes the inner product of these two vectors. Finally, let $\mathcal{S} = \{S_1, \ldots, S_n\}$, and define the cost of each set to be 1.

It is easy to check that each set contains $2^{k-1} = (n+1)/2$ elements, and each element is contained in $(n+1)/2$ sets. Thus, $x_i = 2/(n+1)$, $1 \leq i \leq n$, is a fractional set cover. Its cost is $2n/(n+1)$.

Next, we will show that any integral set cover must pick at least k of the sets. Consider the union of any p sets, where $p < k$. Let $i_1, \ldots, i_p$ be the indices of these p sets, and let $\boldsymbol{A}$ be a $p \times k$ matrix over $GF[2]$ whose rows consist of vectors $\mathbf{i}_1, \ldots, \mathbf{i}_p$, respectively. Since the rank of A is $< k$, the dimension of its null space is ≥ 1, and so the null space contains a nonzero vector, say $\mathbf{j}$. Since $\boldsymbol{A}\mathbf{j} = \mathbf{0}$, the element e_j is not in any of the p sets. Hence the p sets do not form a cover.

Therefore, any integral set cover has cost at least $k = \log_2(n+1)$. Hence, the lower bound on the integrality gap established by this example is

$$\left(\frac{n+1}{2n}\right) \cdot \log_2(n+1) > \frac{\log_2 n}{2}.$$

□

13.2 Generalizations of set cover

The greedy algorithm and its analysis using dual fitting extend naturally to several generalizations of the set cover problem (see Exercise 13.4).

- **Set multicover:** Each element, e, needs to be covered a specified integer number, r_e, of times. The objective again is to cover all elements up to their coverage requirements at minimum cost. We will assume that the cost of picking a set S k times is $kc(S)$.
- **Multiset multicover:** We are given a collection of multisets, rather than sets, of U. A *multiset* contains a specified number of copies of each element. Let $M(S, e)$ denote the multiplicity of element e in set S. The instance satisfies the condition that the multiplicity of an element in a set is at most its coverage requirement, i.e., $\forall S, e \; M(S, e) \leq r_e$. The objective is the same as before.
- **Covering integer programs:** These are integer programs of the form

$$\begin{aligned} &\text{minimize} && \boldsymbol{c} \cdot \boldsymbol{x} \\ &\text{subject to} && \boldsymbol{A}\boldsymbol{x} \geq \boldsymbol{b}, \end{aligned}$$

  where all entries in $\boldsymbol{A}, \boldsymbol{b}, \boldsymbol{c}$ are nonnegative and $\boldsymbol{x}$ is required to be nonnegative and integral.

13.2.1 Dual fitting applied to constrained set multicover

In this section, we will present an H_n factor approximation algorithm for set multicover with the additional constraint that each set can be picked at most once. Let us call this the *constrained set multicover problem*. One interesting feature of this problem is that its linear relaxation and dual contain negative coefficients and thus do not form a covering-packing pair of LP's.

Let $r_e \in \mathbf{Z}^+$ be the coverage requirement for each element $e \in U$. The integer programming formulation of constrained set multicover is not very different from that of set cover.

$$\begin{aligned} &\text{minimize} && \sum_{S \in \mathcal{S}} c(S) x_S && && (13.4) \\ &\text{subject to} && \sum_{S:\, e \in S} x_S \geq r_e, && e \in U \\ & && x_S \in \{0, 1\}, && S \in \mathcal{S} \end{aligned}$$

Notice, however, that in the LP-relaxation, the constraints $x_S \leq 1$ are no longer redundant. If we drop them, then a set may be picked multiple times to satisfy the coverage requirement of the elements. Thus, the LP-relaxation looks different from that for set cover. In particular, because of the negative numbers in the constraint matrix and the right-hand side, it is not even a covering linear program. The analysis given below deals with this added complexity.

$$\begin{aligned} \text{minimize} \quad & \sum_{S \in \mathcal{S}} c(S) x_S && && (13.5) \\ \text{subject to} \quad & \sum_{S:\ e \in S} x_S \geq r_e, && e \in U \\ & -x_S \geq -1, && S \in \mathcal{S} \\ & x_S \geq 0, && S \in \mathcal{S} \end{aligned}$$

The additional constraints in the primal lead to new variables, z_S, in the dual. The dual also has negative numbers in the constraint matrix and is not a packing program. Now, a set S can be overpacked with the y_e's. However, this can be done only if we raise z_S to ensure feasibility, which in turn decreases the objective function value. Overall, overpacking may still be advantageous, since the y_e's appear with coefficients of r_e in the objective function.

$$\begin{aligned} \text{maximize} \quad & \sum_{e \in U} r_e y_e - \sum_{S \in \mathcal{S}} z_S && && (13.6) \\ \text{subject to} \quad & \left(\sum_{e:\ e \in S} y_e \right) - z_S \leq c(S), && S \in \mathcal{S} \\ & y_e \geq 0, && e \in U \\ & z_S \geq 0, && S \in \mathcal{S} \end{aligned}$$

The algorithm is again greedy. Let us say that element e is *alive* if it occurs in fewer than r_e of the picked sets. In each iteration, the algorithm picks, from amongst the currently unpicked sets, the most cost-effective set, where the *cost-effectiveness* of a set is defined to be the average cost at which it covers alive elements. The algorithm halts when there are no more alive elements, i.e., each element has been covered to the extent of its requirement.

When a set S is picked, its cost is distributed equally among the alive elements it covers as follows: if S covers e for the jth time, we set price(e, j) to the current cost-effectiveness of S. Clearly, the cost-effectiveness of sets picked is nondecreasing. Hence, for each element e, price$(e, 1) \leq$ price$(e, 2) \leq \ldots \leq$ price(e, r_e).

At the end of the algorithm, the dual variables are set as follows: For each $e \in U$, let $\alpha_e = \text{price}(e, r_e)$. For each $S \in \mathcal{S}$ that is picked by the algorithm, let

$$\beta_S = \sum_{e \text{ covered by } S} (\text{price}(e, r_e) - \text{price}(e, j_e)),$$

where j_e is the copy of e that is covered by S. Notice that since $\text{price}(e, j_e) \leq \text{price}(e, r_e)$, β_S is nonnegative. If S is not picked by the algorithm, β_S is defined to be 0.

Lemma 13.5 *The multicover picked by the algorithm is fully paid for by the dual solution* $(\boldsymbol{\alpha}, \boldsymbol{\beta})$.

Proof: Since the cost of the sets picked by the algorithm is distributed among the covered elements, it follows that the total cost of the multicover produced by the algorithm is

$$\sum_{e \in U} \sum_{j=1}^{r_e} \text{price}(e, j).$$

The objective function value of the dual solution $(\boldsymbol{\alpha}, \boldsymbol{\beta})$ is

$$\sum_{e \in U} r_e \alpha_e - \sum_{S \in \mathcal{S}} \beta_S = \sum_{e \in U} \sum_{j=1}^{r_e} \text{price}(e, j).$$

The lemma follows. □

The dual solution defined above is, in general, infeasible. We will show that when scaled by a factor of H_n, a feasible solution results. Define for each element $e \in U$ and each set $S \in \mathcal{S}$,

$$y_e = \frac{\alpha_e}{H_n} \quad \text{and} \quad z_S = \frac{\beta_S}{H_n}.$$

Lemma 13.6 *The pair* $(\boldsymbol{y}, \boldsymbol{z})$ *is a feasible solution for the dual program (13.6).*

Proof: Consider a set $S \in \mathcal{S}$ consisting of k elements. Number its elements in the order in which their requirements are fulfilled, i.e., the order in which they stopped being alive. Let the ordered elements be $e_1, \ldots, e_k$.

First, assume that S is not picked by the algorithm. When the algorithm is about to cover the last copy of e_i, S contains at least $k-i+1$ alive elements, so

$$\text{price}(e_i, r_{e_i}) \leq \frac{c(S)}{k-i+1}.$$

Since z_S is zero, we get

$$\begin{aligned}\left(\sum_{i=1}^{k} y_{e_i}\right) - z_S &= \frac{1}{H_n}\sum_{i=1}^{k} \text{price}(e_i, r_{e_i}) \\ &\leq \frac{c(S)}{H_n}\cdot\left(\frac{1}{k}+\frac{1}{k-1}+\cdots+\frac{1}{1}\right) \leq c(S)\,.\end{aligned}$$

Next, assume that S is picked by the algorithm, and before this happens, $k' \geq 0$ elements of S are already completely covered. Then

$$\begin{aligned}&\left(\sum_{i=1}^{k} y_{e_i}\right) - z_S \\ &= \frac{1}{H_n}\cdot\left[\sum_{i=1}^{k}\text{price}(e_i, r_{e_i}) - \sum_{i=k'+1}^{k}(\text{price}(e_i, r_{e_i}) - \text{price}(e_i, j_i))\right] \\ &= \frac{1}{H_n}\cdot\left[\sum_{i=1}^{k'}\text{price}(e_i, r_{e_i}) + \sum_{i=k'+1}^{k}\text{price}(e_i, j_i)\right],\end{aligned}$$

where S covers the j_ith copy of e_i, for each $i \in \{k'+1, \ldots, k\}$. But $\sum_{i=k'+1}^{k}\text{price}(e_i, j_i) = c(S)$, since the cost of S is equally distributed among the copies it covers. Finally consider elements e_i, $i \in \{1, \ldots, k'\}$. When the last copy of e_i is being covered, S is not yet picked and covers at least $k-i+1$ alive elements. Thus, $\text{price}(e_i, r_{e_i}) \leq c(S)/(k-i+1)$. Therefore,

$$\left(\sum_{i=1}^{k} y_{e_i}\right) - z_S \leq \frac{c(S)}{H_n}\cdot\left(\frac{1}{k}+\cdots+\frac{1}{k-k'+1}+1\right) \leq c(S).$$

Hence, $(\boldsymbol{y}, \boldsymbol{z})$ is feasible for the dual program. □

Theorem 13.7 *The greedy algorithm achieves an approximation guarantee of H_n for the constrained set multicover problem.*

Proof: By Lemmas 13.5 and 13.6, the total cost of the multicover produced by the algorithm is

$$\sum_{e\in U} r_e\alpha_e - \sum_{S\in\mathcal{S}}\beta_S = H_n\cdot\left[\sum_{e\in U} r_e y_e - \sum_{S\in\mathcal{S}} z_S\right] \leq H_n\cdot\text{OPT}.$$

□

Observe that as a corollary of Theorem 13.7 we get that the integrality gap of LP (13.5) is bounded by H_n. In contrast, the integrality gap of the corresponding LP for multiset multicover, with the restriction that each set be picked at most once, is not bounded by any function of n (see Exercise 13.5).

13.3 Exercises

13.1 Show that the dual-fitting-based analysis for the greedy set cover and constrained set multicover algorithms actually establishes an approximation guarantee of H_k, where k is size of the largest set in the given instance. (Notice the ease with which this can be established using the LP-duality approach; compare with Exercise 2.8.)

13.2 Give an example in which the dual solution, price(e), for each element e, computed by Algorithm 2.2 overpacks some sets, S, by a factor of essentially $H_{|S|}$.

13.3 Give examples to show that the lower bound used by Algorithm 2.2, $\boldsymbol{y}$, can be smaller than OPT by a factor of $O(\log n)$.

13.4 Give the following approximation algorithms.

1. H_n factor for set multicover.
2. H_m factor for multiset multicover, where m is the size of the largest multiset in the given instance (the size of a multiset counts elements with multiplicity).
3. $O(\log n)$ factor for covering integer programs.

Hint: For H_m factor algorithm for multiset multicover, set the dual variables according to the average price for covering elements, i.e.,

$$y_e = \frac{1}{H_m} \sum_{i=1}^{r_e} \text{price}(e,i)/r_e.$$

Use scaling and rounding to reduce covering integer programs to multiset multicover, with m polynomially bounded in n, at the expense of a small error (which goes into the approximation factor).

13.5 Show that the integrality gap of the relaxation for the following two variants of multiset multicover, based on LP (13.2), is not bounded by any function of n.

1. Remove the restriction that $M(S, e) \leq r_e$.

2. Impose the constraint that each set can be picked at most once.

What is the best approximation guarantee you can establish for the greedy algorithm for the second variant. Why does the proof of factor H_n given in Section 13.2 not extend to this case?

13.6 (Mihail [214]) Consider the following variant on the set multicover problem. Let U be the universal set, $|U| = n$, and $\mathcal{S}$ a collection of subsets of U. For each $S \in \mathcal{S}$, its cost is given as a function of time, $t \in \{1, \ldots, T\}$. Each of these cost functions is nonincreasing with time. In addition, for each element in U, a coverage requirement is specified, again as a function of time; these functions are nondecreasing with time. The problem is to pick sets at a minimum total cost so that the coverage requirements are satisfied for each element at each time. A set can be picked any number of times; the cost of picking a set depends on the time at which it is picked. Once picked, the set remains in the cover for all future times at no additional cost. Give an H_n factor algorithm for this problem. (An $H_{(n \cdot T)}$ factor algorithm is straightforward.)

13.7 In many realistic situations, the cost of picking an item a multiple number of times does not grow linearly. Instead it is given by a concave function. The following variant of the set multicover problem models this situation. For each set S_i we are given a concave function f_i specifying the cost of picking this set multiple times. The problem again is to satisfy all coverage requirements of elements at minimum cost. Give a factor H_n algorithm for this problem.
Hint: Reduce the problem to a multiset multicover problem. For each set S_i, construct sets S_i^j, $j \geq 1$. Set S_i^j contains each element of S_i with multiplicity j and has a cost of $f_i(j)$. The greedy algorithm run on this instance achieves the required factor. Next show that there is no need to explicitly construct all the sets S_i^j. In each iteration of the greedy algorithm, the most cost-effective set can be computed directly in polynomial time, even if the requirements are exponentially large.

13.4 Notes

The dual-fitting-based analysis of set cover is due to Lovász [199] and Chvátal [50]. The analysis of constrained set multicover is due to Rajagopalan and Vazirani [235]. For algorithms for covering integer programs, see Dobson [65] and Rajagopalan and Vazirani [235].

14 Rounding Applied to Set Cover

We will introduce the technique of LP-rounding by using it to design two approximation algorithms for the set cover problem, Problem 2.1. The first is a simple rounding algorithm achieving a guarantee of f, where f is the frequency of the most frequent element. The second algorithm, achieving an approximation guarantee of $O(\log n)$, illustrates the use of randomization in rounding.

Consider the polyhedron defined by feasible solutions to an LP-relaxation. For some problems, one can find special properties of extreme point solutions of this polyhedron, which can yield rounding-based algorithms. One such property is *half-integrality*, i.e., in each extreme point solution, every coordinate is $0, 1$, or $1/2$. In Section 14.3 we will show that the vertex cover problem possesses this remarkable property. This directly gives a factor 2 algorithm for weighted vertex cover; namely, find an optimal extreme point solution and round all the halves to 1. A more general property, together with an enhanced rounding algorithm, called iterated rounding, is introduced in Chapter 23.

14.1 A simple rounding algorithm

A linear programming relaxation for the set cover problem is given in LP(13.2). One way of converting a solution to this linear program into an integral solution is to round up all nonzero variables to 1. It is easy to construct examples showing that this could increase the cost by a factor of $\Omega(n)$ (see Example 14.3). However, this simple algorithm does achieve the desired approximation guarantee of f (see Exercise 14.1). Let us consider a slight modification of this algorithm that is easier to prove and picks fewer sets in general:

Algorithm 14.1 (Set cover via LP-rounding)

1. Find an optimal solution to the LP-relaxation.
2. Pick all sets S for which $x_S \geq 1/f$ in this solution.

Theorem 14.2 *Algorithm 14.1 achieves an approximation factor of f for the set cover problem.*

Proof: Let $\mathcal{C}$ be the collection of picked sets. Consider an arbitrary element e. Since e is in at most f sets, one of these sets must be picked to the extent of at least $1/f$ in the fractional cover. Thus, e is covered by $\mathcal{C}$, and hence $\mathcal{C}$ is a valid set cover. The rounding process increases x_S, for each set $S \in \mathcal{C}$, by a factor of at most f. Therefore, the cost of $\mathcal{C}$ is at most f times the cost of the fractional cover, thereby proving the desired approximation guarantee. □

The set cover instance arising from a vertex cover problem has $f = 2$. Therefore, Algorithm 14.1 gives a factor 2 approximation algorithm for the weighted vertex cover problem, thus matching the approximation guarantee established in Theorem 2.7.

Example 14.3 Let us give a tight example for Algorithm 14.1. For simplicity, we will view a set cover instance as a hypergraph: sets correspond to vertices and elements correspond to hyperedges (this is a generalization of the transformation that helped us view a set cover instance with each element having frequency 2 as a vertex cover instance).

Let $V_1, \ldots, V_k$ be k disjoint sets of cardinality n each. The hypergraph has vertex set $V = V_1 \cup \ldots \cup V_k$, and n^k hyperedges; each hyperedge picks one vertex from each V_i. In the set cover instance, elements correspond to hyperedges and sets correspond to vertices. Once again, inclusion corresponds to incidence. Each set has cost 1. Picking each set to the extent of $1/k$ gives an optimal fractional cover of cost n. Given this fractional solution, the rounding algorithm will pick all nk sets. On the other hand, picking all sets corresponding to vertices in V_1 gives a set cover of cost n. □

14.2 Randomized rounding

A natural idea for rounding an optimal fractional solution is to view the fractions as probabilities, flip coins with these biases and round accordingly. Let us show how this idea leads to an $O(\log n)$ factor randomized approximation algorithm for the set cover problem.

First, we will show that each element is covered with constant probability by the sets picked by this process. Repeating this process $O(\log n)$ times, and picking a set if it is chosen in any of the iterations, we get a set cover with high probability, by a standard coupon collector argument. The expected cost of cover picked in this manner is $O(\log n)\cdot\mathrm{OPT}_f \le O(\log n)\cdot\mathrm{OPT}$, where OPT_f is the cost of an optimal solution to the LP-relaxation. Applying Markov's Inequality, we convert this into a high probability statement. We provide details below.

Let $\boldsymbol{x} = \boldsymbol{p}$ be an optimal solution to the linear program. For each set $S \in \mathcal{S}$, pick S with probability p_S, the entry corresponding to S in $\boldsymbol{p}$. Let $\mathcal{C}$ be the collection of sets picked. The expected cost of $\mathcal{C}$,

$$\mathbf{E}[c(\mathcal{C})] = \sum_{S \in \mathcal{S}} \mathbf{Pr}[S \text{ is picked}] \cdot c(S) = \sum_{S \in \mathcal{S}} p_S \cdot c(S) = \mathrm{OPT}_f.$$

Next, let us compute the probability that an element $a \in U$ is covered by $\mathcal{C}$. Suppose that a occurs in k sets of $\mathcal{S}$. Let the probabilities associated with these sets be $p_1, \ldots, p_k$. Since a is fractionally covered in the optimal solution, $p_1 + p_2 + \cdots + p_k \geq 1$. Using elementary calculus, it is easy to show that under this condition, the probability that a is covered by $\mathcal{C}$ is minimized when each of the p_i's is $1/k$. Thus,

$$\mathbf{Pr}[a \text{ is covered by } \mathcal{C}] \geq 1 - \left(1 - \frac{1}{k}\right)^k \geq 1 - \frac{1}{e},$$

where e is the base of natural logarithms. Hence each element is covered with constant probability by $\mathcal{C}$.

To get a complete set cover, independently pick $d \log n$ such subcollections, and compute their union, say $\mathcal{C}'$, where d is a constant such that

$$\left(\frac{1}{e}\right)^{d \log n} \leq \frac{1}{4n}.$$

Now,

$$\mathbf{Pr}[a \text{ is not covered by } \mathcal{C}'] \leq \left(\frac{1}{e}\right)^{d \log n} \leq \frac{1}{4n}.$$

Summing over all elements $a \in U$, we get

$$\mathbf{Pr}[\mathcal{C}' \text{ is not a valid set cover}] \leq n \cdot \frac{1}{4n} \leq \frac{1}{4}.$$

Clearly, $\mathbf{E}[c(\mathcal{C}')] \leq \mathrm{OPT}_f \cdot d \log n$. Applying Markov's Inequality (see Section B.2) with $t = \mathrm{OPT}_f \cdot 4d \log n$, we get

$$\mathbf{Pr}[c(\mathcal{C}') \geq \mathrm{OPT}_f \cdot 4d \log n] \leq \frac{1}{4}.$$

The probability of the union of the two undesirable events is $\leq 1/2$. Hence,

$$\mathbf{Pr}[\mathcal{C}' \text{ is a valid set cover and has cost} \leq \mathrm{OPT}_f \cdot 4d \log n] \geq \frac{1}{2}.$$

Observe that we can verify in polynomial time whether C' satisfies both these conditions. If not, we repeat the entire algorithm. The expected number of repetitions needed is at most 2.

14.3 Half-integrality of vertex cover

Consider the vertex cover problem with arbitrary weights. Let $c : V \to \mathbf{Q}^+$ be the function assigning nonnegative weights to the vertices. The integer program for this problem is:

$$\begin{aligned} \text{minimize} \quad & \sum_{v\in V} c(v)x_v && && (14.1)\\ \text{subject to} \quad & x_u + x_v \geq 1, && (u,v)\in E \\ & x_v \in \{0,1\}, && v \in V \end{aligned}$$

The LP-relaxation of this integer program is:

$$\begin{aligned} \text{minimize} \quad & \sum_{v\in V} c(v)x_v && && (14.2)\\ \text{subject to} \quad & x_u + x_v \geq 1, && (u,v)\in E \\ & x_v \geq 0, && v \in V \end{aligned}$$

Recall that an *extreme point solution* of a set of linear inequalities is a feasible solution that cannot be expressed as convex combination of two other feasible solutions. A *half-integral solution* to LP (14.2) is a feasible solution in which each variable is 0, 1, or 1/2.

Lemma 14.4 *Let x be a feasible solution to LP (14.2) that is not half-integral. Then, x is the convex combination of two feasible solutions and is therefore not an extreme point solution for the set of inequalities in LP (14.2).*

Proof: Consider the set of vertices for which solution x does not assign half-integral values. Partition this set as follows.

$$V_+ = \left\{ v \;\middle|\; \frac{1}{2} < x_v < 1 \right\}, \quad V_- = \left\{ v \;\middle|\; 0 < x_v < \frac{1}{2} \right\}.$$

For $\varepsilon > 0$, define the following two solutions.

$$y_v = \begin{cases} x_v + \varepsilon, & x_v \in V_+ \\ x_v - \varepsilon, & x_v \in V_- \\ x_v, & \textit{otherwise} \end{cases}, \qquad z_v = \begin{cases} x_v - \varepsilon, & x_v \in V_+ \\ x_v + \varepsilon, & x_v \in V_- \\ x_v, & \textit{otherwise.} \end{cases}$$

By assumption, $V_+ \cup V_- \neq \emptyset$, and so $\boldsymbol{x}$ is distinct from $\boldsymbol{y}$ and $\boldsymbol{z}$. Furthermore, $\boldsymbol{x}$ is a convex combination of $\boldsymbol{y}$ and $\boldsymbol{z}$, since $\boldsymbol{x} = \frac{1}{2}(\boldsymbol{y} + \boldsymbol{z})$. We will show, by choosing $\varepsilon > 0$ small enough, that $\boldsymbol{y}$ and $\boldsymbol{z}$ are both feasible solutions for LP (14.2), thereby establishing the lemma.

Ensuring that all coordinates of $\boldsymbol{y}$ and $\boldsymbol{z}$ are nonnegative is easy. Next, consider the edge constraints. Suppose $x_u + x_v > 1$. Clearly, by choosing ε small enough, we can ensure that $\boldsymbol{y}$ and $\boldsymbol{z}$ do not violate the constraint for such an edge. Finally, for an edge such that $x_u + x_v = 1$, there are only three possibilities: $x_u = x_v = \frac{1}{2}$; $x_u = 0, x_v = 1$; and $u \in V_+, v \in V_-$. In all three cases, for any choice of ε,

$$x_u + x_v = y_u + y_v = z_u + z_v = 1.$$

The lemma follows. □

This leads to:

Theorem 14.5 *Any extreme point solution for the set of inequalities in LP (14.2) is half-integral.*

Theorem 14.5 directly leads to a factor 2 approximation algorithm for weighted vertex cover: find an extreme point solution, and pick all vertices that are set to half or one in this solution.

14.4 Exercises

14.1 Modify Algorithm 14.1 so that it picks all sets that are nonzero in the fractional solution. Show that the algorithm also achieves a factor of f.
Hint: Use the primal complementary slackness conditions to prove this.

14.2 Consider the collection of sets, $\mathcal{C}$, picked by the randomized rounding algorithm. Show that with some constant probability, $\mathcal{C}$ covers at least half the elements at a cost of at most $O(\mathrm{OPT})$.

14.3 Give $O(\log n)$ factor randomized rounding algorithms for the set multicover and multiset multicover problems (see Section 13.2).

14.4 Give a (non-bipartite) tight example for the half-integrality-based algorithm for weighted vertex cover.

14.5 (J. Cheriyan) Give a polynomial time algorithm for the following problem. Given a graph G with nonnegative vertex weights and a valid, though not necessarily optimal, coloring of G, find a vertex cover of weight $\leq (2-\frac{2}{k})\mathrm{OPT}$, where k is the number of colors used.

14.6 Give a counterexample to the following claim. A set cover instance in which each element is in exactly f sets has a $(1/f)$-integral optimal fractional solution (i.e., in which each set is picked an integral multiple of $1/f$).

14.7 This exercise develops a combinatorial algorithm for finding an optimal half integral vertex cover. Given undirected graph $G = (V, E)$ and a non-negative cost function c on vertices, obtain bipartite graph $H(V', V'', E')$ as follows. Corresponding to each vertex $v \in V$, there are vertices $v' \in V'$ and $v'' \in V''$ each of cost $c(v)/2$. Corresponding to each edge $(u, v) \in E$, there are two edges $(u', v''), (u'', v') \in E'$. Show that a vertex cover in H can be mapped to a half-integral vertex cover in G preserving total cost and vice versa. Use the fact that an optimal vertex cover in a bipartite graph can be found in polynomial time to obtain an optimal half-integral vertex cover in G.

14.8 Consider LP (12.8), introduced in Exercise 12.8, for a non-bipartite graph $G = (V, E)$.

1. Show that it is not an exact relaxation for the maximum matching problem in G.
2. Show that this LP always has a half-integral optimal solution.

14.9 In an attempt to improve the running time of the algorithm obtained in Exercise 9.7 for bin packing, consider going to the LP-relaxation of the integer program and using LP-rounding. What guarantee can you establish for bin packing through this method?

14.5 Notes

Algorithm 14.1 is due to Hochbaum [132]. For a more sophisticated randomized rounding algorithm for set cover, see Srinivasan [252]. Theorem 14.5 is due to Nemhauser and Trotter [221].

15 Set Cover via the Primal–Dual Schema

As noted in Section 12.3, the primal–dual schema is the method of choice for designing approximation algorithms since it yields combinatorial algorithms with good approximation factors and good running times. We will first present the central ideas behind this schema and then use it to design a simple f factor algorithm for set cover, where f is the frequency of the most frequent element.

The primal–dual schema has its origins in the design of exact algorithms. In that setting, this schema yielded the most efficient algorithms for some of the cornerstone problems in **P**, including matching, network flow, and shortest paths. These problems have the property that their LP-relaxations have integral optimal solutions. By Theorem 12.3 we know that optimal solutions to linear programs are characterized by fact that they satisfy all the complementary slackness conditions. In fact, the primal–dual schema for exact algorithms is driven by these conditions. Starting with initial feasible solutions to the primal and dual programs, it iteratively starts satisfying complementary slackness conditions. When they are all satisfied, both solutions must be optimal. During the iterations, the primal is always modified integrally, so that eventually we get an integral optimal solution.

Consider an LP-relaxation for an **NP**-hard problem. In general, the relaxation will not have an optimal solution that is integral. Does this rule out a complementary slackness condition driven approach? Interestingly enough, the answer is 'no'. It turns out that the algorithm can be driven by a suitable relaxation of these conditions! This is the most commonly used way of designing primal–dual based approximation algorithms – but not the only way.

15.1 Overview of the schema

Let us consider the following primal program, written in standard form.

$$\text{minimize} \quad \sum_{j=1}^{n} c_j x_j$$

$$\begin{array}{lll} \text{subject to} & \displaystyle\sum_{j=1}^{n} a_{ij}x_j \geq b_i, & i = 1, \dots, m \\ & x_j \geq 0, & j = 1, \dots, n \end{array}$$

where a_{ij}, b_i, and c_j are specified in the input. The dual program is:

$$\begin{array}{lll} \text{maximize} & \displaystyle\sum_{i=1}^{m} b_i y_i & \\ \text{subject to} & \displaystyle\sum_{i=1}^{m} a_{ij}y_i \leq c_j, & j = 1, \dots, n \\ & y_i \geq 0, & i = 1, \dots, m \end{array}$$

Most known approximation algorithms using the primal–dual schema run by ensuring one set of conditions and suitably relaxing the other. In the following description we capture both situations by relaxing both conditions. Eventually, if primal conditions are ensured, we set $\alpha = 1$, and if dual conditions are ensured, we set $\beta = 1$.

Primal complementary slackness conditions
Let $\alpha \geq 1$.
For each $1 \leq j \leq n$: either $x_j = 0$ or $c_j/\alpha \leq \sum_{i=1}^{m} a_{ij}y_i \leq c_j$.

Dual complementary slackness conditions
Let $\beta \geq 1$.
For each $1 \leq i \leq m$: either $y_i = 0$ or $b_i \leq \sum_{j=1}^{n} a_{ij}x_j \leq \beta \cdot b_i$,

Proposition 15.1 *If x and y are primal and dual feasible solutions satisfying the conditions stated above then*

$$\sum_{j=1}^{n} c_j x_j \leq \alpha \cdot \beta \cdot \sum_{i=1}^{m} b_i y_i.$$

Proof: The proof uses the local paying mechanism alluded to in Exercise 12.4. Let us state this mechanism formally. We will assume that the "money" possessed by dual variable y_i is $\alpha\beta b_i y_i$. Hence, the entire dual is worth the r.h.s. of the inequality to be proven and the cost of the primal solution is the l.h.s. Recall that dual variables correspond to primal constraints and vice versa – we will use this correspondence below.

Dual variables pay to buy the primal solution as follows: y_i pays $\alpha y_i a_{ij} x_j$ towards primal variable x_i. The total amount paid by y_i is

$$\alpha y_i \sum_{j=1}^{n} a_{ij}x_j \leq \alpha\beta b_i y_i,$$

where the inequality follows from the relaxed dual complementary slackness conditions. Thus the total amount paid by y_i is bounded by the amount of money it has.

The total amount collected by x_j is

$$\alpha x_j \sum_{i=1}^{m} a_{ij} y_i \geq c_j x_j,$$

where the inequality follows from the relaxed primal complementary slackness conditions. Hence the total amount collected by the primal variables covers the cost of the primal solution. □

The algorithm starts with a primal infeasible solution and a dual feasible solution; these are usually the trivial solutions $\boldsymbol{x} = \mathbf{0}$ and $\boldsymbol{y} = \mathbf{0}$. It iteratively improves the feasibility of the primal solution and the optimality of the dual solution, ensuring that in the end a primal feasible solution is obtained and all conditions stated above, with a suitable choice of α and β, are satisfied. The primal solution is always extended integrally, thus ensuring that the final solution is integral. Finally, the cost of the dual solution is used as a lower bound on OPT, and by Proposition 15.1, the approximation guarantee of the algorithm is $\alpha\beta$.

The improvements to the primal and the dual go hand-in-hand: the current primal solution suggests how to improve the dual, and vice versa. The improvements are "local", much in the spirit of the local paying mechanism outlined in Proposition 15.1. This is also the paradigm underlying the schema: two processes making local improvements relative to each other and achieving global optimality.

15.2 Primal–dual schema applied to set cover

Let us obtain a factor f algorithm for the set cover problem using the primal–dual schema. For this algorithm, we will choose $\alpha = 1$ and $\beta = f$. We will work with the primal and dual pair of LP's given in (13.2) and (13.3), respectively. The complementary slackness conditions are:

Primal conditions

$$\forall S \in \mathcal{S} : x_S \neq 0 \Rightarrow \sum_{e:\ e \in S} y_e = c(S).$$

Set S will be said to be *tight* if $\sum_{e:\ e \in S} y_e = c(S)$. Since we will increment the primal variables integrally, we can state the conditions as: *Pick only tight sets in the cover.*

Clearly, in order to maintain dual feasibility, we are not allowed to overpack any set.

Dual conditions

$$\forall e : y_e \neq 0 \Rightarrow \sum_{S:\ e \in S} x_S \leq f$$

Since we will find a 0/1 solution for $\boldsymbol{x}$, these conditions are equivalent to: *Each element having a nonzero dual value can be covered at most f times.* Since each element is in at most f sets, this condition is trivially satisfied for all elements.

The two sets of conditions naturally suggest the following algorithm:

Algorithm 15.2 (Set cover – factor f)

1. **Initialization:** $\boldsymbol{x} \leftarrow \mathbf{0}$; $\boldsymbol{y} \leftarrow \mathbf{0}$
2. Until all elements are covered, do:
   Pick an uncovered element, say e, and raise y_e until some set goes tight.
   Pick all tight sets in the cover and update $\boldsymbol{x}$.
   Declare all the elements occurring in these sets as "covered".
3. Output the set cover $\boldsymbol{x}$.

Theorem 15.3 *Algorithm 15.2 achieves an approximation factor of f.*

Proof: Clearly there will be no uncovered elements and no overpacked sets at the end of the algorithm. Thus, the primal and dual solutions will both be feasible. Since they satisfy the relaxed complementary slackness conditions with $\alpha = 1$ and $\beta = f$, by Proposition 15.1 the approximation factor is f. □

Example 15.4 A tight example for this algorithm is provided by the following set system:

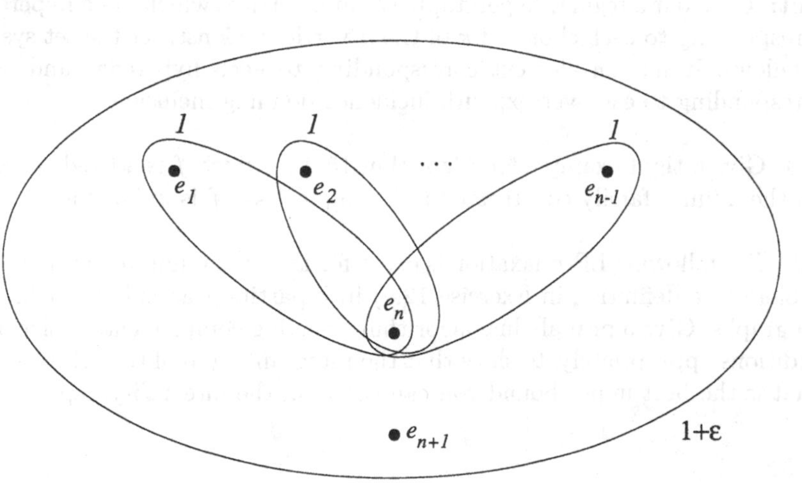

Here, $\mathcal{S}$ consists of $n-1$ sets of cost 1, $\{e_1, e_n\}, \ldots, \{e_{n-1}, e_n\}$, and one set of cost $1+\varepsilon$, $\{e_1, \ldots, e_{n+1}\}$, for a small $\varepsilon > 0$. Since e_n appears in all n sets, this set system has $f = n$.

Suppose the algorithm raises y_{e_n} in the first iteration. When y_{e_n} is raised to 1, all sets $\{e_i, e_n\}$, $i = 1, \ldots, n-1$ go tight. They are all picked in the cover, thus covering the elements $e_1, \ldots, e_n$. In the second iteration, $y_{e_{n+1}}$ is raised to ε and the set $\{e_1, \ldots, e_{n+1}\}$ goes tight. The resulting set cover has a cost of $n+\varepsilon$, whereas the optimum cover has cost $1+\varepsilon$. □

15.3 Exercises

15.1 This extends Exercise 12.4. Show that if $\boldsymbol{x}$ and $\boldsymbol{y}$ are primal and dual feasible solutions and satisfy the conditions of Section 15.2 with $\alpha = 1$ and $\beta = f$, then $\boldsymbol{y}$ pays for $\boldsymbol{x}$ at a rate of f, i.e.,

$$\sum_S c(S) x_S \leq f \cdot \sum_e y_e .$$

15.2 Remove the scaffolding of linear programming from Algorithm 15.2 to obtain a purely combinatorial factor f algorithm for set cover.
Hint: See the algorithm in Exercise 2.11.

15.3 Let k be a fixed constant, and consider instances of set cover whose maximum frequency, f, is bounded by k. Algorithm 15.2 shows that the integrality gap of LP (13.2) is upper bounded by k for these instances. Provide examples to show that this bound is essentially tight.
Hint: Consider a regular hypergraph, G, on n vertices which has a hyperedge corresponding to each choice of k of the n vertices. Construct the set system as follows. It has an element corresponding to each hyperedge and a set corresponding to each vertex, with incidence defining inclusion.

15.4 Give a tight example for Algorithm 15.2 in which f is a fixed constant (for the infinite family constructed in Example 15.4, f is unbounded).

15.5 The following LP-relaxation is exact for the maximum weight matching problem (see definition in Exercise 12.9) in bipartite graphs but not in general graphs. Give a primal–dual algorithm, relaxing complementary slackness conditions appropriately, to show that the integrality gap of this LP is $\geq 1/2$. What is the best upper bound you can place on the integrality gap?

$$\begin{array}{lll} \text{maximize} & \sum_e w_e x_e & \\ \text{subject to} & \sum_{e:\ e \text{ incident at } v} x_e \le 1, & v \in V \\ & x_e \ge 0, & e \in E \end{array} \tag{15.1}$$

15.6 (Chudak, Goemans, Hochbaum, and Williamson [48]) Interpret the layering-based algorithms obtained for set cover and feedback vertex set problems in Chapters 2 and 6 as primal–dual schema based algorithms. How are the complementary slackness conditions being relaxed?

15.4 Notes

Kuhn [186] gave the first primal–dual algorithm – for the weighted bipartite matching problem – however, he used the name "Hungarian Method" to describe his algorithm. Dantzig, Ford, and Fulkerson [63] used this method for giving another means of solving linear programs and called it the primal–dual method. Although the schema was not very successful for solving linear programs, it soon found widespread use in combinatorial optimization.

Algorithm 15.2 is due to Bar-Yehuda and Even [21]. Although it was not originally stated as a primal–dual algorithm, in retrospect, this was the first use of the schema in approximation algorithms. The works of Agrawal, Klein, and Ravi [1] and Goemans and Williamson [111] revived the use of this schema in the latter setting, and introduced the powerful idea of growing duals in a synchronized manner (see Chapter 22). The mechanism of relaxing complementary slackness conditions was first formalized in Williamson, Goemans, Mihail, and Vazirani [266]. For further historical information, see Goemans and Williamson [113].

16 Maximum Satisfiability

The maximum satisfiability problem has been a classical problem in approximation algorithms. More recently, its study has led to crucial insights in the area of hardness of approximation (see Chapter 29). In this chapter, we will use LP-rounding, with randomization, to obtain a 3/4 factor approximation algorithm. We will derandomize this algorithm using the *method of conditional expectation.*

Problem 16.1 (Maximum satisfiability (MAX-SAT)) Given a conjunctive normal form formula f on Boolean variables $x_1, \ldots, x_n$, and non-negative weights, w_c, for each clause c of f, find a truth assignment to the Boolean variables that maximizes the total weight of satisfied clauses. Let $\mathcal{C}$ represent the set of clauses of f, i.e., $f = \bigwedge_{c \in \mathcal{C}} c$. Each clause is a disjunction of literals; each literal being either a Boolean variable or its negation. Let size(c) denote the *size* of clause c, i.e., the number of literals in it. We will assume that the sizes of clauses in f are arbitrary.

For any positive integer k, we will denote by MAX-kSAT the restriction of MAX-SAT to instances in which each clause is of size at most k. MAX-SAT is **NP**-hard; in fact, even MAX-2SAT is **NP**-hard (in contrast, 2SAT is in **P**). We will first present two approximation algorithms for MAX-SAT, having guarantees of $1/2$ and $1-1/e$, respectively. The first performs better if the clause sizes are large, and the seconds performs better if they are small. We will then show how an appropriate combination of the two algorithms achieves the promised approximation guarantee.

In the interest of minimizing notation, let us introduce common terminology for all three algorithms. Random variable W will denote the total weight of satisfied clauses. For each clause c, random variable W_c denotes the weight contributed by clause c to W. Thus, $W = \sum_{c \in \mathcal{C}} W_c$ and

$$\mathbf{E}[W_c] = w_c \cdot \mathbf{Pr}[c \text{ is satisfied}].$$

(Strictly speaking, this is abuse of notation, since the randomization used by the three algorithms is different.)

16.1 Dealing with large clauses

The first algorithm is straightforward. Set each Boolean variable to be True independently with probability 1/2 and output the resulting truth assignment, say τ. For $k \geq 1$, define $\alpha_k = 1 - 2^{-k}$.

Lemma 16.2 *If* $\text{size}(c) = k$, *then* $\mathbf{E}[W_c] = \alpha_k w_c$.

Proof: Clause c is not satisfied by τ iff all its literals are set to False. The probability of this event is 2^{-k}. □

For $k \geq 1, \alpha_k \geq 1/2$. By linearity of expectation,

$$\mathbf{E}[W] = \sum_{c \in \mathcal{C}} \mathbf{E}[W_c] \geq \frac{1}{2} \sum_{c \in \mathcal{C}} w_c \geq \frac{1}{2}\text{OPT},$$

where we have used a trivial upper bound on OPT – the total weight of clauses in $\mathcal{C}$.

Instead of converting this into a high probability statement, with a corresponding loss in guarantee, we show how to derandomize this procedure. The resulting algorithm deterministically computes a truth assignment such that the weight of satisfied clauses is $\geq \mathbf{E}[W] \geq \text{OPT}/2$.

Observe that α_k increases with k and the guarantee of this algorithm is 3/4 if each clause has two or more literals. (The next algorithm is designed to deal with unit clauses more effectively.)

16.2 Derandomizing via the method of conditional expectation

We will critically use the self-reducibility of SAT (see Section A.5). Consider the self-reducibility tree T for formula f. Each internal node at level i corresponds to a setting for Boolean variables $x_1, \ldots, x_i$, and each leaf represents a complete truth assignment to the n variables. Let us label each node of T with its conditional expectation as follows. Let $a_1, \ldots, a_i$ be a truth assignment to $x_1, \ldots, x_i$. The node corresponding to this assignment will be labeled with $\mathbf{E}[W|x_1 = a_1, \ldots, x_i = a_i]$. If $i = n$, this is a leaf node and its conditional expectation is simply the total weight of clauses satisfied by its truth assignment.

Lemma 16.3 *The conditional expectation of any node in* T *can be computed in polynomial time.*

Proof: Consider a node $x_1 = a_1, \ldots, x_i = a_i$. Let ϕ be the Boolean formula, on variables $x_{i+1}, \ldots, x_n$, obtained for this node via self-reducibility. Clearly, the expected weight of satisfied clauses of ϕ under a random truth assignment to the variables $x_{i+1}, \ldots, x_n$ can be computed in polynomial time. Adding to this the total weight of clauses of f already satisfied by the partial assignment $x_1 = a_1, \ldots, x_i = a_i$ gives the answer. □

Theorem 16.4 *We can compute, in polynomial time, a path from the root to a leaf such that the conditional expectation of each node on this path is* $\geq \mathbf{E}[W]$.

Proof: The conditional expectation of a node is the average of the conditional expectations of its two children, i.e.,

$$\begin{aligned}\mathbf{E}[W|x_1 = a_1, ..., x_i = a_i] = \mathbf{E}[W|x_1 = a_1, ..., x_i = a_i, x_{i+1} = \text{True}]/2 + \\ \mathbf{E}[W|x_1 = a_1, ..., x_i = a_i, x_{i+1} = \text{False}]/2.\end{aligned}$$

The reason, of course, is that x_{i+1} is equally likely to be set to True or False. As a result, the child with the larger value has a conditional expectation at least as large as that of the parent. This establishes the existence of the desired path. As a consequence of Lemma 16.3, it can be computed in polynomial time. □

The deterministic algorithm follows as a corollary of Theorem 16.4. We simply output the truth assignment on the leaf node of the path computed. The total weight of clauses satisfied by it is $\geq \mathbf{E}[W]$.

Let us show that the technique outlined above can, in principle, be used to derandomize more complex randomized algorithms. Suppose the algorithm does not set the Boolean variables independently of each other (for instance, see Remark 16.6). Now,

$$\begin{aligned}&\mathbf{E}[W|x_1 = a_1, ..., x_i = a_i] = \\ &\mathbf{E}[W|x_1 = a_1, ..., x_i = a_i, x_{i+1} = \text{True}] \cdot \mathbf{Pr}[x_{i+1} = \text{True}|x_1 = a_1, ..., x_i = a_i] + \\ &\mathbf{E}[W|x_1 = a_1, ..., x_i = a_i, x_{i+1} = \text{False}] \cdot \mathbf{Pr}[x_{i+1} = \text{False}|x_1 = a_1, ..., x_i = a_i].\end{aligned}$$

The sum of the two conditional probabilities is again 1, since the two events are exhaustive. So, the conditional expectation of the parent is still a convex combination of the conditional expectations of the two children. If we can determine, in polynomial time, which of the two children has a larger value, we can again derandomize the algorithm. However, computing the conditional expectations may not be easy. Observe how critically independence was used in the proof of Lemma 16.3. It was because of independence that we could assume a random truth assignment on Boolean variables $x_{i+1}, \ldots, x_n$ and thereby compute the expected weight of satisfied clauses of ϕ.

In general, a randomized algorithm may pick from a larger set of choices and not necessarily with equal probability. But once again a convex combination of the conditional expectations of these choices, given by the probabilities of picking them, equals the conditional expectation of the parent. Hence there must be a choice that has at least as large a conditional expectation as the parent.

16.3 Dealing with small clauses via LP-rounding

Following is an integer program for MAX-SAT. For each clause $c \in \mathcal{C}$, let S_c^+ (S_c^-) denote the set of Boolean variables occurring nonnegated (negated) in c. The truth assignment is encoded by $\boldsymbol{y}$. Picking $y_i = 1$ ($y_i = 0$) denotes setting x_i to True (False). The constraint for clause c ensures that z_c can be set to 1 only if at least one of the literals occurring in c is set to True, i.e., if clause c is satisfied by the picked truth assignment.

$$\begin{aligned} \text{maximize} \quad & \sum_{c \in \mathcal{C}} w_c z_c && (16.1) \\ \text{subject to} \quad & \forall c \in \mathcal{C}: \sum_{i \in S_c^+} y_i + \sum_{i \in S_c^-} (1 - y_i) \geq z_c \\ & \forall c \in \mathcal{C}: z_c \in \{0, 1\} \\ & \forall i: y_i \in \{0, 1\} \end{aligned}$$

The LP-relaxation is:

$$\begin{aligned} \text{maximize} \quad & \sum_{c \in \mathcal{C}} w_c z_c && (16.2) \\ \text{subject to} \quad & \forall c \in \mathcal{C}: \sum_{i \in S_c^+} y_i + \sum_{i \in S_c^-} (1 - y_i) \geq z_c \\ & \forall c \in \mathcal{C}: 1 \geq z_c \geq 0 \\ & \forall i: 1 \geq y_i \geq 0 \end{aligned}$$

The algorithm is again straightforward. Solve LP (16.2). Let $(\boldsymbol{y}^*, \boldsymbol{z}^*)$ denote the optimal solution. Independently set x_i to True with probability y_i^*, for $1 \leq i \leq n$. Output the resulting truth assignment, say τ.

We will use the random variables W and W_c defined in Section 16.1. For $k \geq 1$, define

$$\beta_k = 1 - \left(1 - \frac{1}{k}\right)^k.$$

Lemma 16.5 *If* size$(c) = k$, *then*

$$\mathbf{E}[W_c] \geq \beta_k w_c z_c^*.$$

Proof: We may assume w.l.o.g. that all literals in c appear nonnegated (if x_i appears negated, we can replace x_i with $\overline{x}_i$ throughout f and modify LP (16.2) accordingly without affecting z_c^* or W_c). Further, by renaming variables, we may assume $c = (x_1 \vee \ldots \vee x_k)$.

Clause c is satisfied if $x_1, \ldots, x_k$ are not all set to False. The probability of this event is

$$1 - \prod_{i=1}^{k}(1 - y_i) \geq 1 - \left(\frac{\sum_{i=1}^{k}(1 - y_i)}{k}\right)^k = 1 - \left(1 - \frac{\sum_{i=1}^{k} y_i}{k}\right)^k$$
$$\geq 1 - \left(1 - \frac{z_c^*}{k}\right)^k,$$

where the first inequality follows from the arithmetic-geometric mean inequality which states that for nonnegative numbers $a_1, \ldots, a_k$,

$$\frac{a_1 + \ldots + a_k}{k} \geq \sqrt[k]{a_1 \times \ldots \times a_k}.$$

The second inequality uses the constraint in LP (16.2) that $y_1 + \ldots + y_k \geq z_c$.

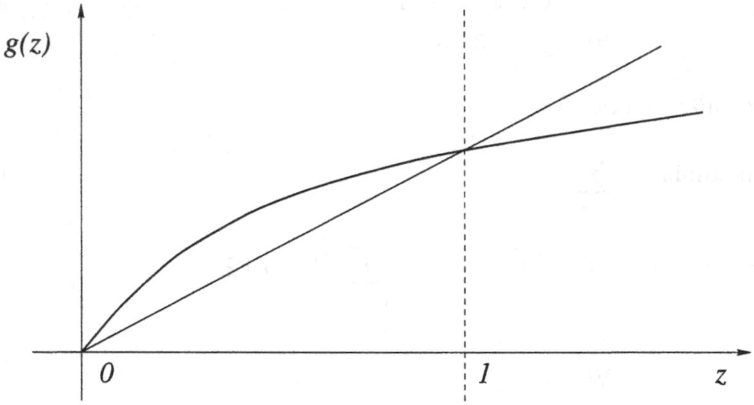

Define function g by:

$$g(z) = 1 - \left(1 - \frac{z}{k}\right)^k.$$

This is a concave function with $g(0) = 0$ and $g(1) = \beta_k$. Therefore, for $z \in [0,1]$, $g(z) \geq \beta_k z$. Hence, $\mathbf{Pr}[c$ is satisfied$] \geq \beta_k z_c^*$. The lemma follows. □

Notice that β_k is a decreasing function of k. Thus, if all clauses are of size at most k,

$$\mathbf{E}[W] = \sum_{c \in \mathcal{C}} \mathbf{E}[W_c] \geq \beta_k \sum_{c \in \mathcal{C}} w_c z_c^* = \beta_k \mathrm{OPT}_f \geq \beta_k \mathrm{OPT},$$

where OPT_f is the value of an optimal solution to LP (16.2). Clearly, $\mathrm{OPT}_f \geq \mathrm{OPT}$. This algorithm can also be derandomized using the method of conditional expectation (Exercise 16.3). Hence, for MAX-SAT instances with clause sizes at most k, it is a β_k factor approximation algorithm. Since

$$\forall k \in \mathbf{Z}^+ : \left(1 - \frac{1}{k}\right)^k < \frac{1}{e},$$

this is a $1 - 1/e$ factor algorithm for MAX-SAT.

16.4 A 3/4 factor algorithm

We will combine the two algorithms as follows. Let b be the flip of a fair coin. If $b = 0$, run the first randomized algorithm, and if $b = 1$, run the second randomized algorithm.

Remark 16.6 Notice that we are effectively setting x_i to True with probability $\frac{1}{4} + \frac{1}{2}y_i^*$; however, the x_i's are *not* set independently!

Let $(\boldsymbol{y}^*, \boldsymbol{z}^*)$ be an optimal solution to LP (16.2) on the given instance.

Lemma 16.7 $\quad \mathbf{E}[W_c] \geq \frac{3}{4} w_c z_c^*.$

Proof: Let size$(c) = k$. By Lemma 16.2,

$$\mathbf{E}[W_c \mid b = 0] = \alpha_k w_c \geq \alpha_k w_c z_c^*,$$

where we have used the fact that $z_c^* \leq 1$. By Lemma 16.5,

$$\mathbf{E}[W_c \mid b = 1] \geq \beta_k w_c z_c^*.$$

Combining we get

$$\mathbf{E}[W_c] = \frac{1}{2}(\mathbf{E}[W_c \mid b = 0] + \mathbf{E}[W_c \mid b = 1]) \geq w_c z_c^* \frac{(\alpha_k + \beta_k)}{2}.$$

Now, $\alpha_1 + \beta_1 = \alpha_2 + \beta_2 = 3/2$, and for $k \geq 3$, $\alpha_k + \beta_k \geq 7/8 + (1 - 1/e) \geq 3/2$. The lemma follows. □

By linearity of expectation,

$$\mathbf{E}[W] = \sum_{c \in \mathcal{C}} \mathbf{E}[W_c] \geq \frac{3}{4} \sum_{c \in \mathcal{C}} w_c z_c^* = \frac{3}{4}\mathrm{OPT}_f \geq \frac{3}{4}\mathrm{OPT}, \tag{16.3}$$

where OPT_f is the optimal solution to LP (16.2). Finally, consider the following deterministic algorithm.

Algorithm 16.8 (MAX-SAT – factor 3/4)

1. Use the derandomized factor 1/2 algorithm to get a truth assignment, τ_1.
2. Use the derandomized factor $1 - 1/e$ algorithm to get a truth assignment, τ_2.
3. Output the better of the two assignments.

Theorem 16.9 *Algorithm 16.8 is a deterministic factor* 3/4 *approximation algorithm for MAX-SAT.*

Proof: By Lemma 16.7, the average of the weights of satisfied clauses under τ_1 and τ_2 is $\geq \frac{3}{4}\mathrm{OPT}$. Hence the better of these two assignments also does at least as well. □

By (16.3), $\mathbf{E}[W] \geq \frac{3}{4}\mathrm{OPT}_f$. The weight of the integral solution produced by Algorithm 16.8 is at least $\mathbf{E}[W]$. Therefore, the integrality gap of LP (16.2) is $\geq 3/4$. Below we show that this is tight.

Example 16.10 Consider the SAT formula $f = (x_1 \vee x_2) \wedge (\overline{x}_1 \vee x_2) \wedge (x_1 \vee \overline{x}_2) \wedge (\overline{x}_1 \vee \overline{x}_2)$, where each clause is of unit weight. It is easy to see that setting $y_i = 1/2$ and $z_c = 1$ for all i and c is an optimal solution to LP (16.2) for any instance having size 2 clauses. Therefore $\mathrm{OPT}_f = 4$. On the other hand $\mathrm{OPT} = 3$, and thus for this instance LP (16.2) has a integrality gap of 3/4. □

Example 16.11 Let us provide a tight example to Algorithm 16.8. Let $f = (x \vee y) \wedge (x \vee \overline{y}) \wedge (\overline{x} \vee z)$, and let the weights of these three clauses be 1, 1, and $2 + \varepsilon$, respectively. By the remark made in Example 16.10, on this instance the factor $1 - 1/e$ algorithm will set each variable to True with probability 1/2 and so will be the same as the factor 1/2 algorithm. During derandomization, suppose variable x is set first. The conditional expectations are $\mathbf{E}[W \mid x = \text{True}] = 3 + \varepsilon/2$ and $\mathbf{E}[W \mid x = \text{False}] = 3 + \varepsilon$. Thus, x will be set to False. But this leads to a total weight of $3 + \varepsilon$, whereas by setting x to True we can get a weight of $4 + \varepsilon$. Clearly, we can get an infinite family of such examples by replicating these 3 clauses with new variables. □

16.5 Exercises

16.1 The algorithm of Section 16.1 achieves an approximation guarantee of α_k if all clauses in the given instance have size at least k. Give a tight example of factor α_k for this algorithm.

16.2 Show that the following is a factor 1/2 algorithm for MAX-SAT. Let τ be an arbitrary truth assignment and τ' be its complement, i.e., a variable is True in τ iff it is False in τ'. Compute the weight of clauses satisfied by τ and τ', then output the better assignment.

16.3 Use the method of conditional expectation to derandomize the $1-1/e$ factor algorithm for MAX-SAT.

16.4 Observe that the randomization used in the 3/4 factor algorithm does not set Boolean variables independently of each other. As remarked in Section 16.2, the algorithm can still, in principle, be derandomized using the method of conditional expectation. Devise a way of doing so. Observe that the algorithm obtained is different from Algorithm 16.8.

16.5 (Goemans and Williamson [110]) Instead of using the solution to LP (16.2), y_i^*, as probability of setting x_i to True, consider the more general scheme of using $g(y_i^*)$, for a suitable function g. Can this lead to an improvement over the factor $1 - 1/e$ algorithm?

16.6 Consider the following randomized algorithm for the maximum cut problem, defined in Exercise 2.1. After the initialization step of Algorithm 2.13, each of the remaining vertices is equally likely to go in sets A or B. Show that the expected size of the cut found is at least OPT/2. Show that the derandomization of this algorithm via the method of conditional expectation is precisely Algorithm 2.13.

16.7 Consider the following generalization of the maximum cut problem.

Problem 16.12 (Linear equations over GF[2]) Given m equations over n GF[2] variables, find an assignment for the variables that maximizes the number of satisfied equations.

1. Show that if $m \leq n$, this problem is polynomial time solvable.
2. In general, the problem is **NP**-hard. Give a factor 1/2 randomized algorithm for it, and derandomize using the method of conditional expectation.

16.8 Consider the obvious randomized algorithm for the MAX k-CUT problem, Problem 2.14 in Exercise 2.3, which assigns each vertex randomly to one of the sets $S_1, \ldots, S_k$. Show that the expected number of edges running between these sets is at least OPT/2. Show that the derandomization of this algorithm, via the method of conditional expectation, gives the greedy algorithm sought in Exercise 2.3.

16.9 Repeat Exercise 16.8 for the maximum directed cut problem, Problem 2.15 in Exercise 2.4, i.e., give a factor 1/4 randomized algorithm, and show that its derandomization gives a greedy algorithm.

16.6 Notes

The factor 1/2 algorithm, which was also the first approximation algorithm for MAX-SAT, is due to Johnson [157]. The first factor 3/4 algorithm was due to Yannakakis [270]. The (simpler) algorithm given here is due to Goemans and Williamson [110]. The method of conditional expectation is implicit in Erdös and Selfridge [80]. Its use for obtaining polynomial time algorithms was pointed out by Spencer [251] (see Raghavan [233] and Alon and Spencer [7] for enhancements to this technique).

17 Scheduling on Unrelated Parallel Machines

LP-rounding has yielded approximation algorithms for a large number of **NP**-hard problems in scheduling theory (see Section 17.6). As an illustrative example, we present a factor 2 algorithm for the problem of scheduling on unrelated parallel machines. We will apply the technique of parametric pruning, introduced in Chapter 5, together with LP-rounding, to obtain the algorithm.

Problem 17.1 (Scheduling on unrelated parallel machines) Given a set J of jobs, a set M of machines, and for each $j \in J$ and $i \in M$, $p_{ij} \in \mathbf{Z}^+$, the time taken to process job j on machine i, the problem is to schedule the jobs on the machines so as to minimize the *makespan*, i.e., the maximum processing time of any machine. We will denote the number of jobs by n and the number of machines by m.

The reason for the name "unrelated" is that we have not assumed any relation between the processing times of a job on the different machines. If each job j has the same running time, say p_j, on each of the machines, then the machines are said to be *identical*. This problem was studied in Chapter 10 under the name minimum makespan scheduling, and we had derived a PTAS for it. A generalization of minimum makespan, that also admits a PTAS, is that of *uniform* parallel machines (see Exercise 17.5). In this case there is a speed s_i associated with each machine i, and the processing time for job j on machine i is p_j/s_i.

17.1 Parametric pruning in an LP setting

An obvious integer program for this problem is the following. In this program x_{ij} is an indicator variable denoting whether job j is scheduled on machine i. The objective is to minimize t, the makespan. The first set of constraints ensures that each job is scheduled on one of the machines, and the second set ensures that each machine has a processing time of at most t.

$$\begin{aligned} \text{minimize} \quad & t \qquad\qquad (17.1) \\ \text{subject to} \quad & \sum_{i \in M} x_{ij} = 1, \qquad j \in J \end{aligned}$$

$$\sum_{j \in J} x_{ij} p_{ij} \le t, \quad i \in M$$

$$x_{ij} \in \{0,1\}, \quad i \in M,\ j \in J$$

We show below that this integer program has unbounded integrality gap.

Example 17.2 Suppose we have only one job, which has a processing time of m on each of the m machines. Clearly, the minimum makespan is m. However, the optimal solution to the linear relaxation is to schedule the job to the extent of $1/m$ on each machine, thereby leading to an objective function value of 1, and giving an integrality gap of m. □

This example is exploiting an "unfair" advantage that we have given to the linear relaxation. The integer program automatically sets x_{ij} to 0 if $p_{ij} > t$. On the other hand, the linear relaxation is allowed to set these variables to nonzero values, and thereby obtain a cheaper solution. The situation could be rectified if we could add the following constraint to the linear relaxation:

$$\forall i \in M,\ j \in J: \text{ if } p_{ij} > t \text{ then } x_{ij} = 0.$$

However, this is not a linear constraint.

We will use the technique of parametric pruning to get around this difficulty. The parameter will be $T \in \mathbf{Z}^+$, which is our guess for a lower bound on the optimal makespan. The parameter will enable us to prune away all job–machine pairs such that $p_{ij} > T$. Define $S_T = \{(i,j) \mid p_{ij} \le T\}$. We will define a family of linear programs, LP(T), one for each value of parameter $T \in \mathbf{Z}^+$. LP(T) uses the variables x_{ij} for $(i,j) \in S_T$ only, and asks if there is a feasible, fractional schedule of makespan $\le T$ using the restricted possibilities.

$$\sum_{i:(i,j) \in S_T} x_{ij} = 1, \quad j \in J$$

$$\sum_{j:(i,j) \in S_T} x_{ij} p_{ij} \le T, \quad i \in M$$

$$x_{ij} \ge 0, \quad (i,j) \in S_T$$

17.2 Properties of extreme point solutions

Via an appropriate binary search, we will find the smallest value of T such that LP(T) has a feasible solution. Let T^* be this value. Clearly, T^* is a lower bound on OPT. The algorithm will round an extreme point solution to

LP(T^*) to find a schedule having makespan $\leq 2T^*$. Extreme point solutions to LP(T) have several useful properties.

Lemma 17.3 *Any extreme point solution to* LP(T) *has at most* $n + m$ *nonzero variables.*

Proof: Let $r = |S_T|$ represent the number of variables on which LP(T) is defined. Recall that a feasible solution to LP(T) is an extreme point solution iff it corresponds to setting r linearly independent constraints of LP(T) to equality. Of these r linearly independent constraints, at least $r - (n + m)$ must be chosen from the third set of constraints (of the form $x_{ij} \geq 0$). The corresponding variables are set to 0. So, any extreme point solution has at most $n + m$ nonzero variables. □

Let $\boldsymbol{x}$ be an extreme point solution to LP(T). We will say that job j is *integrally set* in $\boldsymbol{x}$ if it is entirely assigned to one machine. Otherwise, we will say that job j is *fractionally set.*

Corollary 17.4 *Any extreme point solution to* LP(T) *must set at least* $n-m$ *jobs integrally.*

Proof: Let $\boldsymbol{x}$ be an extreme point solution to LP(T), and let α and β be the number of jobs that are integrally and fractionally set by $\boldsymbol{x}$, respectively. Each job of the latter kind is assigned to at least 2 machines and therefore results in at least 2 nonzero entries in $\boldsymbol{x}$. Hence we get

$$\alpha + \beta = n \text{ and } \alpha + 2\beta \leq n + m.$$

Therefore, $\beta \leq m$ and $\alpha \geq n - m$. □

The LP-rounding algorithm is based on several interesting combinatorial properties of extreme point solutions to LP(T). Some of these are established in Section 17.4. Corresponding to an extreme point solution $\boldsymbol{x}$ to LP(T), define $G = (J, M, E)$ to be the bipartite graph on vertex set $J \cup M$ such that $(j, i) \in E$ iff $x_{ij} \neq 0$. Let $F \subset J$ be the set of jobs that are fractionally set in $\boldsymbol{x}$, and let H be the subgraph of G induced on vertex set $F \cup M$. Clearly, (i, j) is an edge in H iff $0 < x_{ij} < 1$. A matching in H will be called a *perfect matching* if it matches every job $j \in F$. The rounding procedure uses the fact that graph H has a perfect matching (see Lemma 17.7).

17.3 The algorithm

The algorithm starts by computing the range in which it finds the right value of T. For this, it constructs the greedy schedule, in which each job is assigned to the machine on which it has the smallest processing time. Let α be the makespan of this schedule. Then the range is $[\alpha/m, \alpha]$.

Algorithm 17.5 (Scheduling on unrelated parallel machines)

1. By a binary search in the interval $[\alpha/m, \alpha]$, find the smallest value of $T \in \mathbf{Z}^+$ for which $\mathrm{LP}(T)$ has a feasible solution. Let this value be T^*.
2. Find an extreme point solution, say $\boldsymbol{x}$, to $\mathrm{LP}(T^*)$.
3. Assign all integrally set jobs to machines as in $\boldsymbol{x}$.
4. Construct graph H and find a perfect matching $\mathcal{M}$ in it (e.g., using the procedure of Lemma 17.7).
5. Assign fractionally set jobs to machines according to matching $\mathcal{M}$.

17.4 Additional properties of extreme point solutions

We will say that a connected graph on vertex set V is a *pseudo-tree* if it contains at most $|V|$ edges. Since the graph is connected, it must have at least $|V| - 1$ edges. So, it is either a tree or a tree plus a single edge. In the latter case it has a unique cycle. Let us say that a graph is a *pseudo-forest* if each of its connected components is a pseudo-tree. Recall that in Section 17.2 we defined two graphs, G and H, corresponding to an extreme point solution $\boldsymbol{x}$ to $\mathrm{LP}(T)$.

Lemma 17.6 *Graph G is a pseudo-forest.*

Proof: We will show that the number of edges in each connected component of G is bounded by the number of vertices in it. Hence, each connected component is a pseudo-tree.

Consider a connected component G_c. Restrict $\mathrm{LP}(T)$ and $\boldsymbol{x}$ to the jobs and machines of G_c only, to obtain $\mathrm{LP}_c(T)$ and $\boldsymbol{x}_c$. Let $\boldsymbol{x}_{\overline{c}}$ represent the rest of $\boldsymbol{x}$. The important observation is that $\boldsymbol{x}_c$ must be an extreme point solution to $\mathrm{LP}_c(T)$. Suppose that this is not the case. Then, $\boldsymbol{x}_c$ is a convex combination of two feasible solutions to $\mathrm{LP}_c(T)$. Each of these, together with $\boldsymbol{x}_{\overline{c}}$, form a feasible solution to $\mathrm{LP}(T)$. Therefore, $\boldsymbol{x}$ is a convex combination of two feasible solutions to $\mathrm{LP}(T)$, leading to a contradiction.

Now, applying Lemma 17.3, we get that G_c is a pseudo-tree. □

Lemma 17.7 *Graph H has a perfect matching.*

Proof: Each job that is integrally set in $\boldsymbol{x}$ has exactly one edge incident at it in G. Remove these jobs, together with their incident edges, from G. Clearly, the remaining graph is H. Since an equal number of edges and vertices were removed, H is also a pseudo-forest.

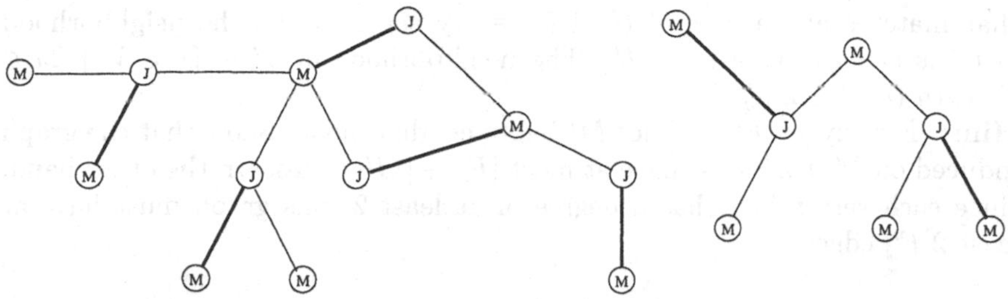

In H, each job has a degree of at least 2. So, all leaves in H must be machines. Keep matching a leaf with the job it is incident to, and remove them both from the graph. (At each stage all leaves must be machines.) In the end we will be left with even cycles (since we started with a bipartite graph). Match off alternate edges of each cycle. This gives a perfect matching in H. □

Theorem 17.8 *Algorithm 17.5 achieves an approximation guarantee of factor 2 for the problem of scheduling on unrelated parallel machines.*

Proof: Clearly, $T^* \leq$ OPT, since LP(OPT) has a feasible solution. The extreme point solution, $\boldsymbol{x}$, to LP(T^*) has a fractional makespan of $\leq T^*$. Therefore, the restriction of $\boldsymbol{x}$ to integrally set jobs has a (integral) makespan of $\leq T^*$. Each edge (i, j) of H satisfies $p_{ij} \leq T^*$. The perfect matching found in H schedules at most one extra job on each machine. Hence, the total makespan is $\leq 2T^* \leq 2 \cdot$ OPT. The algorithm clearly runs in polynomial time. □

Example 17.9 Let us provide a family of tight examples. The mth instance consists of $m^2 - m + 1$ jobs that need to be scheduled on m machines. The first job has a processing time of m on all machines, and all the remaining jobs have unit processing time on each machine. The optimal schedule assigns the first job to one machine, and m of the remaining jobs to each of the remaining $m - 1$ machines. Its makespan is m. It is easy to see that LP(T) has no feasible solutions for $T < m$.

Now suppose the following extreme point solution to LP(m) is picked. It assigns $1/m$ of the first job and $m - 1$ other jobs to each of the m machines. Rounding will produce a schedule having a makespan of $2m - 1$. □

17.5 Exercises

17.1 Give an alternative proof of Lemma 17.7 by using Hall's Theorem. (This theorem states that a bipartite graph $G = (U, V, E)$ has a matching

that matches all vertices of U iff for every set $U' \subseteq U$, the neighborhood of U' is at least as large as U'. The neighborhood of U' is $\{v \in V \mid \exists u \in U' \text{ with } (u,v) \in E\}$.)
Hint: For any set $F' \subset F$, let M' be its neighborhood. Show that the graph induced on $F' \cup M'$ must have at most $|F'| + |M'|$ edges. On the other hand, since each vertex in F has a degree of at least 2, this graph must have at least $2|F'|$ edges.

17.2 Prove that the solution given to LP(m) in Example 17.9 is an extreme point solution.

17.3 Does Algorithm 17.5 achieve a better factor than 2 for the special case that the machines are identical?

17.4 Prove the following strengthening of Lemma 17.6. There is an extreme point solution to LP(T) such that its corresponding bipartite graph, G, is a forest.

17.5 (Hochbaum and Shmoys [136]) Give a PTAS for the problem of minimizing makespan on uniform parallel machines. In this problem there is a speed s_i associated with each machine i, and the processing time for job j on machine i is p_j/s_i.

17.6 Notes

The result of this chapter is due to Lenstra, Shmoys, and Tardos [191]. For other LP-rounding based scheduling algorithms, see the survey by Hall [127].

18 Multicut and Integer Multicommodity Flow in Trees

The theory of cuts in graphs occupies a central place not only in the study of exact algorithms, but also approximation algorithms. We will present some key results in the next four chapters. This will also give us the opportunity to develop further the two fundamental algorithm design techniques introduced in Chapters 14 and 15.

In Chapter 15 we used the primal–dual schema to derive a factor 2 algorithm for the weighted vertex cover problem. This algorithm was particularly easy to obtain because the relaxed dual complementary slackness conditions were automatically satisfied in any integral solution. In this chapter, we will use the primal–dual schema to obtain an algorithm for a generalization of this problem (see Exercise 18.1). This time, enforcing relaxed dual complementary slackness conditions will be a nontrivial part of the algorithm. Furthermore, we will introduce the procedure of reverse delete, which will be used in several other primal–dual algorithms.

18.1 The problems and their LP-relaxations

The following is an important generalization of the minimum s–t cut problem. In fact, it also generalizes the multiway cut problem (Problem 4.1).

Problem 18.1 (Minimum multicut) Let $G=(V,E)$ be an undirected graph with nonnegative capacity c_e for each edge $e \in E$. Let $\{(s_1,t_1),\ldots,(s_k,t_k)\}$ be a specified set of pairs of vertices, where each pair is distinct, but vertices in different pairs are not required to be distinct. A *multicut* is a set of edges whose removal separates each of the pairs. The problem is to find a minimum capacity multicut in G.

The minimum s–t cut problem is the special case of multicut for $k = 1$. Problem 18.1 generalizes multiway cut because separating terminals $s_1,\ldots,s_l$ is equivalent to separating all pairs (s_i,s_j), for $1 \le i < j \le l$. This observation implies that the minimum multicut problem is **NP**-hard even for $k = 3$, since the multiway cut problem is **NP**-hard for the case of 3 terminals.

In Chapter 20 we will obtain an $O(\log k)$ factor approximation algorithm for the minimum multicut problem. In this chapter, we will obtain a factor 2 algorithm for the special case when G is restricted to be a tree. Since G is a tree, there is a unique path between s_i and t_i, and the multicut must pick an edge on this path to disconnect s_i from t_i. Although the problem looks deceptively simple, Exercise 18.1 should convince the reader that this is not so. The minimum multicut problem is **NP**-hard even if restricted to trees of height 1 and unit capacity edges.

Since we want to apply LP-duality theory to design the algorithm, let us first give an integer programming formulation for the problem and obtain its LP-relaxation. Introduce a 0/1 variable d_e for each edge $e \in E$, which will be set to 1 iff e is picked in the multicut. Let p_i denote the unique path between s_i and t_i in the tree.

$$\begin{array}{lll} \text{minimize} & \sum_{e \in E} c_e d_e & \\ \text{subject to} & \sum_{e \in p_i} d_e \geq 1, & i \in \{1, \ldots, k\} \\ & d_e \in \{0,1\}, & e \in E \end{array}$$

The LP-relaxation is obtained by replacing the constraint $d_e \in \{0,1\}$ by $d_e \geq 0$. As in the derivation of LP (13.2), there is no need to add the constraint $d_e \leq 1$ explicitly.

$$\begin{array}{lll} \text{minimize} & \sum_{e \in E} c_e d_e & \\ \text{subject to} & \sum_{e \in p_i} d_e \geq 1, & i \in \{1, \ldots, k\} \\ & d_e \geq 0, & e \in E \end{array} \tag{18.1}$$

We can now think of d_e as specifying the fractional extent to which edge e is picked. A solution to this linear program is a *fractional multicut*: on each path p_i, the sum of fractions of edges picked is at least 1. In general, minimum fractional multicut may be strictly cheaper than minimum integral multicut. This is illustrated in Example 18.2.

We will interpret the dual program as specifying a *multicommodity flow* in G, with a separate commodity corresponding to each vertex pair (s_i, t_i). Dual variable f_i will denote the amount of this commodity routed along the unique path from s_i to t_i.

$$\begin{array}{lll} \text{maximize} & \sum_{i=1}^{k} f_i & \\ \text{subject to} & \sum_{i:\ e \in p_i} f_i \leq c_e, & e \in E \\ & f_i \geq 0, & i \in \{1, \ldots, k\} \end{array} \tag{18.2}$$

The commodities are routed concurrently. The object is to maximize the sum of the commodities routed, subject to the constraint that the sum of flows routed through an edge is bounded by the capacity of the edge. Notice that the sum of flows through an edge (u, v) includes flow going in either direction, u to v and v to u.

By the weak duality theorem, a feasible multicommodity flow gives a lower bound on the minimum fractional multicut and hence also on the minimum integral multicut. By the LP-duality theorem, minimum fractional multicut equals maximum multicommodity flow.

Example 18.2 Consider the following graph with unit capacity edges and 3 vertex pairs:

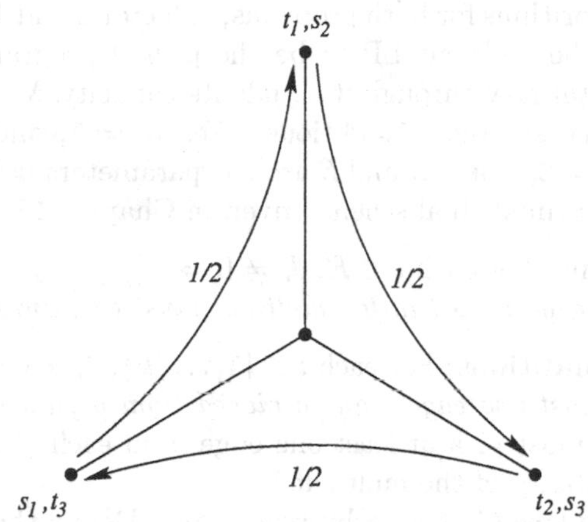

The arrows show how to send 3/2 units of flow by sending 1/2 unit of each commodity. Picking each edge to the extent of 1/2 gives a multicut of capacity 3/2 as well. These must be optimal solutions to the primal and dual programs. On the other hand, any integral multicut must pick at least two of the three edges in order to disconnect all three pairs. Hence, minimum integral multicut has capacity 2. □

Finally, let us state one more problem.

Problem 18.3 (Integer multicommodity flow) Graph G and the source–sink pairs are specified as in the minimum multicut problem; however, the edge capacities are all integral. A separate commodity is defined for each (s_i, t_i) pair. The object is to maximize the sum of the commodities routed, subject to edge capacity constraints and subject to routing each commodity integrally.

Let us consider this problem when G is restricted to be a tree. If in (18.2), the variables are constrained to be nonnegative integers, we would get an integer programming formulation for this problem. Clearly, the objective function value of this integer program is bounded by that of the linear program (18.2).

Furthermore, the best fractional flow may be strictly larger. For instance, in Example 18.2, maximum integral multicommodity flow is 1, since sending 1 unit of any of the three commodities will saturate two of the edges. This problem is **NP**-hard, even for trees of height 3 (though the capacity has to be arbitrary).

18.2 Primal–dual schema based algorithm

We will use the primal–dual schema to obtain an algorithm that simultaneously finds a multicut and an integer multicommodity flow that are within a factor of 2 of each other, provided the given graph is a tree. Hence, we get approximation algorithms for both problems, of factor 2 and 1/2, respectively.

Let us define the multicut LP to be the primal program. An edge e is *saturated* if the total flow through it equals its capacity. We will ensure primal complementary slackness conditions, i.e., $\alpha = 1$, and relax the dual conditions with $\beta = 2$, where α and β are the parameters used in the general description of the primal–dual schema given in Chapter 15.

Primal conditions: For each $e \in E$, $d_e \neq 0 \Rightarrow \sum_{i:\ e \in p_i} f_i = c_e$. Equivalently, *any edge picked in the multicut must be saturated.*

Relaxed dual conditions: For each $i \in \{1, \ldots, k\}$, $f_i \neq 0 \Rightarrow \sum_{e \in p_i} d_e \leq 2$. Equivalently, *at most two edges can be picked from a path carrying nonzero flow.* (Clearly, we must pick at least one edge from each (s_i, t_i) path simply to ensure the feasibility of the multicut.)

Let us root the tree G at an arbitrary vertex. Define the *depth* of vertex v to be the length of the path from v to the root; the depth of the root is 0. For two vertices $u, v \in V$, let $\text{lca}(u, v)$ denote the *lowest common ancestor* of u and v, i.e., the minimum depth vertex on the path from u to v. Let e_1 and e_2 be two edges on a path from a vertex to the root. If e_1 occurs before e_2 on this path, then e_1 is said to be *deeper* than e_2.

The algorithm starts with an empty multicut and flow, and iteratively improves the feasibility of the primal solution and the optimality of the dual solution. In an iteration, it picks the deepest unprocessed vertex, say v, and greedily routes integral flow between pairs that have v as their lowest common ancestor. When no more flow can be routed between these pairs, all edges that were saturated in this iteration are added to the list D in arbitrary order. When all the vertices have been processed, D will be a multicut; however, it may have redundant edges. To remove them, a *reverse delete* step is performed: edges are considered in the reverse of the order in which they were added to D, and if the deletion of edge e from D still gives a valid multicut, e is discarded from D.

Algorithm 18.4 (Multicut and integer multicommodity flow in trees)

1. **Initialization:** $f \leftarrow 0$; $D \leftarrow \emptyset$.
2. **Flow routing:** For each vertex v, in nonincreasing order of depth, do:
   For each pair (s_i, t_i) such that $\text{lca}(s_i, t_i) = v$, greedily route integral flow from s_i to t_i.
   Add to D all edges that were saturated in the current iteration in arbitrary order.
3. Let $e_1, e_2, \ldots, e_l$ be the ordered list of edges in D.
4. **Reverse delete:** For $j = l$ downto 1 do:
   If $D - \{e_j\}$ is a multicut in G, then $D \leftarrow D - \{e_j\}$.
5. Output the flow and multicut D.

Lemma 18.5 *Let (s_i, t_i) be a pair with nonzero flow, and let $\text{lca}(s_i, t_i) = v$. At most one edge is picked in the multicut from each of the two paths, s_i to v and t_i to v.*

Proof: The argument is the same for each path. Suppose two edges e and e' are picked from the s_i–v path, with e being the deeper edge. Clearly, e' must be in D all through reverse delete. Consider the moment during reverse delete when edge e is being tested. Since e is not discarded, there must be a pair, say (s_j, t_j), such that e is the only edge of D on the s_j–t_j path. Let u be the lowest common ancestor of s_j and t_j. Since e' does not lie on the s_j–t_j path, u must be deeper than e', and hence deeper than v. After u has been processed, D must contain an edge from the s_j–t_j path, say e''.

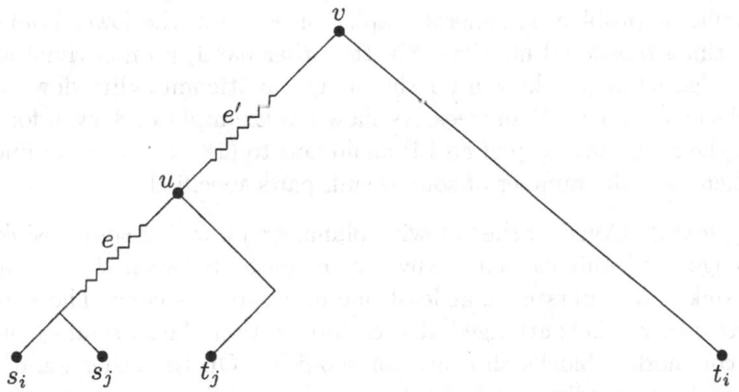

Since nonzero flow has been routed from s_i to t_i, e must be added during or after the iteration in which v is processed. Since v is an ancestor of u, e is added after e''. So e'' must be in D when e is being tested. This contradicts the fact that at this moment e is the only edge of D on the s_j–t_j path. □

Theorem 18.6 *Algorithm 18.4 achieves approximation guarantees of factor 2 for the minimum multicut problem and factor* $1/2$ *for the maximum integer multicommodity flow problem on trees.*

Proof: The flow found at the end of Step 2 is maximal, and since at this point D contains all the saturated edges, D is a multicut. Since the reverse delete step only discards redundant edges, D is a multicut after this step as well. Thus, feasible solutions have been found for both the flow and the multicut.

Since each edge in the multicut is saturated, the primal conditions are satisfied. By Lemma 18.5, at most two edges have been picked in the multicut from each path carrying nonzero flow. Therefore, the relaxed dual conditions are also satisfied. Hence, by Proposition 15.1, the capacity of the multicut found is within twice the flow. Since a feasible flow is a lower bound on the optimal multicut, and a feasible multicut is an upper bound on the optimal integer multicommodity flow, the claim follows. □

Finally, we obtain the following approximate min–max relation from Theorem 18.6:

Corollary 18.7 *On trees with integer edge capacities,*

$$\max_{\textit{int. flow } F} |F| \;\leq\; \min_{\textit{multicut } C} c(C) \;\leq 2 \cdot \max_{\textit{int. flow } F} |F|,$$

where $|F|$ *represents the value of flow function* F *and* $c(C)$ *represents the capacity of multicut* C.

In Chapter 20 we will present an $O(\log k)$ factor algorithm for the minimum multicut problem in general graphs; once again, the lower bound used is an optimal fractional multicut. On the other hand, no nontrivial approximation algorithms are known for the integer multicommodity flow problem in graphs more general than trees. As shown in Example 18.8, even for planar graphs, the integrality gap of an LP analogous to (18.2) is lower bounded by $n/2$, where n is the number of source–sink pairs specified.

Example 18.8 Consider the following planar graph with n source–sink pairs. Every edge is of unit capacity. Any pair of paths between the ith and jth source–sink pairs intersect in at least one unit capacity edge. The magnified part shows how this is arranged at each intersection. Thus, sending one unit of any commodity blocks all other commodities. On the other hand, half a unit of each commodity can be routed simultaneously.

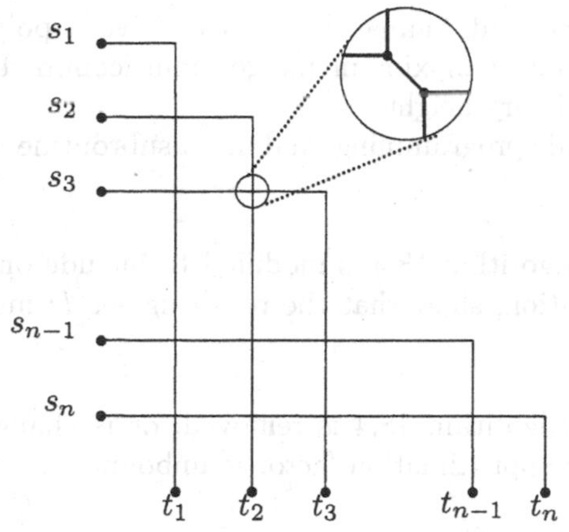

□

18.3 Exercises

18.1 (Garg, Vazirani, and Yannakakis [104]) Give approximation factor preserving reductions between the following pairs of problems:

(a) cardinality vertex cover and minimum multicut in trees of height 1 and unit capacity edges,
(b) vertex cover with arbitrary weights and minimum multicut in trees of height 1 and arbitrary edge capacities.

Hint: Given a vertex cover instance G, construct a height 1 tree that has a leaf corresponding to each vertex of G and a source–sink pair corresponding to each edge of G.

18.2 The following is a well-studied polynomial time solvable generalization of the maximum matching problem. Given an undirected graph $G = (V, E)$ and a function $b : V \rightarrow \mathbf{Z}^+$, a b-matching is a set of edges, $E' \subseteq E$, with associated multiplicities, $m : E' \rightarrow \mathbf{Z}^+$, such that each vertex $v \in V$ has at most $b(v)$ edges incident at it, counting multiplicities. The size of this b-matching is the sum of multiplicities of edges in E'. The *maximum b-matching problem* is that of finding a b-matching of maximum size. Show that the following pairs of problems are polynomial time equivalent:

(a) maximum integer multicommodity flow problem on trees of height 1 and unit capacity edges, and the maximum matching problem,
(b) maximum integer multicommodity flow problem on trees of height 1 and arbitrary capacity edges, and the maximum b-matching problem.

18.3 (Garg, Vazirani, and Yannakakis [104]) Give a polynomial time algorithm for computing a maximum integer multicommodity flow on unit capacity trees of arbitrary height.
Hint: Apply dynamic programming, and use a subroutine for the maximum matching problem.

18.4 If Step 2 of Algorithm 18.4 is modified to include only one saturated edge after each iteration, show that the resulting set D may not even be a multicut.

18.5 If Step 4 in Algorithm 18.4 is removed, or is changed to a forward delete, show that its approximation factor is unbounded.

18.6 Modify step 4 in Algorithm 18.4 to: sort edges in D by decreasing capacity and remove redundant edges in this order. What factor can you prove for the modified algorithm?

18.7 Give tight examples for Algorithm 18.4 for both multicut and integer multicommodity flow.

18.8 Prove that if e and e' are both in D in Step 3 of Algorithm 18.4, and e is deeper than e', then e is added before or in the same iteration as e'.

18.9 Find the best integral and fractional multicut and the best multicommodity flow in the following graph. All capacities are 1, and the specified pairs are $(s_1, t_1), \ldots, (s_5, t_5)$. Notice that the optimal fractional multicut is not half integral. In contrast, the LP-relaxation of the multiway cut problem always has a half-integral optimal solution (see Chapter 19).

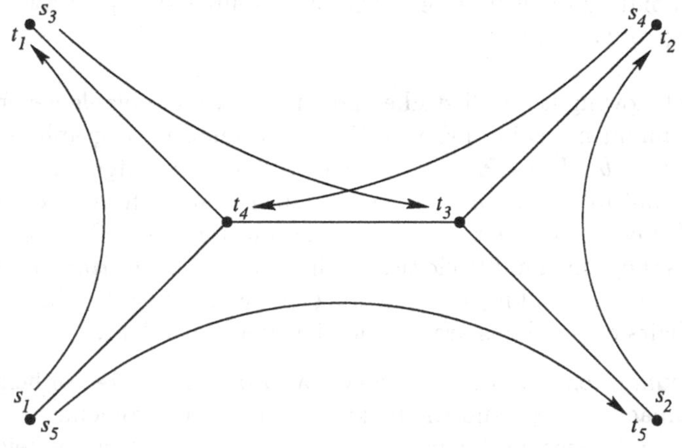

18.4 Notes

Algorithm 18.4 is due to Garg, Vazirani, and Yannakakis [104]. For recent results on the integer multicommodity flow problem, see Guruswami, Khanna, Rajaraman, Shepherd, and Yannakakis [125].

19 Multiway Cut

A simple combinatorial algorithm achieving an approximation factor of $2 - 2/k$ for the multiway cut problem, Problem 4.1, was presented in Chapter 4. In this chapter we will use LP-rounding to improve the factor to 3/2.

In Chapter 14 we mentioned the remarkable property of half-integrality, possessed by LP-relaxations of certain **NP**-hard problems. The multiway cut problem and its generalization, the node multiway cut problem, possess this property. We will present a proof of this fact in Section 19.3. This is the only avenue known for obtaining a constant factor approximation algorithm for the latter problem.

19.1 An interesting LP-relaxation

The usual LP-relaxation for multiway cut has an integrality gap of $2 - 2/k$ (see Exercise 19.2). The key to an improved approximation guarantee is a clever LP-relaxation.

Let Δ_k denote the $k-1$ dimensional simplex. This is the $k-1$ dimensional convex polytope in $\mathbf{R}^k$ defined by $\{\boldsymbol{x} \in \mathbf{R}^k \mid \boldsymbol{x} \geq 0 \text{ and } \sum_i x_i = 1\}$, where x_i is the ith coordinate of point $\boldsymbol{x}$. The simplex Δ_3 is shown below.

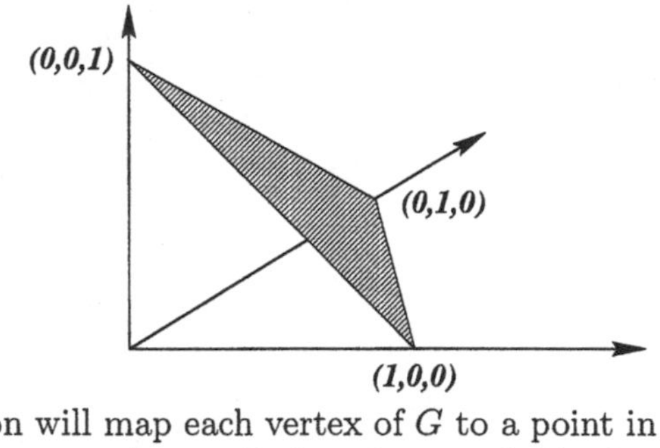

The relaxation will map each vertex of G to a point in Δ_k. Each of the k terminals will be mapped to a distinct vertex of this simplex, i.e., to a unit vector $e_i \in \mathbf{R}^k$. Let $x_v \in \Delta_k$ denote the point to which vertex v is mapped. The length of an edge $(u, v) \in E$ will be defined to be half the ℓ_1 distance between x_u and x_v. The entire relaxation is:

$$\begin{aligned}
\text{minimize} \quad & \sum_{(u,v)\in E} c(u,v)d(u,v) && (19.1)\\
\text{subject to} \quad & d(u,v) = \frac{1}{2}\sum_{i=1}^{k} |x_u^i - x_v^i|, && (u,v)\in E\\
& x_v \in \Delta_k, && v \in V\\
& x_{s_i} = e_i, && s_i \in S
\end{aligned}$$

In Lemma 19.1 we show that this relaxation is really a linear program. An integral solution to this relaxation maps each vertex of G to a vertex of the simplex, respectively. Each edge (u,v) has length either 0 or 1, depending on whether u and v are mapped to the same or different vertices of the simplex. Edges of length 1 form a multiway cut. The cost of this cut is the objective function value of this integral solution. Thus, an optimal integral solution corresponds to an optimal multiway cut.

Lemma 19.1 *Relaxation (19.1) can be expressed as a linear program.*

Proof: For each edge (u,v), replace the first constraint with:

$$x_{uv}^i \geq x_u^i - x_v^i, \quad 1 \leq i \leq k$$

$$x_{uv}^i \geq x_v^i - x_u^i, \quad 1 \leq i \leq k$$

$$d(u,v) = \frac{1}{2}\sum_{i=1}^{k} x_{uv}^i$$

Since the objective function is being minimized, an optimal solution must satisfy $x_{uv}^i = |x_u^i - x_v^i|$. The rest of the constraints are clearly linear. □

Example 19.2 In the example given below, the optimal fractional multiway cut is cheaper than the optimal integral cut. The mapping of vertices to Δ_3 in the optimal fractional solution is shown below; it achieves a cost of 7.5. On the other hand, the optimal integral solution costs 8.

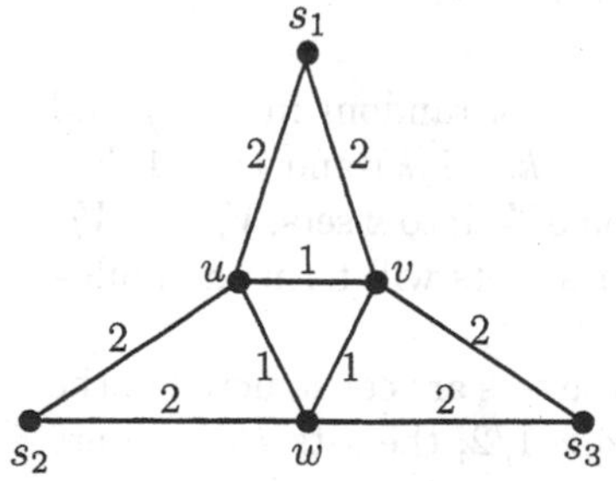

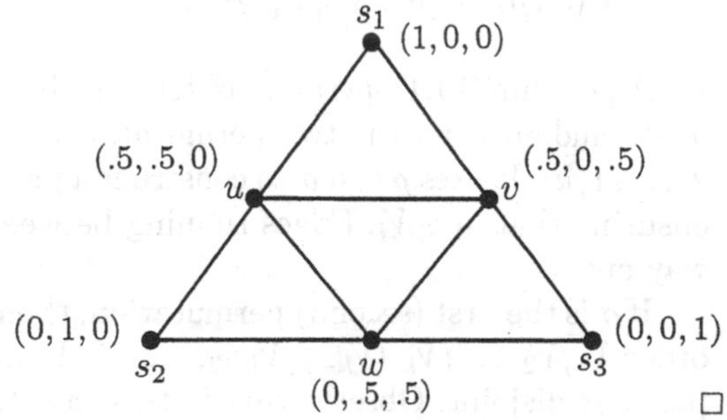

□

The following property will greatly simplify matters:

Lemma 19.3 *Let $\boldsymbol{x}$ be a feasible solution to relaxation (19.1). We may assume w.l.o.g. that for each edge $(u,v) \in E$, x_u and x_v differ in at most two coordinates.*

Proof: We will divide edges by adding new vertices in such a way that this property holds and the cost of the solution remains unchanged.

Suppose that $(u,v) \in E$ and that x_u and x_v differ in more than two coordinates. Replace this edge by two new edges (u,w) and (w,v), where w is a new vertex. Each of the new edges is of the same cost as $c(u,v)$, thereby ensuring that the cost of the integral optimal solution is unchanged. We show below how to enforce $d(u,v) = d(u,w) + d(w,v)$, thereby ensuring that the cost of the fractional solution remains unchanged.

Consider the coordinates in which x_u and x_v differ. Let i be the coordinate in which the difference is minimum. Without loss of generality, assume $x_u^i < x_v^i$. Let $\alpha = x_v^i - x_u^i$. There must be a coordinate j such that $x_u^j \geq x_v^j + \alpha$. We will define point x_w as follows. The ith and jth coordinates of x_w are $x_w^i = x_u^i$ and $x_w^j = x_v^j + \alpha$. The remaining coordinates of x_w are the same as those of x_v. Clearly, $x_w \in \Delta_k$ and $d(u,v) = d(u,w) + d(w,v)$.

Notice that v and w differ in two coordinates and w and u differ in fewer coordinates than u and v. Therefore, each edge of E requires at most $k-2$ such subdivisions to enforce the required property. □

19.2 Randomized rounding algorithm

Let $\boldsymbol{x}$ be an optimal solution to relaxation (19.1) satisfying the property stated in Lemma 19.3, and let OPT_f denote its cost. Let E_i denote the subset of edges whose endpoints differ in coordinate i, i.e., $E_i = \{(u,v) \in E \mid x_u^i \neq x_v^i\}$. Clearly, each edge e with $d(e) > 0$ will lie in two of these sets. Let $W_i = \sum_{e \in E_i} c(e)d(e)$. Renumber the terminals so that W_k is the largest of $W_1, \ldots, W_k$. For $\rho \in (0,1)$, define

$$B(s_i, \rho) = \{v \in V \mid x_v^i \geq \rho\}.$$

Algorithm 19.4 operates as follows. It picks ρ at random in $(0,1)$ and σ at random from the two permutations $(1,2,\ldots,k-1,k)$ and $(k-1,k-2,\ldots,1,k)$. It uses ρ and σ to construct a partition of V into k sets, $V_1, \ldots, V_k$, ensuring that $s_i \in V_i$. Edges running between these sets will form the multiway cut.

If σ is the first (second) permutation, then these sets are constructed in the order $V_1, V_2, \ldots, V_k$ $(V_{k-1}, V_{k-2}, \ldots, V_1, V_k)$. If $\rho > 1/2$, the sets $B(s_i, \rho)$ are pairwise disjoint. Observe that in this case the partition is not affected by σ, because V_i is simply $B(s_i, \rho)$ for $1 \le i \le k-1$, and $V_k = V - (V_1 \cup \cdots \cup V_{k-1})$. If $\rho \le 1/2$, the sets $B(s_i, \rho)$ overlap and σ plays a role, as illustrated in the figure below for $k = 3$.

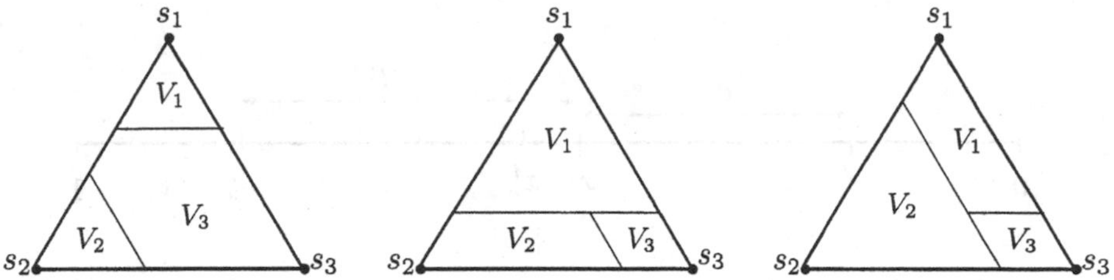

$\rho \ge 1/2$ $\qquad$ $\rho < 1/2, \sigma = (1, 2, 3)$ $\qquad$ $\rho < 1/2, \sigma = (2, 1, 3)$

Algorithm 19.4 (Multiway cut)

1. Compute an optimal solution, $\boldsymbol{x}$, to relaxation (19.1).
2. Renumber the terminals so that W_k is largest among $W_1, \ldots, W_k$.
3. Pick uniformly at random $\rho \in (0, 1)$ and
   $\sigma \in \{(1, 2, \ldots, k-1, k), (k-1, k-2, \ldots, 1, k)\}$.
4. For $i = 1$ to $k - 1$: $V_{\sigma(i)} \leftarrow B(s_i, \rho) - \bigcup_{j<i} V_{\sigma(j)}$.
5. $V_k \leftarrow V - \bigcup_{i<k} V_i$.
6. Let C be the set of edges that run between sets in the partition $V_1, \ldots, V_k$. Output C.

We will show that the expected cost of the multiway cut produced by the algorithm, $\mathbf{E}[c(C)]$, is at most $(1.5 - 1/k) \cdot \mathrm{OPT}_f$. The following lemma will be critical.

Lemma 19.5 *If* $e \in E - E_k$, $\mathbf{Pr}[e \in C] \le 1.5\, d(e)$, *and if* $e \in E_k$, $\mathbf{Pr}[e \in C] \le d(e)$.

Proof: Suppose $e \in E - E_k$. Let $e = (u, v)$, and let i and j be the coordinates in which x_u and x_v differ. There are two cases: the intervals $[x_u^i, x_v^i]$ and $[x_v^j, x_u^j]$ either overlap or they are disjoint. These two cases are shown below. Note that in either case the two intervals have the same length since $x_v^i - x_u^i = x_u^j - x_v^j = d(e)$. Intervals α and β are defined in the figure below for the two cases.

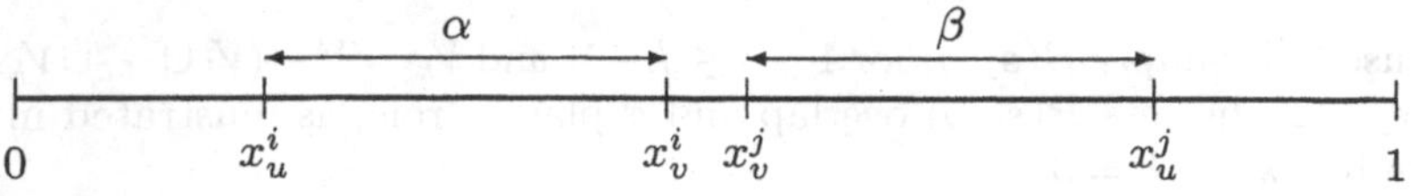

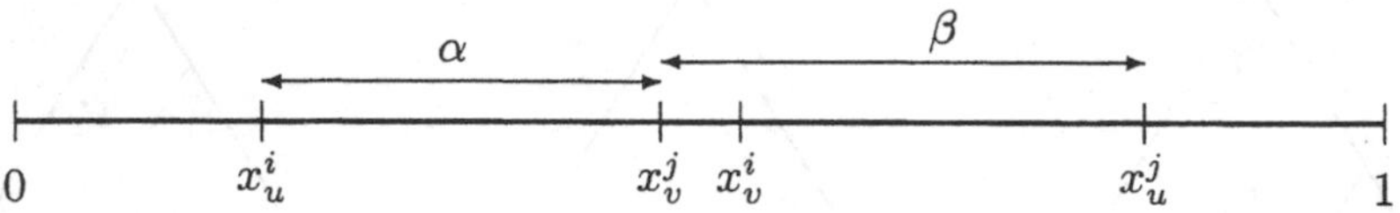

Observe that the vertices u and v can end up in one of three sets, V_i, V_j, or V_k. Furthermore, if $\rho \in [0,1] - (\alpha \cup \beta)$, then both vertices will end up in the same set, and edge e will not be in the cut. Clearly, $\mathbf{Pr}[\rho \in (\alpha \cup \beta)] = |\alpha| + |\beta| \leq 2d(e)$.

The critical observation that leads to the desired bound is that in the event $\rho \in \alpha$ and $\sigma(j) < \sigma(i)$, u and v will both be put in the set V_j, and thus e will not be in the cut. Clearly, the probability of this event is $|\alpha|/2$. Therefore

$$\mathbf{Pr}[e \in C] = |\beta| + |\alpha|/2 \leq 1.5\, d(e).$$

Next, suppose that $e \in E_k$, and that its endpoints differ in coordinates i and k. In this case $\sigma(i) < \sigma(k)$, and u and v will end up in different sets only if ρ falls between x_u^i and x_v^i. The probability of this is $d(e)$. □

Lemma 19.6 *The multiway cut, C, output by Algorithm 19.4 satisfies*

$$\mathbf{E}[c(C)] \leq (1.5 - 1/k)\mathrm{OPT}_f.$$

Proof: Clearly, C forms a multiway cut. Now, $\mathrm{OPT}_f = \sum_e c(e)d(e)$. Since each edge with nonzero length is in two of the sets E_i, $\sum_{i=1}^k W_i = 2 \cdot \mathrm{OPT}_f$. Since k was chosen so that W_k is the largest of these sets,

$$W_k = \sum_{e \in E_k} c(e)d(e) \geq \frac{2}{k} \cdot \mathrm{OPT}_f.$$

Therefore

$$\begin{aligned}\mathbf{E}[c(C)] &= \sum_{e \in E} c(e)\mathbf{Pr}[e \in C] = \sum_{e \in E - E_k} c(e)\mathbf{Pr}[e \in C] + \sum_{e \in E_k} c(e)\mathbf{Pr}[e \in C] \\ &\leq 1.5 \sum_{e \in E - E_k} c(e)d(e) + \sum_{e \in E_k} c(e)d(e)\end{aligned}$$

$$= 1.5 \sum_{e \in E} c(e)d(e) - 0.5 \sum_{e \in E_k} c(e)d(e)$$
$$\leq (1.5 - 1/k) \cdot \mathrm{OPT}_f$$

where the first inequality follows from Lemma 19.5. □

Lemma 19.6 places an upper bound of $1.5 - 1/k$ on the integrality gap of relaxation 19.1 (see the notes in Section 19.5 for references to a slightly better result). The best lower bound know on the integrality gap is $8/(7 + \frac{1}{k-1})$; Example 19.2 places a lower bound of 16/15.

The bound on the expected weight of the multiway cut established in Lemma 19.6 can be converted into a high probability statement using standard techniques (see Exercises 1.10 and 19.4). Hence we get

Theorem 19.7 *There is a 3/2 factor randomized approximation algorithm for the multiway cut problem.*

19.3 Half-integrality of node multiway cut

The following is a generalization of the multiway cut problem, in the sense that there is an approximation factor preserving reduction from the multiway cut problem to it (see Exercise 19.13).

Problem 19.8 (Node multiway cut) Given a connected, undirected graph $G = (V, E)$ with an assignment of costs to vertices, $c : V \to \mathbf{R}^+$, and a set of terminals $S = \{s_1, s_2, \ldots, s_k\} \subseteq V$ that form an independent set in G, a *node multiway cut* is a subset of $V - S$ whose removal disconnects the terminals from each other. The node multiway cut problem asks for the minimum cost such subset.

We will show that the relaxation to the following integer program always has a half-integral optimal solution. A factor $2-2/k$ approximation algorithm will follow from this fact (see Exercise 19.11). In this program we have introduced a 0/1 variable d_v for each vertex $v \in V - S$, which indicates whether vertex v has been picked. Let $\mathcal{P}$ denote the set of all paths running between distinct terminals. There is a constraint for each path p in $\mathcal{P}$ – it ensures that at least one vertex is picked from each path.

$$\begin{aligned}
\text{minimize} \quad & \sum_{v \in V-S} c_v d_v \\
\text{subject to} \quad & \sum_{v \in p} d_v \geq 1, \qquad p \in \mathcal{P} \\
& d_v \in \{0, 1\}, \qquad v \in V - S
\end{aligned}$$

The LP-relaxation is given below. As before, we will interpret d_v's as distance labels. With respect to an assignment to these distance labels, let us define the *length* of a path to be the sum of distance labels of nonterminals on this path. The *distance* between a pair of vertices will be the length of the shortest path between them. A solution, $\boldsymbol{d}$, is feasible only if the distance between every pair of terminals is at least 1.

$$\begin{aligned}
\text{minimize} \quad & \sum_{v \in V-S} c_v d_v & & \qquad (19.2)\\
\text{subject to} \quad & \sum_{v \in p} d_v \geq 1, & p \in \mathcal{P} & \\
& d_v \geq 0, & v \in V - S &
\end{aligned}$$

As in Chapter 18, the dual will be interpreted as seeking a maximum multicommodity flow. The commodities flow between distinct terminals, and the constraint is that the total amount of flow through a vertex be bounded by its cost.

$$\begin{aligned}
\text{maximize} \quad & \sum_{p \in \mathcal{P}} f_p & & \qquad (19.3)\\
\text{subject to} \quad & \sum_{p: v \in p} f_p \leq c_v, & v \in V - S & \\
& f_p \geq 0, & p \in \mathcal{P} &
\end{aligned}$$

Let $\boldsymbol{d}$ be an optimal solution to LP (19.2). We will show how to obtain, efficiently, a half-integral optimal solution from $\boldsymbol{d}$. For the purposes of proof, let $\boldsymbol{f}$ be an optimal solution to the dual LP. Complementary slackness conditions give:

Primal conditions: For each $v \in V - S$, if $d_v > 0$ then v must be saturated.

Dual conditions: For each path p, if $f_p > 0$ then the length of p is exactly 1.

Consider graph G with distance labels on vertices $v \in V - S$ specified by $\boldsymbol{d}$. For each terminal s_i, define its *region* S_i to be the set of vertices reachable from s_i by paths of length zero (we will assume that $s_i \in S_i$). Define the *boundary*, B_i, of S_i to be all vertices that are adjacent to S_i, i.e., $B_i = \{v \in \overline{S_i} \mid \text{for some } u \in S_i, (u,v) \in E\}$. The feasibility of $\boldsymbol{d}$ ensures that the k regions are disjoint and the boundaries do not contain any terminals.

Claim 19.9 *Suppose $v \in B_i \cap B_j$ for $i \neq j$. Then $d_v = 1$.*

Proof: Clearly there is a path from s_i to s_j on which v is the only vertex having a positive distance label. The claim follows from the feasibility of $\boldsymbol{d}$. □

Let $M = \bigcup_{i=1}^{k} B_i$ be the set of boundary vertices. Partition this into two sets: M^{int} being boundary vertices that occur in two or more boundary sets, and M^{disj} being the rest; each vertex in M^{disj} is in a unique boundary set. By Claim 19.9, each vertex in M^{int} has distance label of 1.

Lemma 19.10 *Let p be a path between two distinct terminals such that $f_p > 0$. Then, from the vertices in M, p uses either exactly one vertex of M^{int} or exactly two vertices of M^{disj}.*

Proof: By the dual complementary slackness condition, the length of p must be exactly 1. Thus, if p uses a vertex of M^{int}, then it cannot have any other vertices of M on it.

Suppose p uses three or more vertices of M^{disj}. Assume that p runs from s_i to s_j and that u and w are the first and last vertices of M^{disj} on p, respectively. Let v be any intermediate vertex of M^{disj} on p. Since $v \in M^{\text{disj}}$, v must be in a unique boundary, say B_k; $k = i$ or $k = j$ are possible.

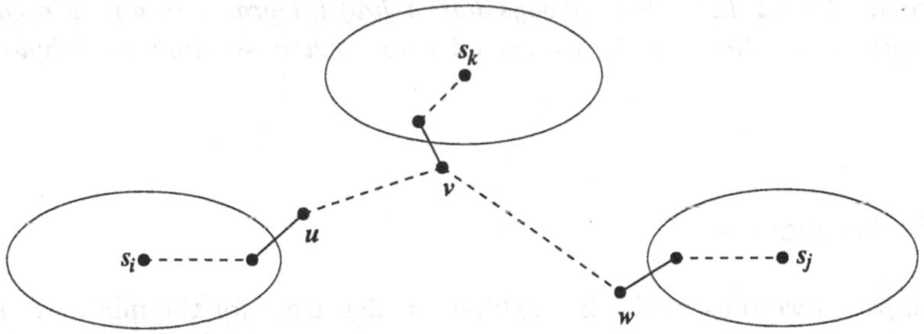

Let q be a path connecting v to s_k via vertices in S_k; such a path must exist since $v \in B_k$. Now consider the following two paths: the first consists of the part of the path p from s_i to v followed by q, and the second consists of the reverse of q followed by the part of p from v to s_j. At least one of these is a valid path running between distinct terminals (even if $k = i$ or $k = j$). Moreover, since it is missing at least one of the positive distance label vertices of p, it must have length strictly less than 1. This contradicts the feasibility of $\boldsymbol{d}$. The lemma follows. □

Let $\boldsymbol{h}$ be a solution to LP (19.2) that assigns distance labels of 1 to each vertex in M^{int}, 1/2 to each vertex in M^{disj}, and 0 to all remaining vertices.

Lemma 19.11 *$\boldsymbol{h}$ is an optimal solution to LP (19.2).*

Proof: Any valid path, p, from terminal s_i to s_j must use vertices of both boundary sets B_i and B_j. Suppose it uses $v \in B_i \cap B_j$. By definition $v \in M^{\text{int}}$, and so $h_v = 1$. Otherwise, it uses two vertices of M^{disj}. In either case the length of p is at least 1, thus showing that $\boldsymbol{h}$ is a feasible solution.

Next we will show that the objective function value of $\boldsymbol{h}$ is the same as that of flow $\boldsymbol{f}$, thereby showing that $\boldsymbol{h}$ is optimal. Partition paths carrying nonzero flow in $\boldsymbol{f}$ into two sets: $\mathcal{P}_1$ consists of paths that use one vertex of M^{int} and $\mathcal{P}_2$ consists of paths that use two vertices of M^{disj}. By Lemma 19.10 these are the only two possibilities. By the primal complementary slackness conditions and the optimality of $\boldsymbol{d}$, each vertex in M is saturated by $\boldsymbol{f}$. Therefore, the total flow carried by paths in $\mathcal{P}_1$ is $\sum_{v \in M^{\text{int}}} c_v$ and by paths in $\mathcal{P}_2$ is $\frac{1}{2}\sum_{v \in M^{\text{disj}}} c_v$. Hence the total flow is

$$\sum_{v \in M^{\text{int}}} c_v + \frac{1}{2} \sum_{v \in M^{\text{disj}}} c_v = \sum_{v \in V-S} h_v c_v.$$

This proves the lemma. □

Clearly $\boldsymbol{h}$ can be obtained from an optimal solution, $\boldsymbol{d}$, to LP (19.2) in polynomial time. This gives:

Theorem 19.12 *LP (19.2) always has a half-integral solution. Moreover, any optimal solution can be converted into such a solution in polynomial time.*

19.4 Exercises

In Chapter 4 we presented a $2 - 2/k$ factor algorithm for the minimum multiway cut problem by comparing the solution found to the integral optimal solution. In the next two exercises we develop an algorithm with the same guarantee using LP-duality.

19.1 Given terminals $s_1, \ldots, s_k$, consider the multicommodity flow problem in which each pair of terminals can form a source–sink pair. Thus there are $\binom{k}{2}$ commodities. Give an LP for maximizing this multicommodity flow and obtain the dual LP. The dual seeks a distance label assignment for edges satisfying the triangle inequality and ensures that the distance between any two terminals is at least 1. An optimal solution to the dual can be viewed as a fractional multiway cut.

19.2 Consider the following algorithm for finding a multiway cut. Solve the dual LP to obtain an optimal fractional multiway cut. This gives a distance label assignment, say d. Pick ρ at random in the interval $[0, \frac{1}{2}]$. An edge (u, v) is picked iff for some terminal s, $d(u, s) \leq \rho \leq d(v, s)$. Prove that the expected cost of the cut picked is at most twice the optimal fractional multiway cut. Derandomize this algorithm, and give a modification to make it a factor $2 - 2/k$ algorithm.

Hint: Show that for each edge (u, v), the probability that it is picked is bounded by $2 \cdot d(u, v)$.

19.3 In an attempt to improve the factor of the previous algorithm, suppose we choose ρ at random in the interval $[0, 1]$. What goes wrong? How is this rectified in Algorithm 19.4?

19.4 Derive Theorem 19.7 from Lemma 19.6.
Hint: Lemma 19.6 implies that $\mathbf{Pr}[c(C) \leq 1.5 \cdot \mathrm{OPT}_f] \geq 2/k \geq 2/n$. Run Algorithm 19.4 polynomially many times and output the best cut.

19.5 How does the approximation guarantee of the algorithm change if σ is picked to be a random permutation from S_k?

19.6 (Y. Rabani) For the case $k = 3$, replace the randomized rounding procedure of Algorithm 19.4 with the following. Pick ρ_1 and ρ_2 independently and uniformly from $(0, 1)$. Pick one of the three dimensions at random, say i. Merge with s_i all nonterminals v satisfying $x_v^i \geq \rho_1$. Arbitrarily pick one of the remaining two dimensions, say j, and denote the third dimension by k. Merge with s_j all remaining nonterminals v satisfying $x_v^j + x_v^i/2 \geq \rho_2$. Finally, merge with s_k all remaining nonterminals. Show that this modified algorithm achieves an approximation guarantee of $7/6$ for the 3-way cut problem.

19.7 In this exercise, we introduce another LP-relaxation for the multiway cut problem (see also Chapter 30). Given an undirected graph $G = (V, E)$ with costs on edges, obtain the directed graph H by replacing each edge (u, v) of G by two directed edges $(u \to v)$ and $(v \to u)$, each having the same cost as (u, v). Assign a 0/1 indicator variable d_e to each edge e in H. Suppose the terminals are numbered $s_1, \ldots, s_k$ in some order. Let $\mathcal{P}$ be the collection of all simple paths from a lower-numbered terminal to a higher-numbered terminal. Consider the following *bidirected* integer programming formulation for the multiway cut problem.

$$
\begin{aligned}
\text{minimize} \quad & \sum_{e \in H} c(e) d_e && (19.4) \\
\text{subject to} \quad & \sum_{e \in p} d_e \geq 1, \quad && p \in \mathcal{P} \\
& d_e \in \{0, 1\}, \quad && e \in H
\end{aligned}
$$

1. Show that an optimal solution to IP (19.4) yields an optimal solution to the multiway cut problem.
2. Obtain the LP-relaxation and dual program. Give a good physical interpretation of the dual.

3. Show that the graph given in Example 19.2 has an integrality gap of 16/15 for this relaxation as well (by showing a primal and dual solution of cost 7.5).

19.8 Consider Algorithm 4.3 for the multiway cut problem. Show that the analogous algorithm for the node multiway cut problem, based on isolating cuts, does not achieve a constant factor. What is the best factor you can prove for this algorithm?

19.9 The multiway cut problem also possesses the half-integrality property. Give an LP for the multiway cut problem similar to LP (19.2), and prove this fact.

19.10 Show that the lower bound on OPT given by LP (19.2) can be smaller by a factor of $2 - 2/k$ by giving a graph in which the optimal node multiway cut is $2 - 2/k$ times bigger than the maximum flow.

19.11 Theorem 19.12 leads directly to a factor 2 approximation algorithm for the node multiway cut problem, by rounding up the halves in a half-integral solution. Obtain a factor $2 - 2/k$ algorithm, and give a family of tight examples for this algorithm.
Hint: Not all vertices of M^{disj} are required for obtaining a multiway cut. For the tight example, consider the following graph.

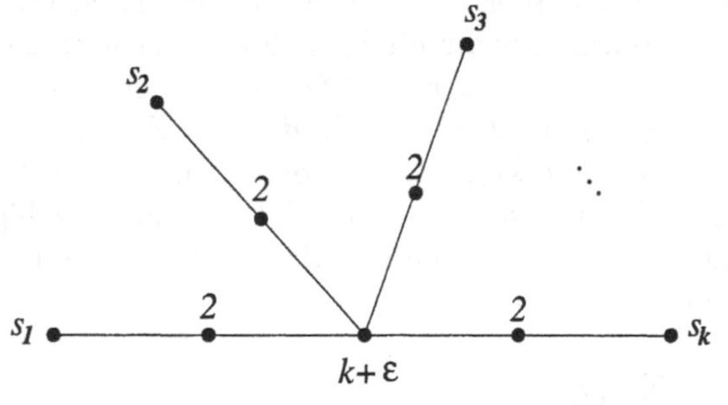

19.12 Consider the following problem.

Problem 19.13 (Directed multiway cut) Given a directed graph $G = (V, E)$ with an assignment of capacities to edges, $c : E \to \mathbf{R}^+$, and a set of terminals $S = \{s_1, s_2, \ldots, s_k\} \subseteq V$, a *directed multiway cut* is a set of edges whose removal ensures that the remaining graph has no path from s_i to s_j for each pair of distinct terminals s_i and s_j. The directed multiway cut problem asks for the minimum cost such set.

Obtain an LP-relaxation for this problem similar to LP (19.2). The dual can be interpreted as a directed multicommodity flow LP. Find the optimal fractional directed multiway cut and flow in the following example:

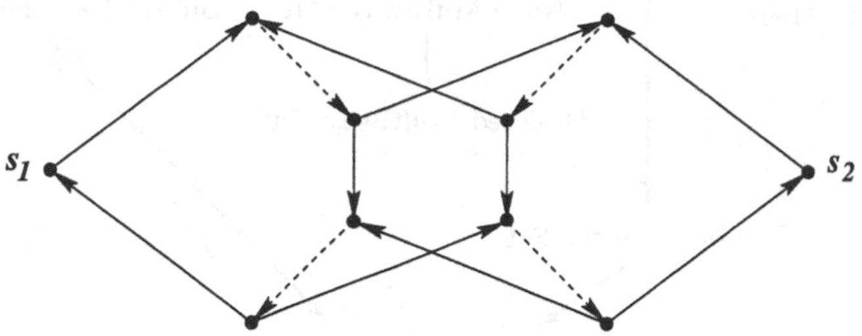

Notice that unlike LP (19.2), this relaxation does not always have an optimal half-integral solution.

19.13 Let us define the following two problems:

Problem 19.14 (Subset feedback edge set) Given a connected, undirected graph $G = (V, E)$ with an assignment of weights to edges, $w : E \to \mathbf{R}^+$, and a set of special vertices $S = \{s_1, s_2, \ldots, s_k\} \subseteq V$, a *subset feedback edge set* is a set of edges whose removal ensures that the remaining graph has no cycle containing a special vertex. The subset feedback edge set problem asks for the minimum weight such set.

Problem 19.15 (Subset feedback vertex set) Given a connected, undirected graph $G = (V, E)$ with an assignment of weights to vertices, $w : V \to \mathbf{R}^+$, and a set of special vertices $S = \{s_1, s_2, \ldots, s_k\} \subseteq V$, a *subset feedback vertex set* is a subset of $V - S$ whose removal ensures that the remaining graph has no cycle containing a special vertex. The subset feedback vertex set problem asks for the minimum weight such set.

These and previously introduced problems are related by approximation factor preserving reductions given in the following figure (each arrow represents such a reduction). Give these reductions. For a definition of such reductions, see Section A.3.1.

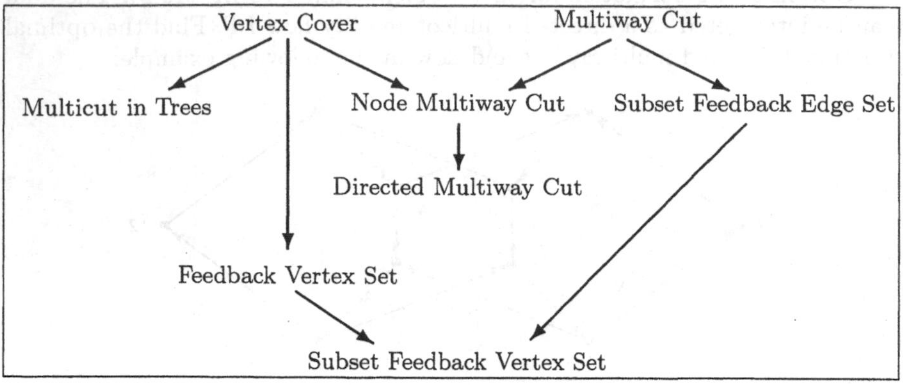

The current best factors known for multiway cut and subset feedback vertex set are 1.34 and 8, respectively. For the rest of the problems, the current best factor is 2.

19.5 Notes

Algorithm 19.4 is due to Calinescu, Karloff, and Rabani [37]. The current best guarantee known for the multiway cut problem is 1.3438, due to Karger, Klein, Stein, Thorup, and Young [164]. This is also the best upper bound known on the integrality gap of the relaxation used. Freund and Karloff [93] give a family of instances achieving a lower bound of $8/(7 + \frac{1}{k-1})$ on the integrality gap; Example 19.2 is from their paper. Theorem 19.12 is due to Garg, Vazirani, and Yannakakis [102]. For currently best approximation algorithms known for directed multiway cut, subset feedback edge set, and subset feedback vertex set, see Naor and Zosin [218], Even, Naor, Schieber, and Zosin [82], and Even, Naor, and Zosin [83], respectively.

20 Multicut in General Graphs

The importance of min–max relations to combinatorial optimization was mentioned in Chapter 1. Perhaps the most useful of these is the celebrated max-flow min-cut theorem. Indeed, much of flow theory, and the theory of cuts in graphs, has been built around this theorem. It is not surprising, therefore, that a concerted effort was made to obtain generalizations of this theorem to the case of multiple commodities.

There are two such generalizations. In the first one, the objective is to maximize the sum of the commodities routed, subject to flow conservation and capacity constraints. In the second generalization, a demand dem(i) is specified for each commodity i, and the objective is to maximize f, called throughput, such that for each i, $f \cdot \text{dem}(i)$ amount of commodity i can be routed simultaneously. We will call these *sum multicommodity flow* and *demands multicommodity flow* problems, respectively. Clearly, for the case of a single commodity, both problems are the same as the maximum flow problem.

Each of these generalizations is associated with a fundamental **NP**-hard cut problem, the first with the minimum multicut problem, Problem 18.1, and the second with the sparsest cut problem, Problem 21.2. In each case an approximation algorithm for the cut problem gives, as a corollary, an approximate max-flow min-cut theorem. In this chapter we will study the first generalization; the second is presented in Chapter 21. We will obtain an $O(\log k)$ factor approximation algorithm for the minimum multicut problem, where k is the number of commodities. A factor 2 algorithm for the special case of trees was presented in Chapter 18.

20.1 Sum multicommodity flow

Problem 20.1 (Sum multicommodity flow) Let $G = (V, E)$ be an undirected graph with nonnegative capacity c_e for each edge $e \in E$. Let $\{(s_1, t_1), \ldots, (s_k, t_k)\}$ be a specified set of pairs of vertices where each pair is distinct, but vertices in different pairs are not required to be distinct. A separate commodity is defined for each (s_i, t_i) pair. For convenience, we will think of s_i as the source and t_i as the sink of this commodity. The objective is to maximize the sum of the commodities routed. Each commodity must satisfy flow conservation at each vertex other than its own source and sink. Also, the sum of flows routed through an edge, in both directions combined, should not exceed the capacity of this edge.

Let us first give a linear programming formulation for this problem. For each commodity i, let P_i denote the set of all paths from s_i to t_i in G, and let $P = \bigcup_{i=1}^{k} P_i$. The LP will have a variable f_p for each $p \in P$, which will denote the flow along path p. The endpoints of this path uniquely specify the commodity that flows on this path. The objective is to maximize the sum of flows routed on these paths, subject to edge capacity constraints. Notice that flow conservation constraints are automatically satisfied in this formulation. The program has exponentially many variables; however, that is not a concern since we will use it primarily to obtain a clean formulation of the dual program.

$$\begin{aligned} \text{maximize} \quad & \sum_{p \in P} f_p && (20.1) \\ \text{subject to} \quad & \sum_{p:e \in p} f_p \le c_e, \qquad e \in E \\ & f_p \ge 0, \qquad p \in P \end{aligned}$$

Let us obtain the dual of this program. For this, let d_e be the dual variable associated with edge e. We will interpret these variables as distance labels of edges.

$$\begin{aligned} \text{minimize} \quad & \sum_{e \in E} c_e d_e && (20.2) \\ & \sum_{e \in p} d_e \ge 1, \qquad p \in P \\ & d_e \ge 0, \qquad e \in E \end{aligned}$$

The dual program tries to find a distance label assignment to edges so that on each path $p \in P$, the distance labels of edges add up to at least 1. Equivalently, a distance label assignment is feasible iff for each commodity i, the shortest path from s_i to t_i has length at least 1.

Notice that the programs (18.2) and (18.1) are special cases of the two programs presented above for the restriction that G is a tree.

The following remarks made in Chapter 18 hold for the two programs presented above as well: an optimal integral solution to LP (20.2) is a minimum multicut, and an optimal fractional solution can be viewed as a minimum fractional multicut. By the LP-duality theorem, minimum fractional multicut equals maximum multicommodity flow and, as shown in Example 18.2, it may be strictly smaller than minimum integral multicut.

This naturally raises the question whether the ratio of minimum multicut and maximum multicommodity flow is bounded. Equivalently, is the integrality gap of LP (20.2) bounded? In the next section we present an algorithm for finding a multicut within an $O(\log k)$ factor of the maximum flow, thereby showing that the gap is bounded by $O(\log k)$.

20.2 LP-rounding-based algorithm

First notice that the dual program (20.2) can be solved in polynomial time using the ellipsoid algorithm, since there is a simple way of obtaining a separation oracle for it: simply compute the length of a minimum s_i–t_i path, for each commodity i, w.r.t. the current distance labels. If all these lengths are ≥ 1, we have a feasible solution. Otherwise, the shortest such path provides a violated inequality. Alternatively, the LP obtained in Exercise 20.1 can be solved in polynomial time. Let d_e be the distance label computed for each edge e, and let $F = \sum_{e \in E} c_e d_e$.

Our goal is to pick a set of edges of small capacity, compared to F, that is a multicut. Let D be the set of edges with positive distance labels, i.e., $D = \{e \mid d_e > 0\}$. Clearly, D is a multicut; however, its capacity may be very large compared to F (Exercises 20.3 and 20.4). How do we pick a small capacity subset of D that is still a multicut? Since the optimal fractional multicut is the most cost-effective way of disconnecting all source–sink pairs, edges with large distance labels are more important than those with small distance labels for this purpose. The algorithm described below indirectly gives preference to edges with large distance labels.

The algorithm will work on graph $G = (V, E)$ with edge lengths given by d_e. The *weight of edge* e is defined to be $c_e d_e$. Let $\text{dist}(u, v)$ denote the length of the shortest path from u to v in this graph. For a set of vertices $S \subset V$, $\delta(S)$ denotes the set of edges in the cut $(S, \overline{S})$, $c(S)$ denotes the capacity of this cut, i.e., the total capacity of edges in $\delta(S)$, and $\text{wt}(S)$ denotes the *weight* of set S, which is roughly the sum of weights of all edges having both endpoints in S (a more precise definition is given below).

The algorithm will find disjoint sets of vertices, $S_1, \ldots, S_l, l \leq k$, in G, called *regions*, such that:

- No region contains any source–sink pair, and for each i, either s_i or t_i is in one of the regions.
- For each region S_i, $c(S_i) \leq \varepsilon \, \text{wt}(S_i)$, where ε is a parameter that will be defined below.

By the first condition, the union of the cuts of these regions, i.e., $M = \delta(S_1) \cup \delta(S_2) \cup \ldots \cup \delta(S_l)$, is a multicut, and by the second condition, its capacity $c(M) \leq \varepsilon F$. (When we give the precise definition of $\text{wt}(S)$, this inequality will need to be modified slightly.)

20.2.1 Growing a region: the continuous process

The sets $S_1, \ldots, S_l$ are found through a region growing process. Let us first present a continuous process to clarify the issues. For the sake of time efficiency, the algorithm itself will use a discrete process (see Section 20.2.2).

Each region is found by growing a set starting from one vertex, which is the source or sink of a pair. This will be called the *root* of the region. Suppose the root is s_1. The process consists of growing a ball around the root. For each radius r, define $S(r)$ to be the set of vertices at a distance $\leq r$ from s_1, i.e., $S(r) = \{v \mid \text{dist}(s_1, v) \leq r\}$. $S(0) = \{s_1\}$, and as r increases continuously from 0, at discrete points, $S(r)$ grows by adding vertices in increasing order of their distance from s_1.

Lemma 20.2 *If the region growing process is terminated before the radius becomes* $1/2$, *then the set* S *that is found contains no source–sink pair.*

Proof: The distance between any pair of vertices in $S(r)$ is $\leq 2r$. Since for each commodity i, $\text{dist}(s_i, t_i) \geq 1$, the lemma follows. □

For technical reasons that will become clear in Lemma 20.3 (see also Exercises 20.5 and 20.6), we will assign a weight to the root, $\text{wt}(s_1) = F/k$. The weight of $S(r)$ is the sum of $\text{wt}(s_1)$ and the sum of the weights of edges, or parts of edges, in the ball of radius r around s_1. Let us state this formally. For edges e having at least one endpoint in $S(r)$, let q_e denote the *fraction* of edge e that is in $S(r)$. If both endpoints of e are in $S(r)$, then $q_e = 1$. Otherwise, suppose $e = (u, v)$ with $u \in S(r)$ and $v \notin S(r)$. For such edges,

$$q_e = \frac{r - \text{dist}(s_1, u)}{\text{dist}(s_1, v) - \text{dist}(s_1, u)}.$$

Define the *weight of region* $S(r)$,

$$\text{wt}(S(r)) = \text{wt}(s_1) + \sum c_e d_e q_e,$$

where the sum is over all edges having at least one endpoint in $S(r)$.

We want to fix ε so that we can guarantee that we will encounter the condition $c(S(r)) \leq \varepsilon \, \text{wt}(S(r))$ for $r < 1/2$. The important observation is that at each point the rate at which the weight of the region is growing is at least $c(S(r))$. Until this condition is encountered,

$$\text{d}\, \text{wt}(S(r)) \geq c(S(r))\, \text{d}r > \varepsilon \, \text{wt}(S(r))\, \text{d}r.$$

Exercise 20.5 will help the reader gain some understanding of such a process.

Lemma 20.3 *Picking* $\varepsilon = 2\ln(k+1)$ *suffices to ensure that the condition* $c(S(r)) \leq \varepsilon \, \text{wt}(S(r))$ *will be encountered before the radius becomes* $1/2$.

Proof: The proof is by contradiction. Suppose that throughout the region growing process, starting with $r = 0$ and ending at $r = 1/2$, $c(S(r)) > \varepsilon\,\mathrm{wt}(S(r))$. At any point the incremental change in the weight of the region is

$$\mathrm{d}\,\mathrm{wt}(S(r)) = \sum_e c_e d_e \,\mathrm{d}q_e.$$

Clearly, only edges having one endpoint in $S(r)$ will contribute to the sum. Consider such an edge $e = (u, v)$ such that $u \in S(r)$ and $v \notin S(r)$. Then,

$$c_e d_e \,\mathrm{d}q_e = c_e \frac{d_e}{\mathrm{dist}(s_1, v) - \mathrm{dist}(s_1, u)} \,\mathrm{d}r.$$

Since $\mathrm{dist}(s_1, v) \leq \mathrm{dist}(s_1, u) + d_e$, we get $d_e \geq \mathrm{dist}(s_1, v) - \mathrm{dist}(s_1, u)$, and hence $c_e d_e \,\mathrm{d}q_e \geq c_e \,\mathrm{d}r$. This gives

$$\mathrm{d}\,\mathrm{wt}(S(r)) \geq c(S(r))\,\mathrm{d}r > \varepsilon\,\mathrm{wt}(S(r))\,\mathrm{d}r.$$

As long as the terminating condition is not encountered, the weight of the region increases exponentially with the radius. The initial weight of the region is F/k and the final weight is at most $F + F/k$. Integrating we get

$$\int_{\frac{F}{k}}^{F+\frac{F}{k}} \frac{1}{\mathrm{wt}(S(r))} \,\mathrm{d}\,\mathrm{wt}(S(r)) > \int_0^{\frac{1}{2}} \varepsilon \,\mathrm{d}r.$$

Therefore, $\ln(k+1) > \frac{1}{2}\varepsilon$. However, this contradicts the assumption that $\varepsilon = 2\ln(k+1)$, thus proving the lemma. □

20.2.2 The discrete process

The discrete process starts with $S = \{s_1\}$ and adds vertices to S in increasing order of their distance from s_1. Essentially, it involves executing a shortest path computation from the root. Clearly, the sets of vertices found by both processes are the same.

The weight of region S is redefined for the discrete process as follows:

$$\mathrm{wt}(S) = \mathrm{wt}(s_1) + \sum_e c_e d_e,$$

where the sum is over all edges that have at least one endpoint in S, and $\mathrm{wt}(s_1) = F/k$. The discrete process stops at the first point when $c(S) \leq \varepsilon\,\mathrm{wt}(S)$, where ε is again $2\ln(k+1)$. Notice that for the same set S, $\mathrm{wt}(S)$ in the discrete process is at least as large as that in the continuous process.

Therefore, the discrete process cannot terminate with a larger set than that found by the continuous process. Hence, the set S found contains no source–sink pair.

20.2.3 Finding successive regions

The first region is found in graph G, starting with any one of the sources as the root. Successive regions are found iteratively. Let $G_1 = G$ and S_1 be the region found in G_1. Consider a general point in the algorithm when regions $S_1, \ldots, S_{i-1}$ have already been found. Now, G_i is defined to be the graph obtained by removing vertices $S_1 \cup \ldots \cup S_{i-1}$, together with all edges incident at them from G.

If G_i does not contain a source–sink pair, we are done. Otherwise, we pick the source of such a pair, say s_j, as the root, define its weight to be F/k, and grow a region in G_i. All definitions, such as distance and weight, are w.r.t. graph G_i. We will denote these with a subscript of G_i. Also, for a set of vertices S in G_i, $c_{G_i}(S)$ will denote the total capacity of edges incident at S in G_i, i.e., the total capacity of edges in $\delta_{G_i}(S)$. As before, the value of ε is $2\ln(k+1)$, and the terminating condition is $c_{G_i}(S_i) \leq \varepsilon \,\mathrm{wt}_{G_i}(S_i)$. Notice that in each iteration the root is the only vertex that is defined to have nonzero weight.

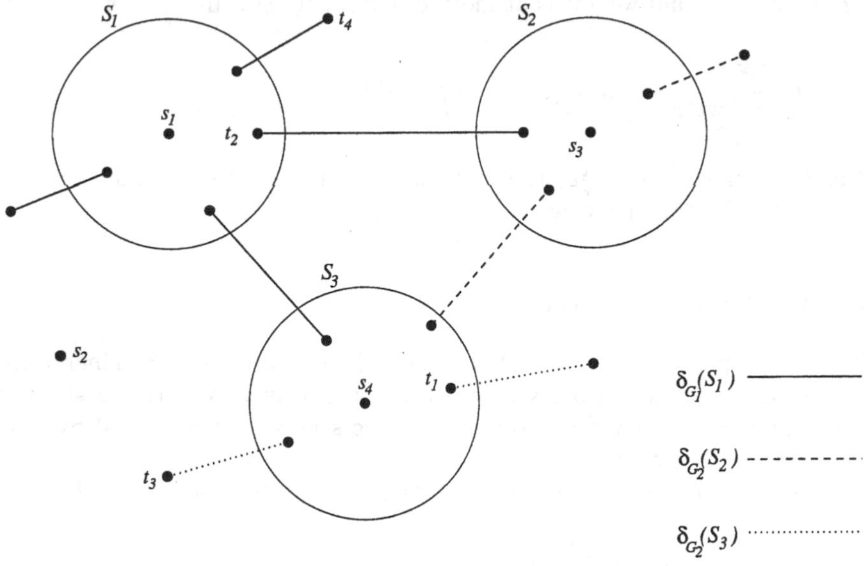

In this manner, we will find regions $S_1, \ldots, S_l$, $l \leq k$, and will output the set $M = \delta_{G_1}(S_1) \cup \ldots \cup \delta_{G_l}(S_l)$. Since edges of each cut are removed from the graph for successive iterations, the sets in this union are disjoint, and $c(M) = \sum_i c_{G_i}(S_i)$.

The algorithm is summarized below. Notice that while a region is growing, edges with large distance labels will remain in its cut for a longer time, and thus are more likely to be included in the multicut found. (Of course, the precise time that an edge remains in the cut is given by the difference between the distances from the root to the two endpoints of the edge.) As promised, the algorithm indirectly gives preference to edges with large distance labels.

Algorithm 20.4 (Minimum multicut)

1. Find an optimal solution to the LP (20.2), thus obtaining distance labels for edges of G.
2. $\varepsilon \leftarrow 2\ln(k+1)$, $H \leftarrow G$, $M \leftarrow \emptyset$;
3. While $\exists$ a source–sink pair in H do:
   Pick such a source, say s_j;
   Grow a region S with root s_j until $c_H(S) \leq \varepsilon\,\mathrm{wt}_H(S)$;
   $M \leftarrow M \cup \delta_H(S)$;
   $H \leftarrow H$ with vertices of S removed;
4. Output M.

Lemma 20.5 *The set M found is a multicut.*

Proof: We need to prove that no region contains a source–sink pair. In each iteration i, the sum of weights of edges of the graph and the weight defined on the current root is bounded by $F + F/k$. By the proof of Lemma 20.3, the continuous region growing process is guaranteed to encounter the terminating condition before the radius of the region becomes $1/2$. Therefore, the distance between a pair of vertices in the region, S_i, found by the discrete process is also bounded by 1. Notice that we had defined these distances w.r.t. graph G_i. Since G_i is a subgraph of G, the distance between a pair of vertices in G cannot be larger than that in G_i. Hence, S_i contains no source–sink pair. □

Lemma 20.6 $c(M) \leq 2\varepsilon F = 4\ln(k+1)F$.

Proof: In each iteration i, by the terminating condition we have $c_{G_i}(S_i) \leq \varepsilon\,\mathrm{wt}_{G_i}(S_i)$. Since all edges contributing to $\mathrm{wt}_{G_i}(S_i)$ will be removed from the graph after this iteration, each edge of G contributes to the weight of at most one region. The total weight of all edges of G is F. Since each iteration helps disconnect at least one source–sink pair, the number of iterations is bounded by k. Therefore, the total weight attributed to source vertices is at most F. Summing gives:

$$c(M) = \sum_i c_{G_i}(S_i) \leq \varepsilon \left(\sum_i \mathrm{wt}_{G_i}(S_i) \right) \leq \varepsilon \left(k\frac{F}{k} + \sum_e c_e d_e \right) = 2\varepsilon F.$$

□

Theorem 20.7 *Algorithm 20.4 achieves an approximation guarantee of* $O(\log k)$ *for the minimum multicut problem.*

Proof: The proof follows from Lemmas 20.5 and 20.6, and from the fact that the value of the fractional multicut, F, is a lower bound on the minimum multicut. □

Exercise 20.6 justifies the choice of $\text{wt}(s_1) = F/k$.

Corollary 20.8 *In an undirected graph with k source–sink pairs,*

$$\max_{m/c \text{ flow } F} |F| \leq \min_{\text{multicut } C} |C| \leq O(\log k) \left(\max_{m/c \text{ flow } F} |F| \right),$$

where $|F|$ *represents the value of multicommodity flow* F*, and* $|C|$ *represents the capacity of multicut* C*.*

20.3 A tight example

Example 20.9 We will construct an infinite family of graphs for which the integrality gap for LP (20.2) is $\Omega(\log k)$, thereby showing that our analysis of Algorithm 20.4 and the approximate max-flow min-multicut theorem presented in Corollary 20.8 are tight within constant factors.

The construction uses expander graphs. An *expander* is a graph $G = (V, E)$ in which every vertex has the same degree, say d, and for any nonempty subset $S \subset V$,

$$|\delta(S)| > \min(|S|, |\overline{S}|),$$

where $\delta(S)$ denotes the set of edges in the cut $(S, \overline{S})$, i.e., edges that have one endpoint in S and the other in $\overline{S}$. Standard probabilistic arguments show that almost every constant degree graph, with $d \geq 3$, is an expander (see Section 20.6). Let H be such a graph containing k vertices.

Source–sink pairs are designated in H as follows. Consider a breadth first search tree rooted at some vertex v. The number of vertices within distance $\alpha - 1$ of vertex v is at most $1 + d + d^2 + \ldots + d^{\alpha-1} < d^\alpha$. Picking $\alpha = \lfloor \log_d k/2 \rfloor$ ensures that at least $k/2$ vertices are at a distance $\geq \alpha$ from v. Let us say that a pair of vertices are a source–sink pair if the distance between them is at least α. Therefore, we have chosen $\Theta(k^2)$ pairs of vertices as source–sink pairs.

Each edge in H is of unit capacity. Thus, the total capacity of edges of H is $O(k)$. Since the distance between each source–sink pair is $\Omega(\log k)$, any flow path carrying a unit of flow uses up $\Omega(\log k)$ units of capacity. Therefore, the value of maximum multicommodity flow in H is bounded by $O(k/\log k)$. Next we will prove that a minimum multicut in H, say M, has capacity $\Omega(k)$, thereby proving the claimed integrality gap. Consider the connected components obtained by removing M from H.

Claim 20.10 *Each connected component has at most $k/2$ vertices.*

Proof: Suppose a connected component has strictly more than $k/2$ vertices. Pick an arbitrary vertex v in this component. By the argument given above, the number of vertices that are within distance $\alpha - 1$ of v in the entire graph H is $< d^\alpha \leq k/2$. Thus, there is a vertex u in the component such that the distance between u and v is at least α, i.e., u and v form a source–sink pair. Thus removal of M has failed to disconnect a source–sink pair, leading to a contradiction. □

By Claim 20.10, and the fact that H is an expander, each component S has $|\delta(S)| \geq |S|$. Since each vertex of H is in one of the components, $\sum_S |\delta(S)| \geq k$, where the sum is over all connected components. Since an edge contributes to the cuts of at most two components, the number of edges crossing components is $\Omega(k)$. This gives the desired lower bound on the minimum multicut.

Next, let us ensure that the number of source–sink pairs defined in the graph is not related to the number of vertices in it. Notice that replacing an edge of H by a path of unit capacity edges does not change the value of maximum flow or minimum multicut. Using this operation we can construct from H a graph G having n vertices, for arbitrary $n \geq k$. The integrality gap of LP (20.2) for G is $\Omega(\log k)$.

□

20.4 Some applications of multicut

We will obtain an $O(\log n)$ factor approximation algorithm for the following problem by reducing to the minimum multicut problem. See Exercise 20.7 for further applications.

Problem 20.11 (2CNF≡ clause deletion) A 2CNF≡ formula consists of a set of clauses of the form $(u \equiv v)$, where u and v are literals. Let F be such a formula, and wt be a function assigning nonnegative rational weights to its clauses. The problem is to delete a minimum weight set of clauses of F so that the remaining formula is satisfiable.

Given a 2CNF≡ formula F on n Boolean variables, let us define graph $G(F)$ with edge capacities as follows: The graph has $2n$ vertices, one corresponding to each literal. Corresponding to each clause $(p \equiv q)$ we include the two edges (p, q) and $(\overline{p}, \overline{q})$, each having capacity equal to the weight of the clause $(p \equiv q)$.

Notice that the two clauses $(p \equiv q)$ and $(\overline{p} \equiv \overline{q})$ are equivalent. We may assume w.l.o.g. that F does not contain two such equivalent clauses, since we can merge their weights and drop one of these clauses. With this assumption each clause corresponds to two distinct edges in $G(F)$.

Lemma 20.12 *Formula F is satisfiable iff no connected component of $G(F)$ contains a variable and its negation.*

Proof: If (p, q) is an edge in $G(F)$ then the literals p and q must take the same truth value in every satisfying truth assignment. Thus, all literals of a connected component of $G(F)$ are forced to take the same truth value. Therefore, if F is satisfiable, no connected component in $G(F)$ contains a variable and its negation.

Conversely, notice that if literals p and q occur in the same connected component, then so do their negations. If no connected component contains a variable and its negation, the components can be paired so that in each pair, one component contains a set of literals and the other contains the complementary literals. For each pair, set the literals of one component to true and the other to false to obtain a satisfying truth assignment. □

For each variable and its negation, designate the corresponding vertices in $G(F)$ to be a source–sink pair, thus defining n source–sink pairs. Let M be a minimum multicut in $G(F)$ and C be a minimum weight set of clauses whose deletion makes F satisfiable. In general, M may have only one of the two edges corresponding to a clause.

Lemma 20.13 $\mathrm{wt}(C) \leq c(M) \leq 2 \cdot \mathrm{wt}(C)$.

Proof: Delete clauses corresponding to edges of M from F to get formula F'. The weight of clauses deleted is at most $c(M)$. Since $G(F')$ does not contain any edges of M, it does not have any component containing a variable and its negation. By Lemma 20.12, F' is satisfiable, thus proving the first inequality.

Next, delete from $G(F)$ the two edges corresponding to each clause in C. This will disconnect all source–sink pairs. Since the capacity of edges deleted is $2\mathrm{wt}(C)$, this proves the second inequality. □

Since we can approximate minimum multicut to within an $O(\log n)$ factor, we get:

Theorem 20.14 *There is an $O(\log n)$ factor approximation algorithm for Problem 20.11.*

20.5 Exercises

20.1 By defining for each edge e and commodity i a flow variable $f_{e,i}$, give an LP that is equivalent to LP (20.1) and has polynomially many variables.

Obtain the dual of this program and show that it is equivalent to LP (20.2); however, unlike LP (20.2), it has only polynomially many constraints.

20.2 Let $\boldsymbol{d}$ be an optimal solution to LP (20.2). Show that $\boldsymbol{d}$ must satisfy the triangle inequality.

20.3 Intuitively, our goal in picking a multicut is picking edges that are bottlenecks for multicommodity flow. In this sense, D is a very good starting point: prove that D is precisely the set of edges that are saturated in *every* maximum multicommodity flow.
Hint: Use complementary slackness conditions.

20.4 Give an example to show that picking all of D gives an $\Omega(n)$ factor for multicut.

20.5 Consider the following growth process. $W(t)$ denotes the weight at time t. Assume that the initial weight is $W(0) = W_0$, and that at each point the rate of growth is proportional to the current weight, i.e.,

$$\mathrm{d}W(t) = \varepsilon W(t)\mathrm{d}t.$$

Give the function $W(t)$. Next, assume that $W_0 = F/k$ and that $W(1/2) = F + F/k$. What is ε?
Hint: $W(t) = W_0 e^{\varepsilon t}$ and $\varepsilon = 2\ln(k+1)$.

20.6 This exercise justifies the choice of $\mathrm{wt}(s_1)$, which was fixed to be F/k. Suppose we fix it at W_0. Clearly, ε is inversely related to W_0 (see Lemma 20.3). However, the approximation factor of the algorithm is given by $\varepsilon(F + kW_0)$ (see Lemma 20.6). For what value of W_0 is the approximation factor minimized?

20.7 Consider the following problem, which has applications in VLSI design.

Problem 20.15 (Graph bipartization by edge deletion) Given an edge weighted undirected graph $G = (V, E)$, remove a minimum weight set of edges to leave a bipartite graph.

Obtain an $O(\log n)$ factor approximation algorithm for this problem by reducing it to Problem 20.11.

20.8 (Even, Naor, Schieber, and Rao [81]) This exercise develops an $O(\log^2 n)$ factor algorithm for the following problem.

Problem 20.16 (Minimum length linear arrangement) Given an undirected graph $G = (V, E)$, find a numbering of its vertices from 1 to n, $h : V \to \{1, \dots, n\}$, so as to minimize

$$\sum_{(u,v)\in E} |h(u) - h(v)|.$$

1. Show that the following is an LP-relaxation of this problem. This LP has a variable d_e for each edge $e \in E$, which we will interpret as a distance label. For any distance label assignment $\boldsymbol{d}$ to the edges of G, define $\text{dist}_{\boldsymbol{d}}(u, v)$ to be the length of the shortest path from u to v in G. Give a polynomial time separation oracle for this LP, thereby showing that it can be solved in polynomial time.

$$\begin{aligned} \text{minimize} \quad & \sum_{e\in E} d_e && (20.3) \\ & \sum_{u\in S} \text{dist}_{\boldsymbol{d}}(u,v) \geq \frac{1}{4}(|S|^2 - 1), && S \subseteq V, v \in S \\ & d_e \geq 0, && e \in E \end{aligned}$$

2. Let $\boldsymbol{d}$ be an optimal solution to LP (20.3). Show that for any $S \subseteq V, v \in S$ there is a vertex $u \in S$ such that $\text{dist}_{\boldsymbol{d}}(u, v) \geq (|S| + 1)/4$.
3. For $S \subseteq V$, define $\text{wt}(S)$ to be the sum of distance labels of all edges having both endpoints in S. Also, define $c(S, \overline{S})$ to be the number of edges in the cut $(S, \overline{S})$. Give a region growing process similar to that described in Section 20.2.1 that finds a cut $(S, \overline{S})$ in G with $\text{wt}(S) \leq \text{wt}(\overline{S})$ such that $c(S, \overline{S})$ is $O(\text{wt}(S)(\log n)/n)$.
4. Show that a divide-and-conquer algorithm that recursively finds a numbering for vertices in S from 1 to $|S|$, and a numbering for vertices in $\overline{S}$ from $|S| + 1$ to n achieves an approximation guarantee of $O(\log^2 n)$.
   Hint: Assuming each edge in the cut $(S, \overline{S})$ is of length $n - 1$, write a suitable recurrence for the cost incurred by the algorithm.

20.6 Notes

Theorem 20.7 and its corollary are due to Garg, Vazirani, and Yannakakis [103]. Problem 20.11 was introduced in Klein, Rao, Agrawal, and Ravi [180]. For showing existence of expanders via a probabilistic argument, see Pinsker [228].

21 Sparsest Cut

In this chapter we will obtain an approximation algorithm for the sparsest cut problem using an interesting LP-rounding procedure that employs results on low distortion embeddings of metrics in ℓ_1 spaces. As mentioned in Chapter 20, we will get as a corollary an approximate max-flow min-cut theorem for the demands version of multicommodity flow. Approximation algorithms for several other important problems will also follow.

21.1 Demands multicommodity flow

Problem 21.1 (Demands multicommodity flow) Let $G = (V, E)$ be an undirected graph with a nonnegative capacity c_e for each edge $e \in E$. Let $\{(s_1, t_1), \ldots, (s_k, t_k)\}$ be a specified set of pairs of vertices, where each pair is distinct, but vertices in different pairs are not required to be distinct. A separate commodity is defined for each (s_i, t_i) pair; for convenience, we will think of s_i as the source and t_i as the sink of this commodity. For each commodity i, a nonnegative demand, $\mathrm{dem}(i)$, is also specified. The objective is to maximize f, called *throughput*, such that for each commodity i, $f \cdot \mathrm{dem}(i)$ units of this commodity can be routed simultaneously, subject to flow conservation and capacity constraints, i.e., each commodity must satisfy flow conservation at each vertex other than its own source and sink, and the sum of flows routed through an edge, in both directions combined, should not exceed the capacity of this edge. We will denote the optimal throughput by f^*.

Consider a cut $(S, \overline{S})$ in G. Let $c(S)$ denote the capacity of edges in this cut and $\mathrm{dem}(S)$ denote the total demand separated by this cut, i.e.,

$$\mathrm{dem}(S) = \sum_{i:\ |\{s_i, t_i\} \cap S| = 1} \mathrm{dem}(i).$$

Clearly, the ratio of these quantities places an upper bound on the throughput, i.e., $f^* \leq \frac{c(S)}{\mathrm{dem}(S)}$. This motivates:

Problem 21.2 (Sparsest cut) Let $G = (V, E)$ be an undirected graph with capacities, source–sink pairs, and demands defined as in Problem 21.1.

The *sparsity* of cut $(S, \overline{S})$ is given by $\frac{c(S)}{\text{dem}(S)}$. The problem is to find a cut of minimum sparsity. We will denote the sparsity of this cut by α^*.

Among all cuts, α^* puts the most stringent upper bound on f^*. Is this upper bound tight? Example 21.3 shows that it is not. However, minimum sparsity cannot be arbitrarily larger than maximum throughput; we will show that their ratio is bounded by $O(\log k)$.

Example 21.3 Consider the bipartite graph $K_{3,2}$ with all edges of unit capacity and a unit demand between each pair of nonadjacent vertices – a total of four commodities.

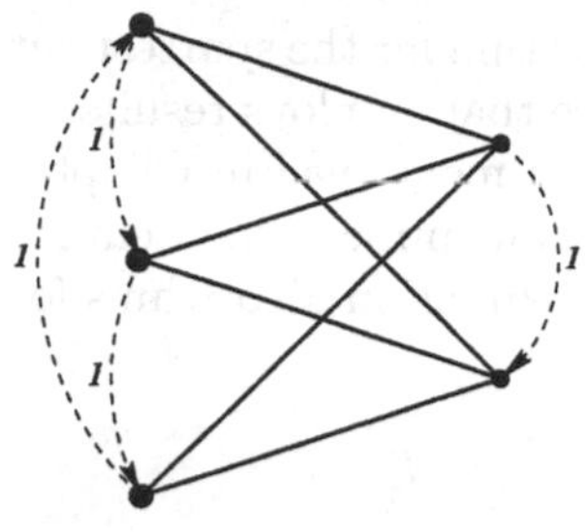

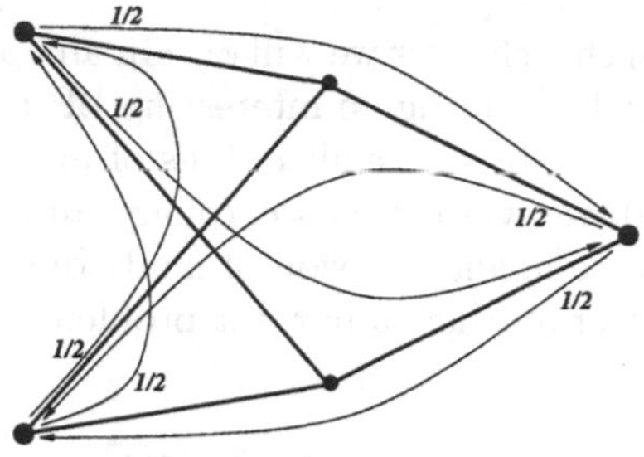

It is easy to check that a sparsest cut of $K_{3,2}$ has sparsity 1. This graph can be viewed as the union of two stars $K_{3,1}$ (the centers of the stars are the vertices on the right side of the bipartition), and, as in Example 18.2, we get the unique way of routing one unit of each of the three commodities having source and sink on the left side of the bipartition. However, this saturates all edges, making it impossible to route the fourth commodity. Hence, throughput is strictly smaller than 1. □

21.2 Linear programming formulation

We start by giving a linear programming formulation of the problem of maximizing throughput, f. Let $\mathcal{P}_i = \{q_j^i\}$ denote the set of all paths between s_i and t_i. Introduce variable f_j^i to denote the flow of commodity i sent along path q_j^i. The first set of constraints ensures that the demand of each commodity is met (with factor f), and the second set are edge capacity constraints.

$$\begin{aligned}
\text{maximize} \quad & f && && (21.1)\\
\text{subject to} \quad & \sum_j f_j^i \geq f \cdot \text{dem}(i), && i = 1, \ldots, k \\
& \sum_{q_j^i : e \in q_j^i} f_j^i \leq c_e, && e \in E \\
& f \geq 0 \\
& f_j^i \geq 0
\end{aligned}$$

Define the graph H with vertex set $V_H = \{s_i, t_i | 1 \le i \le k\}$ and edge set $E_H = \{(s_i, t_i) | 1 \le i \le k\}$ to be the *demand graph*. For each edge $e = (s_i, t_i)$ of H, let $\text{dem}(e) = dem(i)$. We will show that the dual to LP (21.1) yields a metric $(V, \boldsymbol{d})$ satisfying:

Theorem 21.4 *Let f^* denote the optimal throughput. Then,*

$$f^* = \min_{metric\, \boldsymbol{d}} \frac{\sum_{e \in G} c_e d_e}{\sum_{e \in H} \text{dem}(e) d_e}.$$

Let l_i and d_e be dual variables associated with the first and second set of inequalities of LP (21.1). We will interpret d_e's as distance label assignments to the edges of G. The first set of inequalities ensures that for each commodity i, l_i is upper bounded by the length of any path from s_i to t_i w.r.t. the distance label assignment.

$$\begin{aligned} \text{minimize} \quad & \sum_{e \in E} c_e d_e & & (21.2) \\ \text{subject to} \quad & \sum_{e \in q_j^i} d_e \ge l_i, & & q_j^i \in \mathcal{P}_i, i = 1, \ldots, k \\ & \sum_{i=1}^{k} l_i \,\text{dem}(i) \ge 1 & & \\ & d_e \ge 0, & & e \in E \\ & l_i \ge 0, & & i = 1, \ldots, k \end{aligned}$$

Example 21.5 For the instance given in Example 21.3, the optimal throughput is $f^* = 3/4$; this corresponds to routing the four commodities as follows:

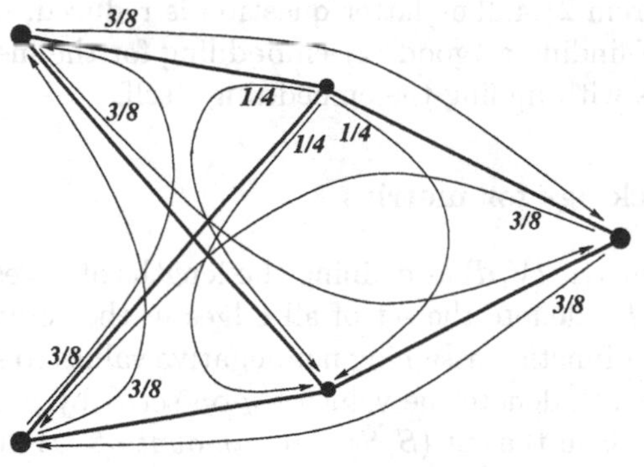

The optimal dual solution is: $d_e = 1/8$ for each edge e and $l_i = 1/4$ for each commodity i. It would be instructive for the reader to verify feasibility and optimality of these solutions. □

Claim 21.6 *There is an optimal distance label assignment* $\boldsymbol{d}$ *for the dual program (21.2) that is a metric on* V*. Furthermore, for each commodity* i*,* $l_i = d_{(s_i,t_i)}$*, and the second inequality holds with equality, i.e.,* $\sum_i d_{(s_i,t_i)}\mathrm{dem}(i) = 1$*.*

Proof: If for some three points u, v, and w, $d_{uw} > d_{uv} + d_{vw}$, then decrease d_{uw} to $d_{uv} + d_{vw}$. Since this does not decrease the shortest path between any s_i–t_i pair, the solution still remains feasible. Moreover, the objective function value cannot increase by this process. Continuing in this manner, we will obtain a metric on V.

Now, the length of a shortest path from s_i to t_i is given by the distance label $d_{(s_i,t_i)}$. Setting $l_i = d_{(s_i,t_i)}$ does not change the feasibility or the objective function value of the solution. Finally, if the second inequality holds strictly, then we can scale down all distance labels without violating feasibility, thus contradicting the optimality of $\boldsymbol{d}$. □

By Claim 21.6, the dual program yields a metric $(V, \boldsymbol{d})$ that minimizes

$$\frac{\sum_{e \in G} c_e d_e}{\sum_{e \in H} \mathrm{dem}(e) d_e}.$$

By the LP-duality theorem, this equals the optimal throughput. This proves Theorem 21.4.

21.3 Metrics, cut packings, and ℓ_1-embeddability

In Section 21.3.1, we will define the notion of a cut packing for a metric and will show that the question of finding a good approximation to the sparsest cut for graph G reduces to that of finding a "good" cut packing for the metric obtained in Theorem 21.4. The latter question is reduced, in Section 21.3.2, to the question of finding a "good" ℓ_1-embedding for the metric. Eventually, Section 21.4 deals with finding the embedding itself.

21.3.1 Cut packings for metrics

Let us think of a metric $(V, \boldsymbol{d})$ as defining the lengths of edges of the complete graph on V. Let E_n denote the set of all edges in the complete graph on n vertices. Let y be a function assigning nonnegative values to subsets of V, i.e., $y : 2^V \rightarrow \mathbf{R}^+$. We will denote the value of y on set S by y_S. Let us say that edge e *feels* y_S if e is in the cut $(S, \overline{S})$. The *amount of cut* that edge e feels is $\sum_{S: e \in \delta(S)} y_S$. Function y is called a *cut packing* for metric $(V, \boldsymbol{d})$ if no edge feels more cut than its length, i.e., for each edge $e \in E_n$, $\sum_{S: e\in\delta(S)} y_S \leq d_e$. If this inequality holds with equality for each edge $e \in E_n$, then y is said to be an *exact cut packing.* The reason for the name "cut packing" is that equivalently, we can think of y as assigning value $y_S + y(\overline{S})$ to each cut $(S, \overline{S})$.

As shown below, in general, there may not be an exact cut packing for metric $(V, \boldsymbol{d})$. Let us relax this notion by allowing edges to be underpacked up to a specified extent. For $\beta \geq 1$, y is said to be a β-approximate cut packing if the amount of cut felt by any edge is at least $1/\beta$ fraction of its length, i.e., for each edge $e \in E_n$, $d_e/\beta \leq \sum_{S: e \in \delta(S)} y_S \leq d_e$. Clearly, the smaller β is, the better the cut packing. The following theorem shows the importance of finding a good cut packing for $(V, \boldsymbol{d})$.

Theorem 21.7 *Let $(V, \boldsymbol{d})$ be the metric obtained in Theorem 21.4, and let y be a β-approximate cut packing for $(V, \boldsymbol{d})$. Among cuts with $y_S \neq 0$, let $(S', \overline{S'})$ be the sparsest. Then, the sparsity of this cut is at most $\beta \cdot f^*$.*

Proof: Let y be a β-approximate cut packing for metric $(V, \boldsymbol{d})$. Then,

$$
\begin{aligned}
f^* = \frac{\sum_{e \in G} c_e d_e}{\sum_{e \in H} \mathrm{dem}(e) d_e} &\geq \frac{\sum_{e \in G} c_e \sum_{S: e \in \delta(S)} y_S}{\sum_{e \in H} \mathrm{dem}(e) \sum_{S: e \in \delta(S)} \beta y_S} \\
&= \frac{\sum_S y_S c(S)}{\beta \sum_S y_S \mathrm{dem}(S)} \\
&\geq \frac{1}{\beta} \cdot \left(\frac{c(S')}{\mathrm{dem}(S')} \right).
\end{aligned}
$$

The first inequality follows using both the upper bound and the lower bound on the amount of cut felt by an edge; the former in the numerator and the latter in the denominator. The equality after that follows by changing the order of summation. The last inequality follows from the well known result stated below. □

Proposition 21.8 *For any nonnegative reals $a_1, \ldots, a_n$ and positive reals $b_1, \ldots, b_n$ and $\alpha_1, \ldots, \alpha_n$,*

$$
\frac{\sum_i \alpha_i a_i}{\sum_i \alpha_i b_i} \geq \min_i \frac{a_i}{b_i}.
$$

Moreover, this inequality holds with equality iff the n values a_i/b_i are all equal.

Corollary 21.9 *If there is an exact cut packing for metric $(V, \boldsymbol{d})$, then every cut $(S, \overline{S})$ with $y_S \neq 0$ has sparsity f^* and thus is a sparsest cut in G.*

Proof: By Theorem 21.7, the minimum sparsity cut with $y_S \neq 0$ has sparsity at most f^* (since $\beta = 1$). Since the sparsity of any cut upper bounds f^*, the sparsity of this cut equals f^*, and this is a sparsest cut in G. But then all inequalities in the proof of Theorem 21.7 must hold with equality. Now, by the second statement in Proposition 21.8, we get that every cut $(S, \overline{S})$ with $y_S \neq 0$ has sparsity f^*. □

The sparsest cut in the instance specified in Example 21.3 has sparsity strictly larger than f^*. By Corollary 21.9, the optimal metric for this instance does not have an exact cut packing. However, it turns out that every metric has an $O(\log n)$-approximate cut packing – we will show this using the notion of ℓ_1-embeddability of metrics.

21.3.2 ℓ_1-embeddability of metrics

A *norm* on the vector space $\mathbf{R}^m$ is a function $\|\cdot\| : \mathbf{R}^m \to \mathbf{R}^+$, such that for any $\boldsymbol{x}, \boldsymbol{y} \in \mathbf{R}^m$, and $\lambda \in \mathbf{R}$:

- $\|\boldsymbol{x}\| = 0$ iff $\boldsymbol{x} = 0$,
- $\|\lambda \boldsymbol{x}\| = |\lambda| \cdot \|\boldsymbol{x}\|$,
- $\|\boldsymbol{x} + \boldsymbol{y}\| \leq \|\boldsymbol{x}\| + \|\boldsymbol{y}\|$.

For $p \geq 1$, the ℓ_p-*norm* is defined by

$$\|\boldsymbol{x}\|_p = \left(\sum_{1 \leq k \leq m} |x_k|^p \right)^{\frac{1}{p}} .$$

The associated ℓ_p-*metric*, denoted by $\boldsymbol{d}_{\ell_p}$, is defined by

$$\boldsymbol{d}_{\ell_p}(\boldsymbol{x}, \boldsymbol{y}) = \|\boldsymbol{x} - \boldsymbol{y}\|_p$$

for all $\boldsymbol{x}, \boldsymbol{y} \in \mathbf{R}^m$. In this section, we will only consider the ℓ_1-norm.

Let σ be a mapping, $\sigma : V \to \mathbf{R}^m$ for some m. Let us say that $\|\sigma(u) - \sigma(v)\|_1$ is the ℓ_1 *length of edge* (u, v) *under* σ. We will say that σ is an *isometric* ℓ_1-*embedding* for metric $(V, \boldsymbol{d})$ if it preserves the ℓ_1 lengths of all edges, i.e.,

$$\forall u, v \in V, d(u, v) = \|\sigma(u) - \sigma(v)\|_1.$$

As shown below, in general, the metric computed by solving the dual program may not be isometrically ℓ_1-embeddable. Thus, we will relax this notion – we will ensure that the mapping does not *stretch* any edge, but we will allow it to *shrink* edges up to a specified factor. For $\beta \geq 1$, we will say that σ is a β-*distortion* ℓ_1-*embedding* for metric $(V, \boldsymbol{d})$ if

$$\forall u, v \in V : \frac{1}{\beta} d(u, v) \leq \|\sigma(u) - \sigma(v)\|_1 \leq d(u, v).$$

Next, we show that the question of finding an approximate cut packing for a metric is intimately related to that of finding a low distortion ℓ_1 embedding for it.

Lemma 21.10 *Let $\sigma : V \rightarrow \mathbf{R}^m$ be a mapping. There is a cut packing $y : 2^V \rightarrow \mathbf{R}^+$ such that each edge feels as much cut under y as its ℓ_1 length under σ. Moreover, the number of nonzero y_S's is at most $m(n-1)$.*

Proof: First consider the case when $m = 1$. Let the n vertices of V be mapped to $u_1 \leq u_2 \leq \cdots \leq u_n$. Assume w.l.o.g. that the vertices are also numbered in this order. For each $i, 1 \leq i \leq n-1$, let $y_{\{v_1,\ldots,v_i\}} = u_{i+1} - u_i$. Clearly, this cut packing satisfies the required condition.

For arbitrary m, we observe that since the ℓ_1-norm is additive, we can define a cut packing for each dimension independently, and the sum of these packings satisfies the required condition. □

Lemma 21.11 *Let $y : 2^V \rightarrow \mathbf{R}^+$ be a cut packing with m nonzero y_S's. There is a mapping $\sigma : V \rightarrow \mathbf{R}^m$ such that for each edge, its ℓ_1 length under σ is the same as the amount of cut it feels under y.*

Proof: We will have a dimension corresponding to each set $S \subseteq V$ such that $y_S \neq 0$. For vertices in S, this coordinate will be 0, and for vertices in $\overline{S}$, this coordinate will be y_S. Thus, this dimension contributes exactly as much to the ℓ_1 length of an edge as the amount of cut felt by this edge due to y_S. Hence this mapping satisfies the required condition. □

Lemmas 21.10 and 21.11 give:

Theorem 21.12 *There exists a β-distortion ℓ_1-embedding for metric $(V, \boldsymbol{d})$ iff there exists a β-approximate cut packing for it. Moreover, the number of nonzero cuts and the dimension of the ℓ_1-embedding are polynomially related.*

Corollary 21.13 *Metric $(V, \boldsymbol{d})$ is isometrically ℓ_1-embeddable iff there exists an exact cut packing for it.*

We have already shown that the metric obtained for the instance in Example 21.3 does not have an exact cut packing. Therefore, it is not isometrically ℓ_1-embeddable. However, we will show that any metric has an $O(\log n)$-distortion ℓ_1-embedding; this fact lies at the heart of the approximation algorithm for the sparsest cut problem.

21.4 Low distortion ℓ_1-embeddings for metrics

First consider the following one-dimensional embedding for metric $(V, \boldsymbol{d})$: pick a set $S \subseteq V$, and define the coordinate of vertex v to be $\sigma(v) = \min_{s \in S} d(s, v)$, i.e., the length of the shortest edge from v to S. This mapping does not stretch any edge:

Lemma 21.14 *For the one-dimensional embedding given above,*

$$\forall u, v \in V, \quad |\sigma(u) - \sigma(v)| \leq d(u, v).$$

Proof: Let s_1 and s_2 be the closest vertices of S to u and v, respectively. Assume w.l.o.g. that $d(s_1, u) \leq d(s_2, v)$. Then, $|\sigma(u) - \sigma(v)| = d(s_2, v) - d(s_1, u) \leq d(s_1, v) - d(s_1, u) \leq d(u, v)$. The last inequality follows by the triangle inequality. □

More generally, consider the following m-dimensional embedding: Pick m subsets of V, $S_1, \ldots, S_m$, and define the ith coordinate of vertex v to be $\sigma_i(v) = \min_{s \in S_i} d(s, v)/m$; notice the scaling factor of m used. The additivity of ℓ_1 metric, together with Lemma 21.14, imply that this mapping also does not stretch any edge.

21.4.1 Ensuring that a single edge is not overshrunk

The remaining task is to choose the sets in such a way that no edge shrinks by a factor of more than $O(\log n)$. It is natural to use randomization for picking the sets. Let us first ensure that a single edge (u, v) is not overshrunk. For this purpose, define the *expected contribution of set* S_i to the ℓ_1 length of edge (u, v) to be $\mathbf{E}[|\sigma_i(u) - \sigma_i(v)|]$.

For simplicity, assume that n is a power of 2; let $n = 2^l$. For $2 \leq i \leq l + 1$, set S_i is formed by picking each vertex of V with probability $1/2^i$. The embedding w.r.t. these sets works for the single edge (u, v) with high probability. The proof of this fact involves cleverly taking into consideration the expected contribution of each set. For different metrics, different sets have a large contribution. In order to develop intuition for the proof, we first illustrate this through a series of examples.

Example 21.15 In the following three metrics, $d(u, v) = 1$, and the n vertices are placed as shown in the figure below.

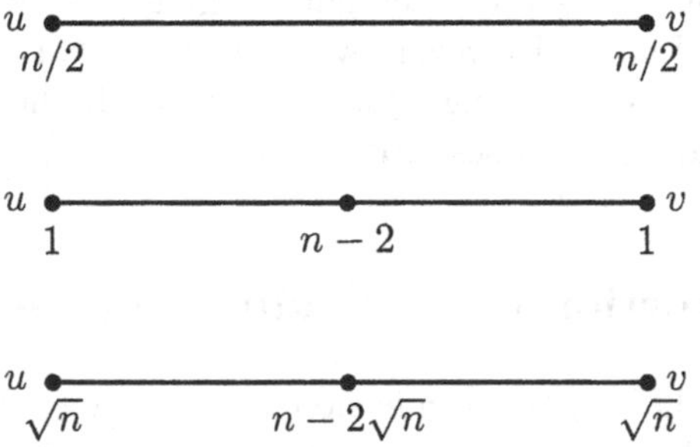

For each metric, the expected contribution of one of the sets is $\Omega(d(u,v)/l)$. For the first metric, this set is S_l, since it will be a singleton with constant probability. For the second metric, this set is S_2, since it will contain exactly one of u and v with constant probability. For the third metric, this set is $S_{\lceil l/2 \rceil}$, since with constant probability, it will contain exactly one vertex of the $2\sqrt{n}$ vertices bunched up with u and v. □

In the next lemma, we encapsulate the basic mechanism for establishing a lower bound on the expected contribution of a set S_i. For any vertex x and nonnegative real r, let $B(x,r)$ denote the *ball* of radius r around x, i.e., $B(x,r) = \{s \in V | d(x,s) \le r\}$.

Lemma 21.16 *If for some choice of $r_1 \ge r_2 \ge 0$, and constant c,*

$$\mathbf{Pr}[(S_i \cap B(u,r_1) = \emptyset) \text{ and } (S_i \cap B(v,r_2) \neq \emptyset)] \ge c,$$

then the expected contribution of S_i is $\ge c(r_1 - r_2)/l$.

Proof: Under the event described, $d(u,S_i) > r_1$ and $d(v,S_i) \le r_2$. If so, $\sigma_i(u) > r_1/l$ and $\sigma_i(v) \le r_2/l$. Therefore, $|\sigma_i(u) - \sigma_i(v)| > (r_1 - r_2)/l$, and the lemma follows. □

The remaining task is to define suitable radii r_1 and r_2 for each set S_i such that the probabilistic statement of Lemma 21.16 holds. We will need the following simple probabilistic fact:

Lemma 21.17 *For $1 \le t \le l-1$, let A and B be disjoint subsets of V, such that $|A| < 2^t$ and $|B| \ge 2^{t-1}$. Form set S by picking each vertex of V independently with probability $p = 1/(2^{t+1})$. Then,*

$$\mathbf{Pr}[(S \cap A = \emptyset) \text{ and } (S \cap B \neq \emptyset)] \ge (1/2)(1 - e^{-1/4}).$$

Proof:

$$\mathbf{Pr}[S \cap A = \emptyset] = (1-p)^{|A|} \ge (1 - p|A|) \ge \frac{1}{2},$$

where the first inequality follows by taking the first two terms of the binomial expansion.

$$\mathbf{Pr}[S \cap B = \emptyset] = (1-p)^{|B|} \le e^{-p|B|} \le e^{-1/4},$$

where we have used the inequality $1 - x \le e^{-x}$. Therefore,

$$\mathbf{Pr}[S \cap B \neq \emptyset] = 1 - (1-p)^{|B|} \ge 1 - e^{-1/4}.$$

Finally, observe that since A and B are disjoint, the two events $[S \cap A = \emptyset]$ and $[S \cap B \neq \emptyset]$ are independent. The lemma follows. □

For convenience, let $c = (1/2)(1 - e^{-1/4})$.

For $0 \leq t \leq l$, define $\rho_t = \min\{\rho \geq 0 : \ |B(u, \rho)| \geq 2^t \text{ and } |B(v, \rho)| \geq 2^t\}$, i.e., ρ_t is the smallest radius such that the ball around u and the ball around v each has at least 2^t vertices. Clearly, $\rho_0 = 0$ and $\rho_l \geq d(u, v)$. Let $\hat{t} = \max\{t : \ \rho_t < d(u, v)/2\}$; clearly, $\hat{t} \leq l - 1$. Finally, for any vertex x and nonnegative real r, let $B^\circ(x, r)$ denote the *open ball* of radius r around x, i.e., $B^\circ(x, r) = \{s \in V | d(x, s) < r\}$.

Lemma 21.18 *For $1 \leq t \leq \hat{t}$, the expected contribution of S_{t+1} is at least $c \cdot \frac{\rho_t - \rho_{t-1}}{l}$, and for $t = \hat{t} + 1$, the expected contribution of S_{t+1} is at least $\frac{c}{l} \cdot \left(\frac{d(u,v)}{2} - \rho_{t-1}\right)$.*

Proof: First consider t such that $1 \leq t \leq \hat{t}$. By the definition of ρ_t, for at least one of the two vertices u and v, the open ball of radius ρ_t contains fewer than 2^t vertices. Assume w.l.o.g. that this happens for vertex u, i.e., $|B^\circ(u, \rho_t)| < 2^t$. Again, by definition, $|B(v, \rho_{t-1})| \geq 2^{t-1}$. Since $\rho_{t-1} \leq \rho_t < d(u, v)/2$, the two sets $B^\circ(u, \rho_t)$ and $B(v, \rho_{t-1})$ are disjoint. Thus, by Lemma 21.17, the probability that S_{t+1} is disjoint from the first set and intersects the second is least c. Now, the first claim follows from Lemma 21.16.

Next, let $t = \hat{t} + 1$. By the definition of $\hat{t}$, for at least one of the two vertices u and v, the open ball of radius $d(u, v)/2$ contains fewer than 2^t vertices. As before, w.l.o.g. assume this happens for vertex u, i.e., $|B^\circ(u, d(u, v)/2)| < 2^t$. Clearly, $|B(v, \rho_{t-1})| \geq 2^{t-1}$. Since $\rho_{t-1} < d(u, v)/2$, the two sets $B^\circ(u, d(u, v)/2)$ and $B(v, \rho_{t-1})$ are disjoint. The rest of the reasoning is the same as before. □

Lemma 21.19 *The expected contribution of all sets $S_2, \ldots, S_{l+1}$ is at least $\frac{c}{2} \cdot \frac{d(u,v)}{l}$.*

Proof: By Lemma 21.18, the expected contribution of all sets $S_2, \ldots, S_{l+1}$ is at least the following telescoping sum:

$$\frac{c}{l} \cdot \left((\rho_1 - \rho_0) + (\rho_2 - \rho_1) + \ldots + \left(\frac{d(u, v)}{2} - \rho_{\hat{t}}\right) \right) = \frac{c}{2} \cdot \frac{d(u, v)}{l}.$$

□

Lemma 21.20 $$\mathbf{Pr}\left[\textit{contribution of all sets is} \geq \frac{c\,d(u,v)}{4l}\right] \geq \frac{c/2}{2 - c/2}.$$

Proof: Denote the probability in question by p. Clearly, the total contribution of all sets $S_2, \ldots, S_{l+1}$ to the ℓ_1 length of edge (u, v) is at most $d(u, v)/l$. This fact and Lemma 21.19 give:

$$p \cdot \frac{d(u,v)}{l} + (1-p) \cdot \frac{c\,d(u,v)}{4l} \geq \frac{c\,d(u,v)}{2l}.$$

Therefore, $p \geq \frac{c/2}{2-c/2}$. □

21.4.2 Ensuring that no edge is overshrunk

The above embedding does not overshrink edge (u, v) with constant probability. In order to ensure that no edge is overshrunk, we will first enhance this probability. The key idea is to repeat the entire process several times independently and use Chernoff bounds to bound the error probability. We will use the following statement of the Chernoff bound: Let $X_1, \ldots, X_n$ be independent Bernoulli trials with $\mathbf{Pr}[X_i = 1] = p$, $0 < p < 1$, and let $X = \sum_{i=1}^{n} X_i$; clearly, $\mathbf{E}[X] = np$. Then, for $0 < \delta \leq 1$,

$$\mathbf{Pr}[X < (1-\delta)\,np] < \exp(-\delta^2 np/2).$$

Pick sets $S_2, \ldots, S_{l+1}$ using probabilities specified above, independently $N = O(\log n)$ times each. Call the sets so obtained S_i^j, $2 \leq i \leq l+1$, $1 \leq j \leq N$. Consider the $N \cdot l = O(\log^2 n)$ dimensional embedding of metric $(V, \boldsymbol{d})$ w.r.t. these $N \cdot l$ sets. We will prove that this is an $O(\log n)$-distortion ℓ_1-embedding for metric $(V, \boldsymbol{d})$.

Lemma 21.21 *For $N = O(\log n)$, this embedding satisfies:*

$$\mathbf{Pr}\left[\|\sigma(u) - \sigma(v)\|_1 \geq \frac{p\,c\,d(u,v)}{8l}\right] \geq 1 - \frac{1}{2n^2},$$

where $p = c/(4-c)$.

Proof: For each j, $1 \leq j \leq N$, we will think of the process of picking sets $S_2^j, \ldots, S_{l+1}^j$ as a single Bernoulli trial; thus, we have N such trials. A trial succeeds if the contribution of all its sets is $\geq (c\,d(u,v))/4l$. By Lemma 21.20, the probability of success is at least p. Using the Chernoff bound with $\delta = 1/2$, the probability that at most $Np/2$ of these trials succeed is at most $\exp(-Np/8)$ which is bounded by $1/2n^2$ for $N = O(\log n)$. If at least $Np/2$ trials succeed, the ℓ_1 length of edge (u, v) will be at least $p\,c\,d(u,v)/8l = d(u,v)/O(\log n)$. The lemma follows. □

Adding the error probabilities for all $n(n-1)/2$ edges, we get:

Theorem 21.22 *The $N\,l = O(\log^2 n)$ dimensional embedding given above is an $O(\log n)$-distortion ℓ_1-embedding for metric $(V, \boldsymbol{d})$, with probability at least $1/2$.*

21.5 LP-rounding-based algorithm

The reader can verify that Claim 21.6 and Theorems 21.7, 21.12, and 21.22 lead to an $O(\log n)$ factor approximation algorithm for the sparsest cut problem. In this section, we will improve the approximation guarantee to $O(\log k)$ where k is the number of source–sink pairs specified.

For this purpose, notice that Theorem 21.7 holds even for the following less stringent approximate cut packing: no edge is allowed to be overpacked, and the edges of the demand graph are not under-packed by more than a β factor (the rest of the edges are allowed to be under-packed to any extent). In turn, such a cut packing can be obtained from an ℓ_1-embedding that does not overshrink edges of the demand graph only. Since these are only $O(k^2)$ in number, where k is the number of source–sink pairs, we can ensure that these edges are not shrunk by a factor of more than $O(\log k)$, thus enabling an improvement in the approximation guarantee.

Let $V' \subseteq V$ be the set of vertices that are sources or sinks, $|V'| \leq 2k$. For simplicity, assume $|V'|$ is a power of 2; let $|V'| = 2^l$. The sets $S_2, \ldots, S_{l+1}$ will be picked from V', and it is easy to verify from the proof of Lemma 21.21 that $N = O(\log k)$ will suffice to ensure that none of the $O(k^2)$ edges of the demand graph is shrunk by more than a factor of $O(\log k)$. The complete algorithm is:

Algorithm 21.23 (Sparsest cut)

1. Solve the dual LP (21.2) to obtain metric $(V, \boldsymbol{d})$.
2. Pick sets S_i^j, $2 \leq i \leq l+1$, $1 \leq j \leq N$, where set S_i^j is formed by picking each vertex of V' independently with probability $1/2^i$.
3. Obtain an ℓ_1-embedding of $(V, \boldsymbol{d})$ in $O(\log^2 k)$-dimensional space w.r.t. these sets.
4. Obtain an approximate cut packing for $(V, \boldsymbol{d})$ from the ℓ_1-embedding.
5. Output the sparsest cut used by the cut packing.

Theorem 21.24 *Algorithm 21.23 achieves an approximation guarantee of $O(\log k)$ for the sparsest cut problem.*

Corollary 21.25 *For a demands multicommodity flow instance with k source–sink pairs,*

$$\frac{1}{O(\log k)}\left(\min_{S \subset V} \frac{c(S)}{\mathrm{dem}(S)}\right) \leq \max_{\textit{throughput } f} f \leq \min_{S \subset V} \frac{c(S)}{\mathrm{dem}(S)}.$$

21.6 Applications

We present below a number of applications of the sparsest cut problem.

21.6.1 Edge expansion

Expander graphs have numerous applications; for instance, see Example 20.9. We will obtain an $O(\log n)$ factor algorithm for the problem of determining the edge expansion of a graph:

Problem 21.26 (Edge expansion) Given an undirected graph $G = (V, E)$, the *edge expansion of a set* $S \subset V$ with $|S| \leq n/2$, is defined to be $|\delta(S)|/|S|$. The problem is to find a minimum expansion set.

Consider the special case of demands multicommodity flow in which we have $n(n-1)/2$ distinct commodities, one for each pair of vertices. This is called the *uniform multicommodity flow problem*. For this problem, the sparsity of any cut $(S, \overline{S})$ is given by

$$\frac{c(S)}{|S| \cdot |\overline{S}|}.$$

Let $(S, \overline{S})$, with $|S| \leq |\overline{S}|$, be the cut found by Algorithm 21.23 when run on G with uniform demands. Notice that $|\overline{S}|$ is known within a factor of 2, since $n/2 \leq |\overline{S}| \leq n$. Thus, S has expansion within an $O(\log n)$ factor of the minimum expansion set in G. Clearly, the generalization of this problem to arbitrary edge costs also has an $O(\log n)$ factor approximation algorithm.

21.6.2 Conductance

The conductance of a Markov chain characterizes its mixing rate, i.e., the number of steps needed to ensure that the probability distribution over states is sufficiently close to its stationary distribution. Let P be the transition matrix of a discrete-time Markov chain on a finite state space X, and let π denote the stationary probability distribution of this chain. We will assume that the chain is aperiodic, connected, and that it satisfies the detailed balance condition, i.e.,

$$\pi(x)P(x,y) = \pi(y)P(y,x) \quad \forall x, y \in X.$$

Define undirected graph $G = (X, E)$ on vertex set X such that $(x, y) \in E$ iff $\pi(x)P(x, y) \neq 0$. The edge weights are defined to be $w(x, y) = \pi(x)P(x, y)$. The *conductance* of this chain is given by

$$\Phi = \min_{S \subset X, 0 < \pi(S) \leq 1/2} \frac{w(S, \overline{S})}{\pi(S)},$$

where $w(S, \overline{S})$ is the sum of weights of all edges in the cut $(S, \overline{S})$. For any set S, the numerator of the quotient defined above is the probability that the chain in equilibrium escapes from set S to $\overline{S}$ in one step. Thus the quotient gives the conditional probability of escape, given that the chain is initially in S and Φ measures the ability of the chain to not get trapped in any small region of the state space.

Theorem 21.24 leads to an $O(\log n)$ factor approximation algorithm for computing conductance. First, observe that it suffices to approximate the following symmetrized variant of Φ:

$$\Phi' = \min_{S \subset X, 0 < \pi(S) \leq 1} \frac{w(S, \overline{S})}{\pi(S)\pi(\overline{S})}, \tag{21.3}$$

since Φ and Φ' are within a factor of 2 of each other (notice that if $0 < \pi(S) \leq 1/2$, then $1/2 \leq \pi(\overline{S}) < 1$).

Next, let us show that computing Φ' is really a special case of the sparsest cut problem. Consider graph $G = (X, E)$ with edge weights as defined above. For each pair of vertices $x, y \in X$, define a distinct commodity with a demand of $\pi(x)\pi(y)$. It is easy to see that the sparsity of a cut $(S, \overline{S})$ for this instance is simply the quotient defined in (21.3). Hence, the sparsity of the sparsest cut is Φ'.

21.6.3 Balanced cut

The following problem finds applications in partitioning problems, such as circuit partitioning in VLSI design. Furthermore, it can be used to perform the "divide" step of the divide-and-conquer algorithms for certain problems; for instance, see the algorithm for Problem 21.29 below.

Problem 21.27 (Minimum b-balanced cut) Given an undirected graph $G = (V, E)$ with nonnegative edge costs and a rational b, $0 < b \leq 1/2$, find a minimum capacity cut $(S, \overline{S})$ such that $b \cdot n \leq |S| < (1 - b) \cdot n$.

A b-balanced cut for $b = 1/2$ is called a *bisection cut*, and the problem of finding a minimum capacity such cut is called the *minimum bisection problem*. We will use Theorem 21.24 to obtain a *pseudo-approximation algorithm* for Problem 21.27 – we will find a (1/3)-balanced cut whose capacity is within an $O(\log n)$ factor of the capacity of a minimum bisection cut (see the notes in Section 21.8 for a true approximation algorithm).

For $V' \subset V$, let $G_{V'}$ denote the subgraph of G induced by V'. The algorithm is: Initialize $U \leftarrow \emptyset$ and $V' \leftarrow V$. Until $|U| \geq n/3$, find a minimum expansion set in $G_{V'}$, say W, then set $U \leftarrow U \cup W$ and $V' \leftarrow V' - W$. Finally, let $S \leftarrow U$, and output the cut $(S, V - S)$.

Claim 21.28 *The cut output by the algorithm is a (1/3)-balanced cut whose capacity is within an $O(\log n)$ factor of the capacity of a minimum bisection cut in G.*

Proof: At the end of the penultimate iteration, $|U| < n/3$. Thus, at the beginning of the last iteration, $|V'| \geq 2n/3$. At most half of these vertices are added to U in the last iteration. Therefore, $|V - S| \geq n/3$ and $n/3 \leq |S| < n/3$. Hence, $(S, V - S)$ is a $(1/3)$-balanced cut.

Let $(T, \overline{T})$ be a minimum bisection cut in G. Since at the beginning of each iteration, $|V'| \geq 2n/3$, each of the sets $T \cap V'$ and $\overline{T} \cap V'$ has at least $n/6$ vertices. Thus, the expansion of a minimum expansion set in $G_{V'}$ in each iteration is at most $\frac{c(T)}{(n/6)}$. Since the algorithm finds a set having expansion within a factor of $O(\log n)$ of optimal in any iteration, the set U found satisfies:

$$\frac{c(U)}{|U|} \leq O(\log n) \cdot \frac{c(T)}{n/6}.$$

Since the final set S has at most $2n/3$ vertices, summing up we get

$$c(S) \leq O(\log n) \cdot \frac{c(T)(2n/3)}{n/6},$$

thereby giving $c(S) \leq O(\log n) \cdot c(T)$. □

21.6.4 Minimum cut linear arrangement

Problem 21.29 (Minimum cut linear arrangement) Given an undirected graph $G = (V, E)$ with nonnegative edge costs, for a numbering of its vertices from 1 to n, define S_i to be the set of vertices numbered at most i, for $1 \leq i \leq n-1$; this defines $n-1$ cuts. The problem is to find a numbering that minimizes the capacity of the largest of these $n-1$ cuts, i.e., it minimizes $\max\{c(S_i) | \ 1 \leq i \leq (n-1)\}$.

Using the pseudo-approximation algorithm obtained above for the $(1/3)$-balanced cut problem, we will obtain a true $O(\log^2 n)$ factor approximation algorithm for this problem. A key observation is that in any arrangement, $S_{n/2}$ is a bisection cut, and thus the capacity of a minimum bisection cut in G, say β, is a lower bound on the optimal arrangement. The reason we get a true approximation algorithm is that the $(1/3)$-balanced cut algorithm compares the cut found to β.

The algorithm is recursive: find a $(1/3)$-balanced cut in G_V, say $(S, \overline{S})$, and recursively find a numbering of S in G_S using numbers from 1 to $|S|$ and a numbering of $\overline{S}$ in $G_{\overline{S}}$ using numbers from $|S| + 1$ to n. Of course, the recursion ends when the set is a singleton, in which case the prescribed number is assigned to this vertex.

Claim 21.30 *The algorithm given above achieves an $O(\log^2 n)$ factor for the minimum cut linear arrangement problem.*

Proof: The following binary tree T (not necessarily complete) encodes the outcomes of the recursive calls made by the algorithm: Each recursive call corresponds to a node of the tree. Suppose recursive call α ends with two further calls, α_1 and α_2, where the first call assigns smaller numbers and the second assigns larger numbers. Then, α_1 will be made the left child of α in T and α_2 will be made the right child of α. If recursive call α was made with a singleton, then α will be a leaf of the tree.

To each nonleaf, we will assign the set of edges in the cut found during this call, and to each leaf we will assign its singleton vertex. Thus, the left to right ordering of leaves gives the numbering assigned by the algorithm to the vertices. Furthermore, the edge sets associated with nonleaf nodes define a partitioning of all edges of G. The cost of edges associated with any nonleaf is $O(\log n)\beta$ by Claim 21.28. Since each recursive call finds a (1/3)-balanced cut, the depth of recursion, and hence the depth of T, is $O(\log n)$.

Following is a crucial observation: Consider any edge (u, v) in G. Let α be the lowest common ancestor of leaves corresponding to u and v in T. Then, (u, v) belongs to the set of edges associated with node α.

With respect to the numbering found by the algorithm, consider a cut $(S_i, \overline{S_i})$, $1 \leq i \leq n-1$. Any edge in this cut connects vertices numbered j and k with $j \leq i$ and $k \geq i+1$. Thus, such an edge must be associated with a node that is a common ancestor of the leaves numbered i and $i+1$. Since the depth of T is $O(\log n)$, there are $O(\log n)$ such common ancestors. Since the cost of edges associated with any node in T is $O(\log n)\beta$, the cost of cut $(S_i, \overline{S_i})$ is bounded by $O(\log^2 n)\beta$. The claim follows since we have already argued that β is a lower bound on the optimal arrangement. □

21.7 Exercises

21.1 For each of the three metrics given in Example 21.15, one of the sets $S_2, \ldots, S_{l+1}$ has an expected contribution of $\Omega(d(u,v)/l)$. Give a metric for which each set has an expected contribution of $\Theta(d(u,v)/l^2)$.

21.2 Show that n points embedded in ℓ_1 space can be an isometric embedding in (a higher dimensional) ℓ_2^2 space (distance between two points in this space is defined to be the square of the ℓ_2 distance between them).
Hint: Since ℓ_1 and ℓ_2^2 are both additive across dimensions, first show that it is sufficient to consider n points in one dimension. Sort these points, and renumber, say $x_1, \ldots, x_n$. Now embed these in $(\mathbf{R}^{n-1}, \ell_2^2)$ as follows. Let $\alpha_i = x_{i+1} - x_i$. Map point x_i to $(\sqrt{\alpha_1}, \ldots, \sqrt{\alpha_{i-1}}, 0, \ldots, 0)$.

21.3 Why can't the pseudo-approximation algorithm given at the beginning of Section 21.6.3 be converted to a true approximation algorithm, i.e., so

that in the end, we compare the (1/3)-balanced cut found to the optimal (1/3)-balanced cut?
Hint: Construct graphs for which the capacity of a minimum bisection cut is arbitrarily higher than that of a (1/3)-balanced cut.

21.4 Show that the above algorithm extends to finding a b-balanced cut that is within an $O(\log n)$ factor of the best b'-balanced cut for $b \leq 1/3$ and $b < b'$. Where in the argument is the restriction $b \leq 1/3$ used?

21.5 Give an approximation factor preserving reduction from the problem of finding a minimum b-balanced cut, for $b < 1/2$, to the minimum bisection problem.

21.6 (Linial, London and Rabinovich [197]) Extend Theorem 21.22 to show that for any $p \geq 1$, there is an $O(\log n)$ distortion ℓ_p-embedding for metric $(V, \boldsymbol{d})$ in $O(\log^2 n)$-dimensional space.
Hint: Map point v to $\frac{d(v,S_i)}{Q^{1/p}}$, for $i = 1, \ldots, Q$, where Q is the dimension of the embedding. Use the fact that $|d(u, S_i) - d(v, S_i)| \leq d(u, v)$ and the monotonicity of ℓ_p-norm.

21.7 (Feige [85]) Consider the following algorithm for:

Problem 21.31 (Bandwidth minimization) Given an undirected graph $G = (V, E)$, number the vertices with distinct integers from 1 to n so that the spread of the longest edge is minimized, where the *spread* of edge (u, v) is the absolute value of the difference of the numbers assigned to u and v.

Algorithm 21.32 (Bandwidth minimization)

1. Define metric $(V, \boldsymbol{d})$, where d_{uv} is the length of the shortest path from u to v in G.
2. Obtain an $O(\log n)$-distortion ℓ_2-embedding of $(V, \boldsymbol{d})$.
3. Pick a line ℓ from a spherically symmetric distribution, and project the n points onto ℓ.
4. Number the vertices from 1 to n according to their ordering on ℓ.
5. Output the numbering.

Remark 21.33 Lemma 26.7 gives an algorithm for picking ℓ.

1. Show that the expected number of pairs of vertices that are within a distance of 1 of each other on ℓ is bounded by

$$O\left(\log n \sum_{u,v} \frac{1}{d_{uv}}\right).$$

2. Show that

$$\sum_{u,v} \frac{1}{d_{uv}} = O(n \log n \cdot \mathrm{OPT}).$$

   Hint: Use the fact that in G, the number of vertices within a distance of k of a vertex v is bounded by $2k \cdot \mathrm{OPT}$.
3. Show that with high probability, the spread of the numbering output is at most $O(\sqrt{n\mathrm{OPT}} \log n)$, i.e., this is an $O(\sqrt{n} \log n)$ factor algorithm.
   Hint: If the spread of the output numbering is s, then the number of pairs of vertices that are within a distance of 1 of each other on ℓ is at least s^2.

21.8 Notes

The seminal work of Leighton and Rao [189] gave the first approximate max-flow min-cut theorem, for the case of uniform multicommodity flow. They also gave a factor $O(\log n)$ approximation algorithm for the associated special case of sparsest cut and a pseudo-approximation algorithm for the b-balanced cut problem. The general version of demands multicommodity flow was first considered by Klein, Agarwal, Ravi, and Rao [180]. Theorem 21.22 is due to Linial, London, and Rabinovich [197], based on a result of Bourgain [33] who showed the existence of such an embedding and gave an exponential time algorithm for finding it. The application of this theorem to the sparsest cut problem, Theorem 21.24, was independently given by Aumann and Rabani [17], and Linial, London, and Rabinovich [197].

An $O(\log^2 n)$ factor algorithm for the minimum bisection problem, and hence for the minimum b-balanced cut problem (see Exercise 21.5), was given by Feige and Krauthgamer [89]. The application of sparsest cut to computing conductance is due to Sinclair [249], and the application of balanced cuts to the minimum cut linear arrangement problem is due to Bhatt and Leighton [27]. See Exercise 26.9 for a semidefinite program for finding an optimal distortion ℓ_2^2-embedding of n points.

22 Steiner Forest

We will obtain a factor 2 approximation algorithm for the Steiner forest problem by enhancing the primal–dual schema with the idea of growing duals in a synchronized manner. The Steiner forest problem generalizes the metric Steiner tree problem, for which a factor 2 algorithm was presented in Chapter 3. Recall, however, that we had postponed giving the lower bounding method behind that algorithm; we will clarify this as well.

As in the Steiner tree problem (Theorem 3.2), the main case of the Steiner forest problem is also the metric case (see Exercise 22.2). However, the primal–dual algorithm remains the same for both cases, so we don't impose this restriction.

Problem 22.1 (Steiner forest) Given an undirected graph $G = (V, E)$, a cost function on edges $c : E \rightarrow \mathbf{Q}^+$, and a collection of disjoint subsets of V, $S_1, \ldots S_k$, find a minimum cost subgraph in which each pair of vertices belonging to the same set S_i is connected.

Let us restate the problem; this will also help generalize it later. Define a *connectivity requirement function* r that maps unordered pairs of vertices to $\{0, 1\}$ as follows:

$$r(u, v) = \begin{cases} 1 \text{ if } u \text{ and } v \text{ belong to the same set } S_i \\ 0 \text{ otherwise} \end{cases}$$

Now, the problem is to find a minimum cost subgraph F that contains a u–v path for each pair (u, v) with $r(u, v) = 1$. In general, the solution will be a forest.

22.1 LP-relaxation and dual

In order to give an integer programming formulation for this problem, let us define a function on all cuts in G, $f : 2^V \rightarrow \{0, 1\}$, which specifies the minimum number of edges that must cross each cut in any feasible solution.

$$f(S) = \begin{cases} 1 \text{ if } \exists\ u \in S \text{ and } v \in \overline{S} \text{ such that } r(u, v) = 1 \\ 0 \text{ otherwise} \end{cases}$$

Let us also introduce a 0/1 variable x_e for each edge $e \in E$; x_e will be set to 1 iff e is picked in the subgraph. The integer program is:

$$\begin{array}{lll} \text{minimize} & \sum_{e \in E} c_e x_e & \qquad (22.1) \\ \text{subject to} & \sum_{e:\ e \in \delta(S)} x_e \geq f(S), & S \subseteq V \\ & x_e \in \{0,1\}, & e \in E \end{array}$$

where $\delta(S)$ denotes the set of edges crossing the cut $(S, \overline{S})$.

Following is the LP-relaxation of (22.1); once again, we have dropped the redundant conditions $x_e \leq 1$.

$$\begin{array}{lll} \text{minimize} & \sum_{e \in E} c_e x_e & \qquad (22.2) \\ \text{subject to} & \sum_{e:\ e \in \delta(S)} x_e \geq f(S), & S \subseteq V \\ & x_e \geq 0, & e \in E \end{array}$$

The dual program is:

$$\begin{array}{lll} \text{maximize} & \sum_{S \subseteq V} f(S) \cdot y_S & \qquad (22.3) \\ \text{subject to} & \sum_{S:\ e \in \delta(S)} y_S \leq c_e, & e \in E \\ & y_S \geq 0, & S \subseteq V \end{array}$$

Notice that the primal and dual programs form a covering and packing pair of LPs (see Section 13.1 for definitions).

22.2 Primal–dual schema with synchronization

We will introduce a new idea in the primal–dual schema for approximation algorithms, setting it apart from the way this schema is used for designing exact algorithms. The later algorithms work *on demand* – in each iteration, we pick one unsatisfied complementary slackness condition, and satisfy it by modifying the primal and dual solutions suitably. The new idea is that of raising duals in a *synchronized manner*. The algorithm is not trying to rectify a *specific* condition. Instead, it tries many possibilities simultaneously, one of which leads to primal improvement.

Some figurative terminology will help describe the algorithm more easily. Let us say that edge e *feels* dual y_S if $y_S > 0$ and $e \in \delta(S)$. Say that set S has been *raised* in a dual solution if $y_S > 0$. Clearly, raising S or $\overline{S}$ has the same effect. Sometimes we will also say that we have raised the cut $(S, \overline{S})$. Further, there is no advantage in raising set S with $f(S) = 0$, since this does not contribute to the dual objective function. Thus, we may assume that such cuts are never raised. Say that edge e is *tight* if the total amount of dual it feels equals its cost. The dual program is trying to maximize the sum of the dual variables y_S subject to the condition that no edge feels more dual than its cost, i.e., no edge is *overtight.*

Next, let us state the primal and relaxed dual complementary slackness conditions. The algorithm will pick edges integrally only. Define the *degree of set S* to be the number of picked edges crossing the cut $(S, \overline{S})$.

Primal conditions: For each $e \in E$, $x_e \neq 0 \Rightarrow \sum_{i:\ e \in \delta(S)} y_S = c_e$. Equivalently, *every picked edge must be tight.*

Relaxed dual conditions: The following relaxation of the dual conditions would have led to a factor 2 algorithm: for each $S \subseteq V, y_S \neq 0 \Rightarrow \sum_{e:\ e \in \delta(S)} x_e \leq 2 \cdot f(S)$, i.e., every raised cut has degree at most 2. However, we do not know how to ensure this condition. Interestingly enough, we can still obtain a factor 2 algorithm – by relaxing this condition further! Raised sets will be allowed to have high degree; however, we will ensure that *on average, raised duals have degree at most 2.* The exact definition of "on average" will be given later.

The algorithm starts with null primal and dual solutions. In the spirit of the primal–dual schema, the current primal solution indicates which cuts need to be raised, and in turn, the current dual solution indicates which edge needs to be picked. Thus, the algorithm iteratively improves the feasibility of the primal, and the optimality of the dual, until a feasible primal is obtained.

Let us describe what happens in an iteration. At any point, the picked edges form a forest. Say that set S is *unsatisfied* if $f(S) = 1$, but there is no picked edge crossing the cut $(S, \overline{S})$. Set S is said to be *active* if it is a minimal (w.r.t. inclusion) unsatisfied set in the current iteration. Clearly, if the currently picked primal solution is infeasible, there must an unsatisfied set and therefore an active set w.r.t. it.

Lemma 22.2 *Set S is active iff it is a connected component in the currently picked forest and $f(S) = 1$.*

Proof: Let S be an active set. Now, S cannot contain part of a connected component because otherwise there will already be a picked edge in the cut $(S, \overline{S})$. Thus, S is a union of connected components. Since $f(S) = 1$, there is a vertex $u \in S$ and $v \in \overline{S}$ such that $r(u, v) = 1$. Let S' be the connected component containing u. Clearly, S' is also unsatisfied, and by the minimality of S, $S = S'$. □

By the characterization of active sets given in Lemma 22.2, it is easy to find all active sets in the current iteration. The dual variables of these sets are raised in a synchronized manner, until some edge goes tight. Any one of the newly tight edges is picked, and the current iteration terminates.

When a primal feasible solution is found, say F, the edge augmentation step terminates. However, F may contain redundant edges, which need to be pruned for achieving the desired approximation factor; this is illustrated in Example 22.4. Formally, edge $e \in F$ is said to be *redundant* if $F - \{e\}$ is also a feasible solution. All redundant edges can be dropped simultaneously from F. Equivalently, only nonredundant edges are retained.

This algorithm is presented below. We leave its efficient implementation as an exercise.

Algorithm 22.3 (Steiner forest)

1. **(Initialization)** $F \leftarrow \emptyset$; for each $S \subseteq V$, $y_S \leftarrow 0$.
2. **(Edge augmentation)** while there exists an unsatisfied set do:
   simultaneously raise y_S for each active set S, until some edge e goes tight;
   $F \leftarrow F \cup \{e\}$.
3. **(Pruning)** return $F' = \{e \in F \mid F - \{e\}$ is primal infeasible$\}$

Example 22.4 Consider a star in which all edges have cost 1, except one edge whose cost is 3.

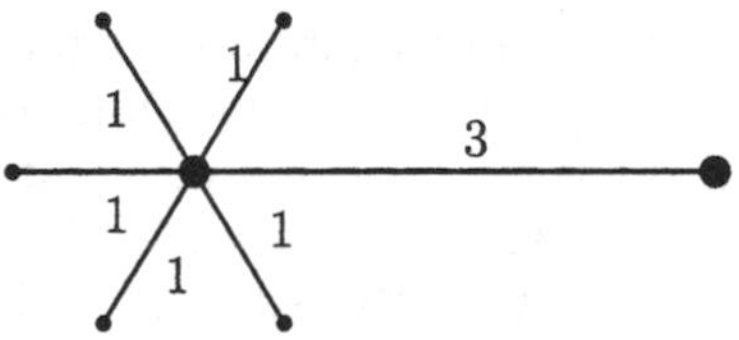

The only requirement is to connect the end vertices of the edge of cost 3. The algorithm will add to F all edges of cost 1 before adding the edge of cost 3. Clearly, at this point, F is not within twice the optimal. However, this will be corrected in the pruning step when all edges of cost 1 will be removed. □

Let us run the algorithm on a nontrivial example to illustrate its finer points.

Example 22.5 Consider the following graph. Costs of edges are marked, and the only nonzero connectivity requirements are $r(u,v) = 1$ and $r(s,t) = 1$. The thick edges indicate an optimal solution of cost 45.

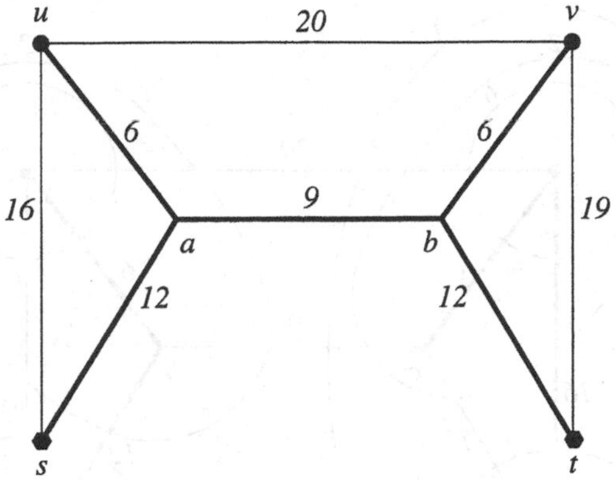

In the first iteration, the following four singleton sets are active: $\{s\}$, $\{t\}$, $\{u\}$, and $\{v\}$. When their dual variables are raised to 6 each, edges (u,a) and (v,b) go tight. One of them, say (u,a) is picked, and the iteration ends. In the second iteration, $\{u,a\}$ replaces $\{u\}$ as an active set. However, in this iteration there is no need to raise duals, since there is already a tight edge, (v,b). This edge is picked, and the iteration terminates. The primal and dual solutions at this point are shown below, with picked edges marked thick:

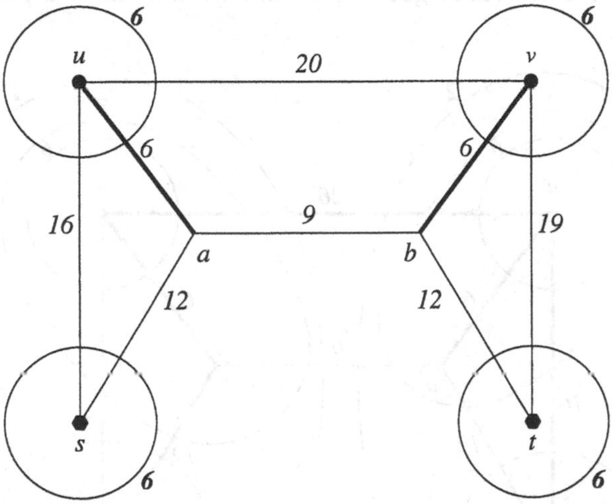

In the third iteration, $\{v,b\}$ replaces $\{v\}$ as an active set. When the active sets are raised by 2 each, edge (u,s) goes tight and is picked. In the fourth iteration, the active sets are $\{u,s,a\}$, $\{v,b\}$ and $\{t\}$. When they are raised by 1 each, edge (b,t) goes tight and is picked. The situation now is:

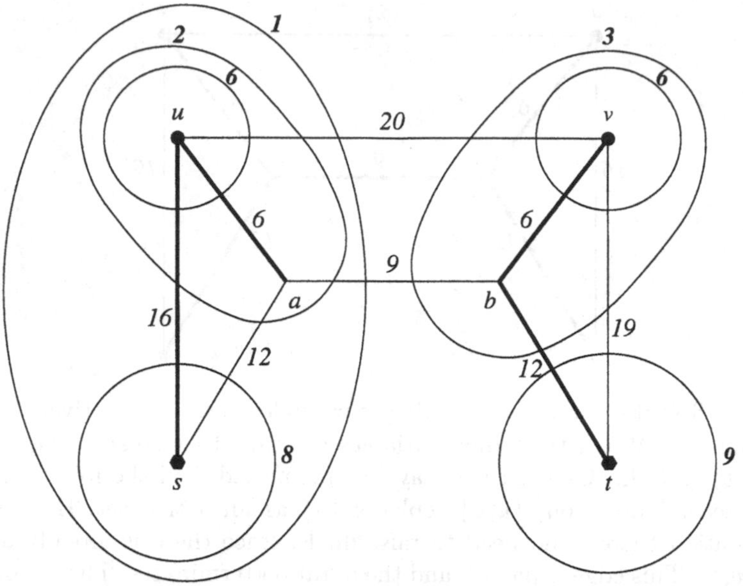

In the fifth iteration, the active sets are $\{a, s, u\}$ and $\{b, v, t\}$. When they are raised by 1 each, (u, v) goes tight, and we now have a primal feasible solution:

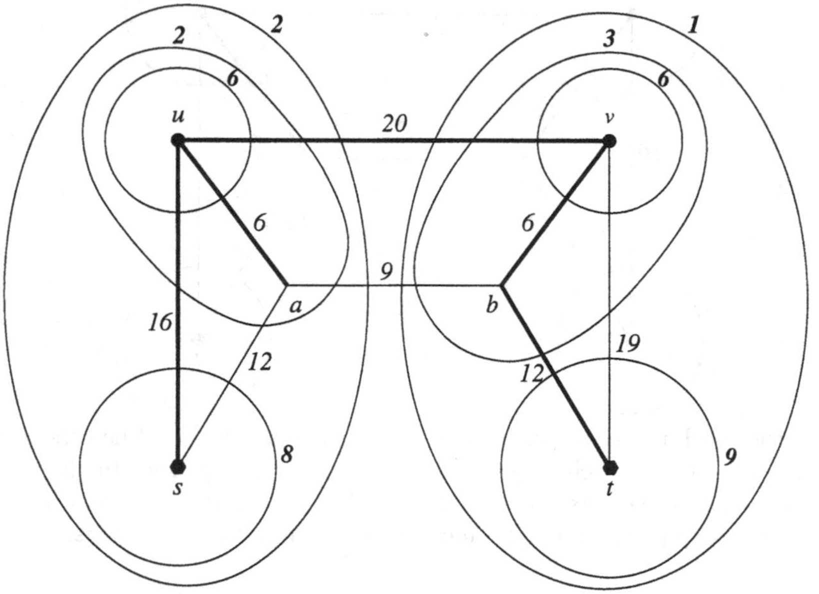

In the pruning step, edge (u, a) is deleted, and we obtain the following solution of cost 54:

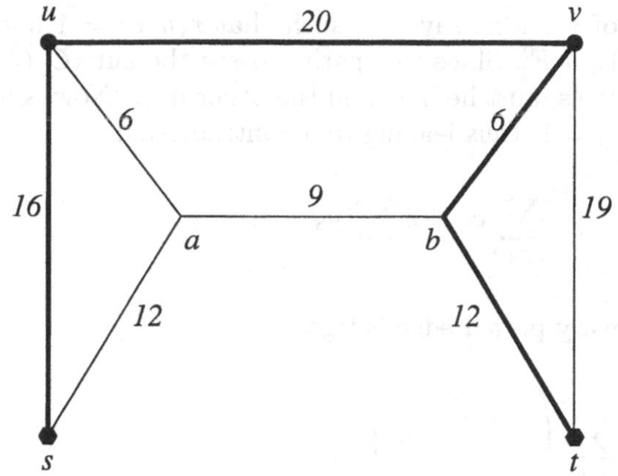

□

22.3 Analysis

In Lemma 22.6 we will show that *simultaneously* deleting all redundant edges still leaves us with a primal feasible solution, i.e., it is never the case that two edges e and f are both redundant individually, but on deletion of e, f becomes nonredundant.

Lemma 22.6 *At the end of the algorithm, F' and $\boldsymbol{y}$ are primal and dual feasible solutions, respectively.*

Proof: At the end of Step 2, F satisfies all connectivity requirements. In each iteration, dual variables of connected components only are raised. Therefore, no edge running within the same component can go tight, and so F is acyclic, i.e., it is a forest. Therefore, if $r(u, v) = 1$, there is a *unique* u–v path in F. Thus, each edge on this path in nonredundant and is not deleted in Step 3. Hence, F' is primal feasible.

When an edge goes tight, the current iteration ends and active sets are redefined. Therefore, no edge is overtightened. Hence, $\boldsymbol{y}$ is dual feasible. □

Let $\deg_{F'}(S)$ denote the number of edges of F' crossing the cut $(S, \overline{S})$. The characterization of degrees of satisfied components established in the next lemma will be crucial in proving the approximation guarantee of the algorithm.

Lemma 22.7 *Consider any iteration of the algorithm, and let C be a component w.r.t. the currently picked edges. If $f(C) = 0$ then $\deg_{F'}(C) \neq 1$.*

Proof: Suppose $\deg_{F'}(C) = 1$, and let e be the unique edge of F' crossing the cut $(C, \overline{C})$. Since e is nonredundant (every edge in F' is nonredundant), there is a pair of vertices, say u, v, such that $r(u,v) = 1$ and e lies on the unique u–v path in F'. Since this path crosses the cut $(C, \overline{C})$ exactly once, one of these vertices must lie in C and the other in $\overline{C}$. Now, since $r(u,v) = 1$, we get that $f(C) = 1$, thus leading to a contradiction. □

Lemma 22.8 $$\sum_{e \in F'} c_e \leq 2 \sum_{S \subseteq V} y_S$$

Proof: Since every picked edge is tight,

$$\sum_{e \in F'} c_e = \sum_{e \in F'} \left(\sum_{S:\ e \in \delta(S)} y_S \right).$$

Changing the order of summation we get:

$$\sum_{e \in F'} c_e = \sum_{S \subseteq V} \left(\sum_{e \in \delta(S) \cap F'} y_S \right) = \sum_{S \subseteq V} \deg_{F'}(S) \cdot y_S.$$

Thus, we need to show that

$$\sum_{S \subseteq V} \deg_{F'}(S) \cdot y_S \leq 2 \sum_{S \subseteq V} y_S. \tag{22.4}$$

We will prove the following stronger claim. In each iteration, the increase in the left-hand side of inequality (22.4) is bounded by the increase in the right-hand side. Consider an iteration, and let Δ be the extent to which active sets were raised in this iteration. Then, we need to show:

$$\Delta \times \left(\sum_{S \text{ active}} \deg_{F'}(S) \right) \leq 2\Delta \times (\# \text{ of active sets}).$$

Notice that the degree w.r.t. F' of any active set S is due to edges that will be picked *during or after* the current iteration. Let us rewrite this inequality as follows:

$$\frac{\sum_{S \text{ active}} \deg_{F'}(S)}{\# \text{ of active sets}} \leq 2. \tag{22.5}$$

Thus, we need to show that in this iteration, the average degree of active sets w.r.t. F' is at most 2. The mechanics of the argument lies in the fact that in a tree, or in general in a forest, the average degree of vertices is at most 2.

Let H be a graph on vertex set V and edge set F'. Consider the set of connected components w.r.t. F at the beginning of the current iteration. In H, shrink the set of vertices of each of these components to a single node to obtain graph H' (we will call the vertices of H' nodes for clarity). Notice that in going from H to H', all edges picked in F before the current iteration have been shrunk. Clearly, the degree of a node in H' is equal to the degree of the corresponding set in H. Let us say that a node of H' corresponding to an active component is an *active node*; any other node will be called *inactive.* Each active node of H' has nonzero degree (since there must be an edge incident to it to satisfy its requirement), and H' is a forest. Now, remove all isolated nodes from H'. The remaining graph is a forest with average degree at most 2. By Lemma 22.7 the degree of each inactive node in this graph is at least 2, i.e., the forest has no inactive leaves. Hence, the average degree of active nodes is at most 2. □

Observe that the proof given above is essentially a charging argument: for each active node of degree greater than 2, there must be correspondingly many active nodes of degree 1, i.e., leaves, in the forest. The exact manner in which the dual conditions have been relaxed must also be clear now: in each iteration, the duals being raised have average degree at most 2. Lemmas 22.6 and 22.8 give:

Theorem 22.9 *Algorithm 22.3 achieves an approximation guarantee of factor 2 for the Steiner forest problem.*

The tight example given for the metric Steiner tree problem, Example 3.4, is also a tight example for this algorithm. Algorithm 22.3 places an upper bound of 2 on the integrality gap of LP-relaxation (22.2) for the Steiner forest problem. Example 22.10 places a lower bound of (essentially) 2 on this LP, even if restricted to the minimum spanning tree problem.

Let us run Algorithm 22.3 on an instance of the metric Steiner tree problem. If the edge costs satisfy the strict triangle inequality, i.e., for any three vertices u, v, w, $c(u,v) < c(u,w) + c(v,w)$, then it is easy to see that the algorithm will find a minimum spanning tree on the required vertices, i.e., it is essentially the algorithm for the metric Steiner tree problem presented in Chapter 3. Even if the triangle inequality is not strictly satisfied, the cost of the solution found is the same as the cost of an MST. Furthermore, if among multiple tight edges, the algorithm always prefers picking edges running between required vertices, it will find an MST. This clarifies the lower bound on which that algorithm was based.

The MST problem is a further special case: every pair of vertices need to be connected. Observe that when run on such an instance, Algorithm 22.3 essentially executes Kruskal's algorithm, i.e., in each iteration, it picks the cheapest edge running between two connected components. Hence it finds an optimal MST. However, as shown in Example 22.10, the dual found may be as small as half the primal.

Example 22.10 Consider a cycle on n vertices, with all edges of cost 1. The cost of an optimal MST is $n-1$. The dual found is $n/2$. Algorithm 22.3 finds a dual of value $n/2$: $1/2$ around each vertex. Indeed, this is an optimal dual solution, since there is a fractional primal solution of the same value: pick each edge to the extent of half. This places a lower bound of (essentially) 2 on the integrality gap of LP (22.2), even if restricted to the minimum spanning tree problem. □

22.4 Exercises

22.1 Show, using the max-flow min-cut theorem, that a subgraph of G has all the required paths iff it does not violate any of the cut requirements in IP (22.1). Use this fact to show that IP (22.1) is an integer programming formulation for the Steiner forest problem.

22.2 Show that there is an approximation-factor-preserving reduction from the Steiner forest problem to the metric Steiner forest problem.

Show that there is no loss of generality in requiring that the edge costs satisfy the triangle inequality for the Steiner forest problem.
Hint: The reasoning is the same as that for the Steiner tree problem.

22.3 How does the feasibility and approximation guarantee of the solution found change if

1. the pruning step of Algorithm 22.3 is replaced with the reverse delete step of Algorithm 18.4.
2. the reverse delete step of Algorithm 18.4 is replaced by the pruning step of Algorithm 22.3.

22.4 Give an example for which some cut raised by Algorithm 22.3 has degree at least 3 w.r.t. the primal solution found.

22.5 Run Algorithm 22.3 on an instance of the minimum spanning tree problem. Pick an arbitrary vertex as the root, and throw away all raised duals containing this vertex. Show that the cost of the tree found is twice the sum of the remaining duals.
Hint: Show that in an iteration which starts with k connected components, and lasts for time Δ, the total increase to the left-hand side of inequality (22.4) is precisely $2(k-1)\Delta$.

22.6 Let us think of running Step 2 of Algorithm 22.3 continuously in time. Thus, in unit time, a dual grows a unit amount. Consider an instance of the

Steiner forest problem, $(G = (V, E), c, S_1, \ldots, S_k)$, and its modification in which one of the vertices from $V - (S_1 \cup \ldots \cup S_k)$ is added to one of the sets. Run Algorithm 22.3 on both these instances. Call these runs R_1 and R_2, respectively.

1. Show that if $k = 1$, i.e., the starting instance was a Steiner tree instance, then the following holds. If at time t two vertices $u, v \in S_1$ are connected by a tight path in run R_1, then they are connected by a tight path at time t in run R_2 as well.
2. Give a counterexample to the previous claim in case $k > 1$.

22.7 (Goemans and Williamson [111]) Algorithm 22.3 actually works for a general class of problems that includes the Steiner forest problem as a special case. A function $f : 2^V \to \{0, 1\}$ is said to be *proper* if it satisfies the following properties:

1. $f(V) = 0$;
2. $f(S) = f(\overline{S})$;
3. If A and B are two disjoint subsets of V and $f(A \cup B) = 1$, then $f(A) = 1$ or $f(B) = 1$.

Notice that function f defined for the Steiner forest problem is a proper function. Consider the integer program (22.1) with f restricted to be a proper function. Show that Algorithm 22.3 is in fact a factor 2 approximation algorithm for this class of integer programs.

22.8 (Goemans and Williamson [111]) Consider the following problem.

Problem 22.11 (Point-to-point connection) Given a graph $G = (V, E)$, a cost function on edges $c : E \to \mathbf{Q}^+$ (not necessarily satisfying the triangle inequality) and two disjoint sets of vertices, S and T, of equal cardinality, find a minimum cost subgraph that has a path connecting each vertex in S to a unique vertex in T.

1. Give a factor 2 approximation algorithm for this problem.
   Hint: Show that this can be formulated as an integer program using (22.1), with f being a proper function.
2. Relax the problem to requiring that each vertex in S be connected to some vertex in T (not necessarily unique). Give a factor 2 approximation algorithm for this problem as well.
   Hint: Reduce to the Steiner tree problem.

22.9 (Goemans and Williamson [111]) Consider the following variant of the metric Steiner tree problem.

Problem 22.12 (Prize-collecting Steiner tree) We are given a complete undirected graph $G = (V, E)$ and a special vertex $r \in V$. Function cost : $E \to$

$\mathbf{Q}^+$ satisfies the triangle inequality, and $\pi : V \to \mathbf{Q}^+$ is the penalty function for vertices. The problem is to find a tree containing r which minimizes the sum of the costs of the edges in the tree and the penalties of vertices not in the tree.

1. Consider the following integer program for this problem. It has a variable, x_e, for each edge e and a variable, Z_T, for each set T of vertices not containing r. Z_T is set to 1 for the set T of vertices that are not included in the optimal tree. Obtain the LP-relaxation and dual for this LP. The dual will have a variable for each set S of vertices not containing r. Let's call this variable y_S.

$$
\begin{array}{lll}
\text{minimize} & \displaystyle\sum_{e\in E} c_e x_e + \sum_{T\subseteq V; r\notin T} \left(Z_T \sum_{v\in T} \pi_v \right) & (22.6) \\
\text{subject to} & \displaystyle\sum_{e\in\delta(S)} x_e + \sum_{T\supseteq S} Z_T \geq 1, & S \subseteq V; r \notin S \\
& x_e \in \{0,1\}, & e \in E \\
& Z_T \in \{0,1\}, & T \subseteq V; r \notin T
\end{array}
$$

2. The following primal–dual algorithm for this problem is along the lines of Algorithm 22.3. Initialize as follows. Each vertex $v \neq r$ is a singleton *active set*, with a *charge* of π_v. Ordered list F is set to $\emptyset$. The dual variables of all active sets are grown in a synchronized manner. As a dual grows, its charge decreases by the same amount. If set S runs out of charge, it is declared *dead* and all of its unmarked vertices are marked with "S". When an edge e goes tight, it is added to F. The rest of the action depends on the following cases.
   - If e connects active set S to r: Set S is deactivated and is declared *connected to* r. All unmarked vertices of S are marked "r".
   - If e connects active set S to a set that is connected to r: Same action as in previous case.
   - If e connects sets S and S' which are either both active or one is active and one is dead: The active sets among S and S' are deactivated. $S \cup S'$ is declared active and is given the sum of the leftover charges of S and S'.

   When there are no more active sets, the algorithm performs a *dynamic reverse delete* on F. This is an enhanced reverse delete procedure in which requirements change dynamically. All vertices marked "r" are labeled *Required.* We will say that F is *feasible* if there is a path from each Required vertex to r using edges of F. Let $e = (u, v) \in F$. Then the *maximal dead set w.r.t. e, containing* v is the maximal set S such that $v \in S, u \notin S$ and S was declared dead by the algorithm. If there is no set satisfying these conditions, then the maximal dead set w.r.t. e, containing v is defined to be $\emptyset$.

Edges $e \in F$ are considered in the reverse order in which they were inserted in F. For each edge e, if $F - e$ is feasible, then e is removed from F. Otherwise, suppose $e = (u, v)$, and let S be the maximal dead set w.r.t. e, containing v. If $S \neq \emptyset$, then declare all vertices marked "S" Required. Repeat for the maximal dead set w.r.t. e containing u.

3. Show that at the beginning of the reverse delete step, F is feasible. Also, show that F is never infeasible, even if the set of Required vertices grows.
4. Show that at the end, F is a tree containing r and satisfying
   a) all vertices marked "r" are included in F, and
   b) if a vertex marked "S" is included in F, then all vertices marked "T", where $T \supseteq S$, are also included in F.
5. The primal solution is constructed as follows. For each edge e in F, set x_e to 1, and set the remaining x_e's to 0. For each maximal dead set T, none of whose vertices are in F, set Z_T to 1. For all the remaining sets, set their Z variable to 0.

   Prove that this is a factor 2 approximation algorithm for the prize-collecting Steiner tree problem, by showing that the primal and dual solutions produced satisfy

$$\sum_{e \in E} c_e x_e + 2 \cdot \sum_{T \subseteq V; r \notin T} \left(Z_T \sum_{v \in T} \pi_v \right) \leq 2 \cdot \sum_{S \subseteq V; r \notin T} y_S .$$

Hint: If $Z_T = 1$, the sum of penalties of vertices in T equals the total dual contained in T. Twice the rest of the duals pays for the cost of the tree. Show the latter by proving, similar to Lemma 22.7, that at any point in the algorithm, there is at most one inactive set of degree 1 (the one containing r).

22.10 Consider the following generalization of the Steiner forest problem to higher connectivity requirements: the specified connectivity requirement function r maps pairs of vertices to $\{0, \ldots, k\}$, where k is part of the input. Assume that multiple copies of any edge can be used; each copy of edge e will cost $c(e)$. Using Algorithm 22.3 as a subroutine, give a factor $2 \cdot (\lfloor \log_2 k \rfloor + 1)$ algorithm for the problem of finding a minimum cost graph satisfying all connectivity requirements.

22.11 We give below the *bidirected cut relaxation* for the Steiner tree problem. This is believed to have a smaller integrality gap than the undirected relaxation (22.2), though there is no proof of this fact yet. From graph G, obtain directed graph H by replacing each edge (u, v) by the two edges $(u \rightarrow v)$ and $(v \rightarrow u)$, each of the same cost as (u, v). Designate an arbitrary required vertex, say r, as the root. Say that $S \subset V$ is *valid* if it contains a required vertex and $r \in \overline{S}$. Let x_e be an indicator variable for each edge $e \in H$. The integer program is:

$$\begin{aligned} \text{minimize} \quad & \sum_{e \in E} c_e x_e && (22.7) \\ \text{subject to} \quad & \sum_{e:\ e \in \delta(S)} x_e \geq 1, && \text{valid set } S \\ & x_e \in \{0,1\}, && e \in H \end{aligned}$$

1. Show that the optimal solution to this integer program is an optimal Steiner tree.
2. Obtain the LP-relaxation and dual for IP (22.7).
3. Show that the cost of the optimal solution to (22.7) and its relaxation is independent of the root chosen.
4. Show that the integrality gap of the relaxation is bounded by 2.
5. (Rajagopalan and Vazirani [234]) Show that the integrality gap of this relaxation for the following graph is 10/9. In this graph, the bold vertices are required and the remaining vertices are Steiner.

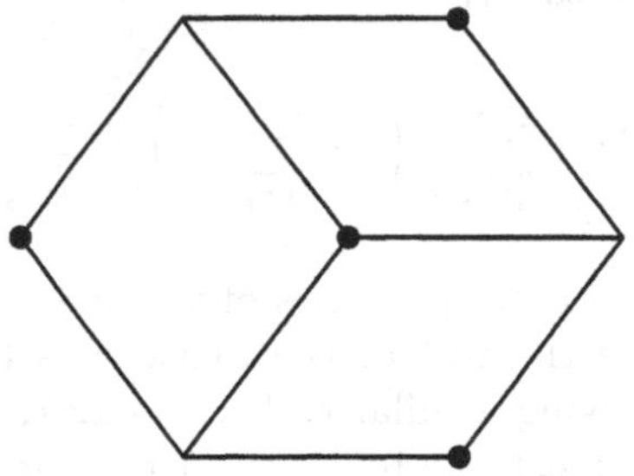

6. (M. Goemans) The following family of graphs puts a lower bound of essentially 8/7 on the integrality gap of relaxation (22.7). This is currently the worst example known.

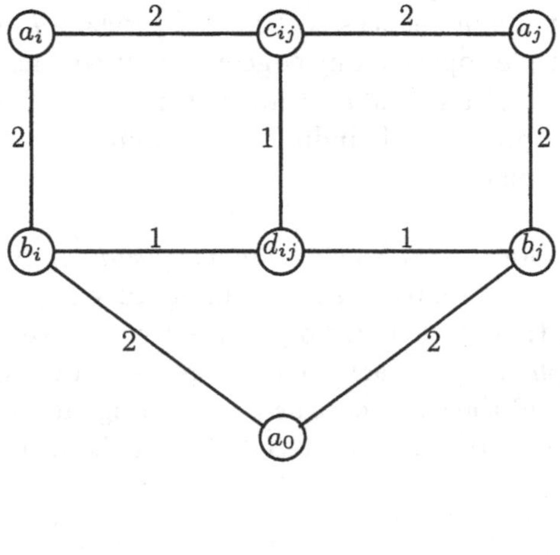

Graph G_n has $n+1$ required vertices $a_0, a_1, \ldots, a_n$, and n^2 Steiner vertices $b_1, \ldots, b_n$ and c_{ij} and d_{ij} for $1 \leq i < j \leq n$. The figure above gives edges and costs. Verify that the optimal Steiner tree has cost $4n$ and the optimal solution to relaxation (22.7) has cost $7n + 1/2$.

7. Construct other graphs for which this relaxation has a gap (it is not easy!).
8. (Edmonds [76]) Consider the special case that there are no Steiner vertices, i.e., we want to find a minimum spanning tree in G. Give a primal–dual algorithm that uses this relaxation to find a tree and a dual of the same cost, thereby showing that this relaxation is exact, i.e., always has an integral optimal solution, for the minimum spanning tree problem. (In contrast, the undirected relaxation has an integrality gap of 2 even for the minimum spanning tree problem.)

22.12 (Prömel and Steger [231]) This exercise develops an algorithm for the Steiner tree problem using the weighted matroid parity problem and the following structural fact. Let us say that a Steiner tree is *3-restricted* if every Steiner vertex used in this tree has exactly three neighbors, all of which are required vertices. The cost of an optimal 3-restricted Steiner tree is within 5/3 of the cost of an optimal Steiner tree (Zelikovsky [271]). Show that an optimal 3-restricted Steiner tree can be found in polynomial time, given an oracle for the weighted matroid parity problem. The latter problem is neither known to be in **P** nor is it known to be **NP**-hard. However, a randomized polynomial time algorithm is known for the case of unary weights. Use this fact, and scaling, to obtain a $5/3 + \varepsilon$ factor algorithm for the Steiner tree problem for any $\varepsilon > 0$.

The *weighted matroid parity problem* is the following. Let $(S, \mathcal{I})$ be a matroid, where S is the ground set and $\mathcal{I}$ is the collection of independent sets. Nonnegative weights are provided for elements of S. Further, a partition of S into pairs $(x_1, x_2), \ldots, (x_{2n-1}, x_{2n})$ is also provided. The problem is to pick a maximum weight collection of pairs so that the picked elements form an independent set.

22.5 Notes

This chapter is based on the work of Goemans and Williamson [111]. The first factor 2 approximation algorithm for the Steiner forest problem was given by Agrawal, Klein, and Ravi [1]. See also the survey by Goemans and Williamson [113].

23 Steiner Network

The following generalization of the Steiner forest problem to higher connectivity requirements has applications in network design and is also known as the *survivable network design problem*. In this chapter, we will give a factor 2 approximation algorithm for this problem by enhancing the LP-rounding technique to *iterated rounding*. A special case of this problem was considered in Exercise 22.10.

Problem 23.1 (Steiner network) We are given an undirected graph $G = (V, E)$, a cost function on edges $c : E \to \mathbf{Q}^+$ (not necessarily satisfying the triangle inequality), a connectivity requirement function r mapping unordered pairs of vertices to $\mathbf{Z}^+$, and a function $u : E \to \mathbf{Z}^+ \cup \{\infty\}$ stating an upper bound on the number of copies of edge e we are allowed to use; if $u_e = \infty$, there is no upper bound for edge e. The problem is to find a minimum cost multigraph on vertex set V that has $r(u, v)$ edge disjoint paths for each pair of vertices $u, v \in V$. Each copy of edge e used for constructing this graph will cost $c(e)$.

23.1 LP-relaxation and half-integrality

In order to give an integer programming formulation for this problem, we will first define a *cut requirement function*, $f : 2^V \to \mathbf{Z}^+$, as we did for the metric Steiner forest problem. For every $S \subseteq V$, $f(S)$ is defined to be the largest connectivity requirement separated by the cut $(S, \overline{S})$, i.e., $f(S) = \max\{r(u, v) | u \in S \text{ and } v \in \overline{S}\}$.

$$
\begin{aligned}
\text{minimize} \quad & \sum_{e \in E} c_e x_e && && (23.1) \\
\text{subject to} \quad & \sum_{e:\ e \in \delta(S)} x_e \geq f(S), && S \subseteq V \\
& x_e \in \mathbf{Z}^+, && e \in E \text{ and } u_e = \infty \\
& x_e \in \{0, 1, \ldots, u_e\}, && e \in E \text{ and } u_e \neq \infty
\end{aligned}
$$

The LP-relaxation is:

$$
\begin{array}{llll}
\text{minimize} & \displaystyle\sum_{e\in E} c_e x_e & & (23.2) \\
\text{subject to} & \displaystyle\sum_{e:\ e\in\delta(S)} x_e \geq f(S), & S \subseteq V & \\
& x_e \geq 0, & e \in E \text{ and } u_e = \infty & \\
& u_e \geq x_e \geq 0, & e \in E \text{ and } u_e \neq \infty &
\end{array}
$$

Since LP (23.2) has exponentially many constraints, we will need the ellipsoid algorithm for finding an optimal solution. Exercise 23.1 develops a polynomial-sized LP.

As shown in Chapters 14 and 19, certain **NP**-hard problems, such as vertex cover and node multiway cut, admit LP-relaxations having the remarkable property that they always have a half-integral optimal solution. Rounding up all halves to 1 in such a solution leads to a factor 2 approximation algorithm. Does relaxation (23.2) have this property? The following lemma shows that the answer is "no".

Lemma 23.2 *Consider the Petersen graph (see Section 1.2) with a connectivity requirement of 1 between each pair of vertices and with each edge of unit cost. Relaxation (23.2) does not have a half-integral optimal solution for this instance.*

Proof: Consider the fractional solution $x_e = 1/3$ for each edge e. Since the Petersen graph is 3-edge connected (in fact, it is 3-vertex connected as well), this is a feasible solution. The cost of this solution is 5. In any feasible solution, the sum of edge variables incident at any vertex must be at least 1, to allow connectivity to other vertices. Therefore, any feasible solution must have cost at least 5 (since the Petersen graph has 10 vertices). Hence, the solution given above is in fact optimal.

Any solution with $x_e = 1$ for some edge e must have cost exceeding 5, since additional edges are required to connect the endpoints of e to the rest of the graph. Therefore, any half-integral solution of cost 5 would have to pick, to the extent of one half each, the edges of a Hamiltonian cycle. Since the Petersen graph has no Hamiltonian cycles, there is no half-integral optimal solution. □

Let us say that an *extreme point solution*, also called a vertex solution or a basic feasible solution, for an LP is a feasible solution that cannot be written as the convex combination of two feasible solutions. The solution $x_e = 1/3$, for each edge e, is not an extreme point solution. An extreme optimal solution is shown in the figure below; thick edges are picked to the extent of 1/2, thin edges to the extent of 1/4, and the missing edge is not picked.

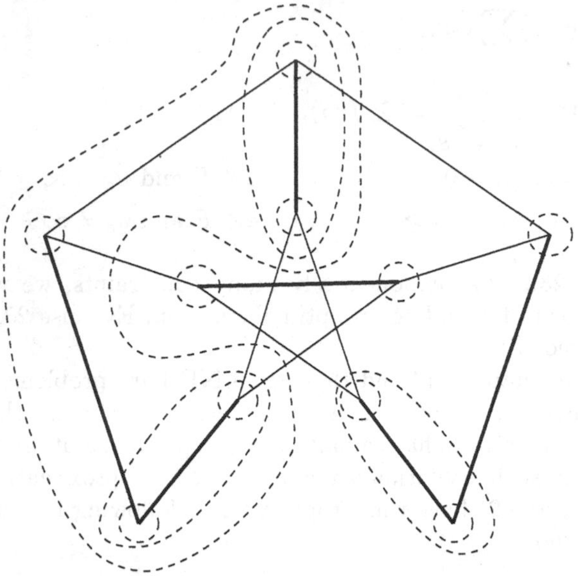

The isomorphism group of the Petersen graph is edge-transitive, and there are 15 related extreme point solutions; the solution $x_e = 1/3$ for each edge e is the average of these.

Notice that although the extreme point solution is not half-integral, it picks some edges to the extent of half. We will show below that in fact this is a property of any extreme point solution to LP (23.2). We will obtain a factor 2 algorithm by rounding up these edges and iterating. Let H be the set of edges picked by the algorithm at some point. Then, the residual requirement of cut $(S, \overline{S})$ is $f'(S) = f(S) - |\delta_H(S)|$, where $\delta_H(S)$ represents the set of edges of H crossing the cut $(S, \overline{S})$. In general, the *residual cut requirement function*, f', may not correspond to the cut requirement function for any set of connectivity requirements. We will need the following definitions to characterize it:

Function $f : 2^V \rightarrow \mathbf{Z}^+$ is said to be *submodular* if $f(V) = 0$, and for every two sets $A, B \subseteq V$, the following two conditions hold:

- $f(A) + f(B) \geq f(A \cap B) + f(A \cup B)$
- $f(A) + f(B) \geq f(A - B) + f(B - A)$.

Remark 23.3 Sometimes submodularity is defined only with the first condition. We will need to work with the stronger definition given above.

Two subsets of V, A and B, are said to *cross* if each of the sets, $A - B$, $B - A$, and $A \cap B$, is nonempty. If A and B don't cross then either they are disjoint or one of these sets is contained in the other.

Lemma 23.4 *For any graph G on vertex set V, the function $|\delta_G(.)|$ is submodular.*

Proof: If sets A and B do not cross, then the two conditions given in the definition of submodular functions hold trivially. Otherwise, edges having one endpoint in $A \cap B$ and the other in $\overline{A \cup B}$ (edge e_1 in the figure below) contribute to $\delta(A)$ and $\delta(B)$ but not to $\delta(A-B)$ or $\delta(B-A)$. Similarly, edge e_2 below does not contribute to $\delta(A \cap B)$ or to $\delta(A \cup B)$. The remaining edges contribute equally to both sides of both conditions. □

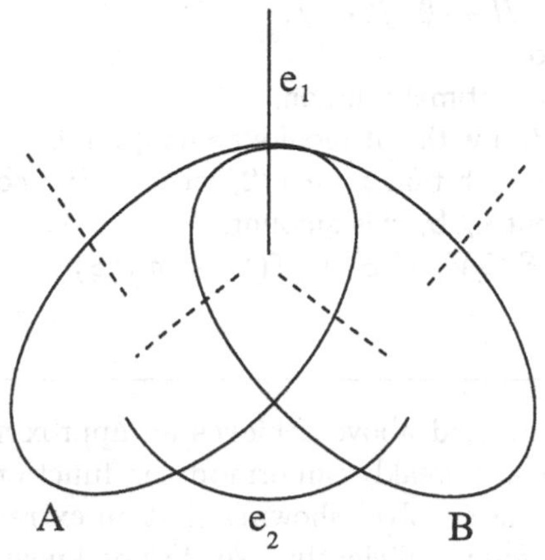

Function $f : 2^V \to \mathbf{Z}$ is said to be *weakly supermodular* if $f(V) = 0$, and for every two sets $A, B \subseteq V$, at least one of the following conditions holds:

- $f(A) + f(B) \leq f(A - B) + f(B - A)$
- $f(A) + f(B) \leq f(A \cap B) + f(A \cup B)$.

It is easy to check that the original cut requirement function is weakly supermodular; by Lemma 23.5, so is the residual cut requirement function.

Lemma 23.5 *Let H be a subgraph of G. If $f : 2^{V(G)} \to \mathbf{Z}^+$ is a weakly supermodular function, then so is the residual cut requirement function f'.*

Proof: Suppose $f(A) + f(B) \leq f(A - B) + f(B - A)$; the proof of the other case is similar. By Lemma 23.4, $|\delta_H(A)| + |\delta_H(B)| \geq |\delta_H(A-B)| + |\delta_H(B-A)|$. Subtracting, we get $f'(A) + f'(B) \leq f'(A - B) + f'(B - A)$. □

We can now state the central polyhedral fact needed for the factor 2 algorithm in its full generality.

Theorem 23.6 *For any weakly supermodular function f, any extreme point solution, x, to LP (23.2) must pick some edge to the extent of at least a half, i.e., $x_e \geq 1/2$ for at least one edge e.*

23.2 The technique of iterated rounding

In this section, we will give an iterated rounding algorithm for the Steiner network problem, using Theorem 23.6.

Algorithm 23.7 (Steiner network)

1. **Initialization:** $H \leftarrow \emptyset$: $f' \leftarrow f$.
2. While $f' \not\equiv \mathbf{0}$, do:
   Find an extreme optimal solution,
   $\boldsymbol{x}$, to LP (23.2) with cut requirements given by f'.
   For each edge e such that $x_e \geq 1/2$, include $\lceil x_e \rceil$ copies of e in H,
   and decrement u_e by this amount.
   Update f': for $S \subseteq V$, $f'(S) \leftarrow f(S) - |\delta_H(S)|$.
3. Output H.

The algorithm presented above achieves an approximation guarantee of factor 2 for an arbitrary weakly supermodular function f. Establishing a polynomial running time involves showing that an extreme optimal solution to LP (23.2) can be found efficiently. We do not know how to do this for an arbitrary weakly supermodular function f. However, if f is the original cut requirement function for some connectivity requirements, then a polynomial time implementation follows from the existence of a polynomial time separation oracle for each iteration.

For the first iteration, a separation oracle follows from a max-flow subroutine. Given a solution $\boldsymbol{x}$, construct a graph on vertex set V with capacity x_e for each edge e. Then, for each pair of vertices $u, v \in V$, check if this graph admits a flow of at least $r(u,v)$ from u to v. If not, we will get a violated cut, i.e., a cut $(S, \overline{S})$ such that $\delta_{\boldsymbol{x}}(S) < f(S)$, where

$$\delta_{\boldsymbol{x}}(S) = \sum_{e:\ e \in \delta(S)} x_e .$$

Let f' be the cut requirement function of a subsequent iteration. Given a solution to LP (23.2) for this function, say $\boldsymbol{x}'$, define $\boldsymbol{x}$ as follows: for each edge e, $x_e = x'_e + e_H$, where e_H is the number of copies of edge e in H. The following lemma shows that a separation oracle for the original function f leads to a separation oracle for f'. Furthermore, this lemma also shows that there is no need to update f' explicitly after each iteration.

Lemma 23.8 *A cut $(S, \overline{S})$ is violated by solution $\boldsymbol{x}'$ under cut requirement function f' iff it is violated by solution $\boldsymbol{x}$ under cut requirement function f.*

Proof: Notice that $\delta_{\boldsymbol{x}}(S) = \delta_{\boldsymbol{x}'}(S) + |\delta_H(S)|$. Since $f(S) = f'(S) + |\delta_H(S)|$, $\delta_{\boldsymbol{x}}(S) \geq f(S)$ iff $\delta_{\boldsymbol{x}'}(S) \geq f'(S)$. □

Lemma 23.8 implies that solution $\boldsymbol{x}'$ is feasible for the cut requirement function f' iff solution $\boldsymbol{x}$ is feasible for f. Assuming Theorem 23.6, whose proof we will provide below, let us show that Algorithm 23.7 achieves an approximation guarantee of 2.

Theorem 23.9 *Algorithm 23.7 achieves an approximation guarantee of 2 for the Steiner network problem.*

Proof: By induction on the number of iterations executed by the algorithm when run with a weakly supermodular cut requirement function f, we will prove that the cost of the integral solution obtained is within a factor of two of the cost of the optimal fractional solution. Since the latter is a lower bound on the cost of the optimal integral solution, the claim follows.

For the base case, if f requires one iteration, the claim follows, since the algorithm rounds up only edges e with $x_e \geq 1/2$.

For the induction step, assume that $\boldsymbol{x}$ is the extreme optimal solution obtained in the first iteration. Obtain $\hat{\boldsymbol{x}}$ from $\boldsymbol{x}$ by zeroing out components that are strictly smaller than $1/2$. By Theorem 23.6, $\hat{\boldsymbol{x}} \neq 0$. Let H be the set of edges picked in the first iteration. Since H is obtained by rounding up nonzero components of $\hat{\boldsymbol{x}}$ and each of these components is $\geq 1/2$, $\text{cost}(H) \leq 2{\cdot}\text{cost}(\hat{\boldsymbol{x}})$.

Let f' be the residual requirement function after the first iteration and H' be the set of edges picked in subsequent iterations for satisfying f'. The key observation is that $\boldsymbol{x} - \hat{\boldsymbol{x}}$ is a feasible solution for f', and thus by the induction hypothesis, $\text{cost}(H') \leq 2 \cdot \text{cost}(\boldsymbol{x} - \hat{\boldsymbol{x}})$. Let us denote by $H + H'$ the edges of H together with those of H'. Clearly, $H + H'$ satisfies f. Now,

$$\begin{aligned}\text{cost}(H + H') &\leq \text{cost}(H) + \text{cost}(H') \\ &\leq 2 \cdot \text{cost}(\hat{\boldsymbol{x}}) + 2 \cdot \text{cost}(\boldsymbol{x} - \hat{\boldsymbol{x}}) \leq 2 \cdot \text{cost}(\boldsymbol{x}).\end{aligned}$$ □

Corollary 23.10 *The integrality gap of LP (23.2) is bounded by 2.*

Notice that previous algorithms obtained using LP-rounding solved the relaxation once and did the entire rounding based on this solution. These algorithms did not exploit the full power of rounding – after part of the solution is rounded, the remaining fractional solution may not be the best solution to continue the rounding process. It may be better to assume integral values for the rounded variables and recompute fractional values for the remaining variables, as is done above. We will call this technique *iterated rounding.*

Example 23.11 The tight example given for the metric Steiner tree problem, Example 3.4, is also a tight example for this algorithm. Observe that after including a subset of edges of the cycle, an extreme optimal solution to the resulting problem picks the remaining edges of the cycle to the extent of one half each. The algorithm finds a solution of cost $2(n - 1)$, whereas the cost of the optimal solution is n. □

23.3 Characterizing extreme point solutions

From polyhedral combinatorics we know that a feasible solution for a set of linear inequalities in $\mathbf{R}^m$ is an extreme point solution iff it satisfies m linearly independent inequalities with equality. Extreme solutions of LP (23.2) satisfy an additional property which leads to a proof of Theorem 23.6.

We will assume that the cut requirement function f in LP (23.2) is an arbitrary weakly supermodular function. Given a solution x to this LP, we will say that an inequality is *tight* if it holds with equality. If this inequality corresponds to the cut requirement of a set S, then we will say that *set S is tight.* Let us make some simplifying assumptions. If $x_e = 0$ for some edge e, this edge can be removed from the graph, and if $x_e \geq 1$, $\lfloor x_e \rfloor$ copies of edge e can be picked and the cut requirement function be updated accordingly. We may assume without loss of generality that an extreme point solution x satisfies $0 < x_e < 1$, for each edge e in graph G. Therefore, each tight inequality corresponds to a tight set. Let the number of edges in G be m.

We will say that a collection, $\mathcal{L}$, of subsets of V forms a *laminar family* if no two sets in this collection cross. The inequality corresponding to a set S defines a vector in $\mathbf{R}^m$: the vector has a 1 corresponding to each edge $e \in \delta_G(S)$, and 0 otherwise. We will call this the *incidence vector* of set S, and will denote it by $\mathcal{A}_S$.

Theorem 23.12 *Corresponding to any extreme point solution to LP (23.2) there is a collection of m tight sets such that*

- *their incidence vectors are linearly independent, and*
- *collection of sets forms a laminar family.*

Example 23.13 The extreme point solution for the Peterson graph assigns nonzero values to 14 of the 15 edges. By Theorem 23.12, there should be 14 tight sets whose incidence vectors are linearly independent. These are marked in figure. □

Fix an extreme point solution, x, to LP (23.2). Let $\mathcal{L}$ be a laminar family of tight sets whose incidence vectors are linearly independent. Denote by span($\mathcal{L}$) the vector space generated by the set of vectors $\{\mathcal{A}_S | S \in \mathcal{L}\}$. Since x is an extreme point solution, the span of the collection of all tight sets is m. We will show that if span($\mathcal{L}$) $< m$, then there is a tight set S whose addition to $\mathcal{L}$ does not violate laminarity and also increases the span. Continuing in this manner, we will obtain m tight sets as required in Theorem 23.12.

We begin by studying properties of crossing tight sets.

Lemma 23.14 *Let A and B be two crossing tight sets. Then, one of the following must hold:*

- *$A - B$ and $B - A$ are both tight and $\mathcal{A}_A + \mathcal{A}_B = \mathcal{A}_{A-B} + \mathcal{A}_{B-A}$*
- *$A \cup B$ and $A \cap B$ are both tight and $\mathcal{A}_A + \mathcal{A}_B = \mathcal{A}_{A \cup B} + \mathcal{A}_{A \cap B}$.*

Proof: Since f is weakly supermodular, either $f(A) + f(B) \leq f(A - B) + f(B - A)$ or $f(A) + f(B) \leq f(A \cup B) + f(A \cap B)$. Let us assume the former holds; the proof for the latter is similar. Since A and B are tight, we have

$$\delta_{\boldsymbol{x}}(A) + \delta_{\boldsymbol{x}}(B) = f(A) + f(B).$$

Since $A - B$ and $B - A$ are not violated,

$$\delta_{\boldsymbol{x}}(A - B) + \delta_{\boldsymbol{x}}(B - A) \geq f(A - B) + f(B - A).$$

Therefore,

$$\delta_{\boldsymbol{x}}(A) + \delta_{\boldsymbol{x}}(B) \leq \delta_{\boldsymbol{x}}(A - B) + \delta_{\boldsymbol{x}}(B - A).$$

As argued in Lemma 23.4 (which established the submodularity of function $|\delta_G(.)|$), edges having one endpoint in $A \cup B$ and the other in $A \cap B$ can contribute only to the left-hand side of this inequality. The rest of the edges must contribute equally to both sides. So, this inequality must be satisfied with equality. Furthermore, since $x_e > 0$ for each edge e, G cannot have any edge having one endpoint in $A \cup B$ and the other in $A \cap B$. Therefore, $\mathcal{A}_A + \mathcal{A}_B = \mathcal{A}_{A-B} + \mathcal{A}_{B-A}$. □

For any set $S \subseteq V$, define its *crossing number* to be the number of sets of $\mathcal{L}$ that S crosses.

Lemma 23.15 *Let S be a set that crosses set $T \in \mathcal{L}$. Then, each of the sets $S - T, T - S, S \cup T$ and $S \cap T$ has a smaller crossing number than S.*

Proof: The figure below illustrates the three ways in which a set $T' \in \mathcal{L}$ can cross one of these four sets without crossing T itself (T' is shown dotted). In all cases, T' crosses S as well. In addition, T crosses S but not any of the four sets. □

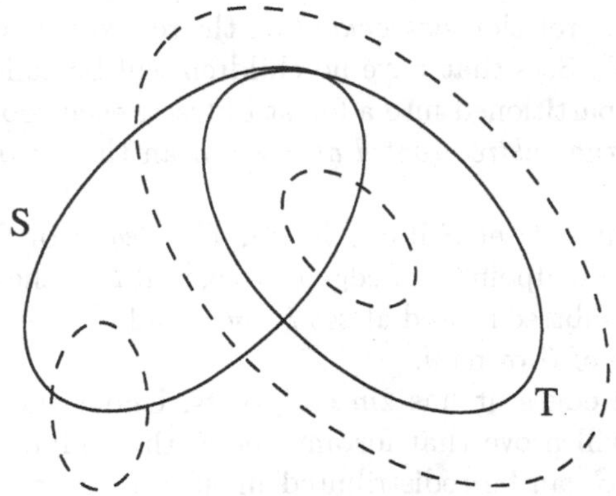

Lemma 23.16 *Let S be a tight set such that $\mathcal{A}_S \notin \text{span}(\mathcal{L})$ and S crosses some set in $\mathcal{L}$. Then, there is a tight set S' having a smaller crossing number than S and such that $\mathcal{A}_{S'} \notin \text{span}(\mathcal{L})$.*

Proof: Let S cross $T \in \mathcal{L}$. Suppose the first possibility established in Lemma 23.14 holds; the proof of the second possibility is similar. Then, $S-T$ and $T-S$ are both tight sets and $\mathcal{A}_S + \mathcal{A}_T = \mathcal{A}_{S-T} + \mathcal{A}_{T-S}$. This linear dependence implies that $\mathcal{A}_{S-T}$ and $\mathcal{A}_{T-S}$ cannot both be in $\text{span}(\mathcal{L})$, since otherwise $\mathcal{A}_S \in \text{span}(\mathcal{L})$. By Lemma 23.15, $S-T$ and $T-S$ both have a smaller crossing number than S. The lemma follows. □

Corollary 23.17 *If $\text{span}(\mathcal{L}) \neq \mathbf{R}^m$, then there is a tight set S such that $\mathcal{A}_S \notin \text{span}(\mathcal{L})$ and $\mathcal{L} \cup \{S\}$ is a laminar family.*

By Corollary 23.17, if $\mathcal{L}$ is a maximal laminar family of tight sets with linearly independent incidence vectors, then $|\mathcal{L}| = m$. This establishes Theorem 23.12.

23.4 A counting argument

The characterization of extreme point solutions given in Theorem 23.12 will yield Theorem 23.6 via a counting argument. Let $\boldsymbol{x}$ be an extreme point solution and $\mathcal{L}$ be the collection of tight sets established in Theorem 23.12. The number of sets in $\mathcal{L}$ equals the number of edges in G, i.e., m. The proof is by contradiction. Suppose that for each edge $e, x_e < 1/2$. Then, we will show that G has more than m edges.

Since $\mathcal{L}$ is a laminar family, it can be viewed as a forest of trees if its elements are ordered by inclusion. Let us make this precise. For $S \in \mathcal{L}$, if S is not contained in any other set of $\mathcal{L}$, then we will say that S is a *root set.* If S is not a root set, we will say that T is the *parent* of S if T is a minimal set in $\mathcal{L}$ containing S; by laminarity of $\mathcal{L}$, T is unique. Further, S will be called a *child* of T. Let the relation *descendent* be the reflexive transitive closure of the relation "child". Sets that have no children will be called *leaves.* In this manner, $\mathcal{L}$ can be partitioned into a forest of trees, each rooted at a root set. For any set S, by the *subtree rooted at S* we mean the set of all descendents of S.

Edge e is *incident* at set S if $e \in \delta_G(S)$. The *degree* of S is defined to be $|\delta_G(S)|$. Set S *owns* endpoint v of edge $e = (u, v)$ if S is the smallest set of $\mathcal{L}$ containing v. The subtree rooted at set S *owns* endpoint v of edge $e = (u, v)$ if some descendent of S owns v.

Since G has m edges, it has $2m$ endpoints. Under the assumption that $\forall e, x_e < 1/2$, we will prove that for any set S, the endpoints owned by the subtree rooted at S can be redistributed in such a way that S gets at least 3 endpoints, and each of its proper descendents gets 2 endpoints. Carrying out this procedure for each of the root sets of the forest, the total number of endpoints in the graph must exceed $2m$, leading to a contradiction.

We have assumed that $\forall e : 0 < x_e < 1/2$. For edge e, define $y_e = 1/2 - x_e$, the *halves complement* of e. Clearly, $0 < y_e < 1/2$. For $S \in \mathcal{L}$ define its *corequirement* to be

$$\text{coreq}(S) = \sum_{e\in\delta(S)} y_e = \frac{1}{2}|\delta_G(S)| - f(S).$$

Clearly, $0 < \text{coreq}(S) < |\delta_G(S)|/2$. Furthermore, since $|\delta_G(S)|$ and $f(S)$ are both integral, $\text{coreq}(S)$ is half-integral. Let us say that $\text{coreq}(S)$ is *semi-integral* if it is not integral, i.e., if $\text{coreq}(S) \in \{1/2, 3/2, 5/2, \ldots\}$. Since $f(S)$ is integral, $\text{coreq}(S)$ is semi-integral iff $|\delta_G(S)|$ is odd.

Sets having a corequirement of 1/2 play a special role in this argument. The following lemma will be useful in establishing that certain sets have this corequirement.

Lemma 23.18 *Suppose S has α children and owns β endpoints, where $\alpha + \beta = 3$. Furthermore, each child of S, if any, has a corequirement of* 1/2. *Then,* $\text{coreq}(S) = 1/2$.

Proof: Since each child of S has corequirement of 1/2, it has odd degree. Using this and the fact that $\alpha + \beta = 3$, one can show that S must have odd degree (see Exercise 23.3). Therefore the corequirement of S is semi-integral. Next, we show that $\text{coreq}(S)$ is strictly smaller than 3/2, thereby proving the lemma. Clearly,

$$\text{coreq}(S) = \sum_{e\in\delta(S)} y_e \leq \sum_{S'} \text{coreq}(S') + \sum_{e} y_e,$$

where the first sum is over all children S' of S, and the second sum is over all edges e having an endpoint in S. Since y_e is strictly smaller than 1/2, if $\beta > 0$, then $\text{coreq}(S) < 3/2$. If $\beta = 0$, all edges incident at the children of S cannot also be incident at S, since otherwise the incidence vectors of these four sets will be linearly dependent. Therefore,

$$\text{coreq}(S) < \sum_{S'} \text{coreq}(S') = 3/2.$$

□

The next two lemmas place lower bounds on the number of endpoints owned by certain sets.

Lemma 23.19 *If set S has only one child, then it must own at least two endpoints.*

Proof: Let S' be the child of S. If S has no endpoint incident at it, the set of edges incident at S and S' must be the same. But then $\mathcal{A}_S = \mathcal{A}_{S'}$, leading to a contradiction. S cannot own exactly one endpoint, because then $\delta_{\boldsymbol{x}}(S)$ and $\delta_{\boldsymbol{x}}(S')$ will differ by a fraction, contradicting the fact that both these sets are tight and have integral requirements. The lemma follows. □

Lemma 23.20 *If set S has two children, one of which has a corequirement of* $1/2$, *then it must own at least one endpoint.*

Proof: Let S' and S'' be the two children of S, with $\text{coreq}(S') = 1/2$. Suppose S does not own any endpoints. Since the three vectors $\mathcal{A}_S, \mathcal{A}_{S'}$, and $\mathcal{A}_{S''}$ are linearly independent, the set of edges incident at S' cannot all be incident at S or all be incident at S''. Let a denote the sum of y_e's of all edges incident at S' and S, and let b denote the sum of y_e's of all edges incident at S' and S''. Thus, $a > 0$, $b > 0$, and $a + b = \text{coreq}(S) = 1/2$.

Since S' has a semi-integral corequirement, it must have odd degree. Therefore, the degrees of S and S'' have different parities, and these two sets have different corequirements. Furthermore, $\text{coreq}(S) = \text{coreq}(S'') + a - b$. Therefore, $\text{coreq}(S) - \text{coreq}(S'') = a - b$. But $-1/2 < a - b < 1/2$. Therefore, S and S'' must have the same corequirement, leading to a contradiction. □

Lemma 23.21 *Consider a tree T rooted at set S. Under the assumption that $\forall e, x_e < 1/2$, the endpoints owned by T can be redistributed in such a way that S gets at least 3 endpoints, and each of its proper descendents gets 2 endpoints. Furthermore, if* $\text{coreq}(S) \neq 1/2$, *then S must get at least 4 endpoints.*

Proof: The proof is by induction on the height of tree T. For the base case, consider a leaf set S. S must have degree at least 3, because otherwise an edge e incident at it will have $x_e \geq 1/2$. If it has degree exactly 3, $\text{coreq}(S)$ is semi-integral. Further, since $\text{coreq}(S) < |\delta_G(S)|/2 = 3/2$, the corequirement of S is $1/2$. Since S is a leaf, it owns an endpoint of each edge incident at it. Therefore, S has the required number of endpoints.

Let us say that a set has a *surplus* of 1 if 3 endpoints have been assigned to it and a surplus of 2 if 4 endpoints have been assigned to it. For the induction step, consider a nonleaf set S. We will prove that by moving the surplus of the children of S and considering the endpoints owned by S itself, we can assign the required number of endpoints to S. There are four cases:

1. If S has 4 or more children, we can assign the surplus of each child to S, thus assigning at least 4 endpoints to S.
2. Suppose S has 3 children. If at least one of them has a surplus of 2, or if S owns an endpoint, we can assign 4 endpoints to S. Otherwise, each child must have a corequirement of half, and by Lemma 23.18, $\text{coreq}(S) = 1/2$ as well. Thus, assigning S the surplus of its children suffices.

3. Suppose S has two children. If each has a surplus of 2, we can assign 4 endpoints to S. If one of them has surplus 1, then by Lemma 23.20, S must own at least one endpoint. If each child has a surplus of 1 and S owns exactly one endpoint, then we can assign 3 endpoints to S, and this suffices by Lemma 23.18. Otherwise, we can assign 4 endpoints to S.
4. If S has one child, say S', then by Lemma 23.19, S owns at least 2 endpoints. If S owns exactly 2 endpoints and S' has surplus of exactly 1, then we can assign 3 endpoints to S; by Lemma 23.18, $\text{coreq}(S) = 1/2$, so this suffices. In all other cases, we can assign 4 endpoints to S.

□

23.5 Exercises

23.1 Give an LP-relaxation for the Steiner network problem, having polynomially many constraints over polynomially many variables.
Hint: Pick a minimum cost set of edges so as to route $\binom{n}{2}$ independent commodities, one for each pair of vertices. Each flow should be at least as large as the connectivity requirement of this pair. The extent to which an edge is picked bounds the amount of each commodity that can flow through this edge.

23.2 Show that a function $f : 2^V \rightarrow \mathbf{Z}^+$ satisfying the following conditions is submodular: $f(V) = 0$, f is symmetric, i.e., for any set $A \subseteq V$ $f(A) = f(V-A)$, and for every two sets $A, B \subseteq V$ $f(A)+f(B) \geq f(A\cap B)+f(A\cup B)$.

23.3 Prove that set S in Lemma 23.18 must have odd degree. (Consider the following possibilities: S owns endpoint v of edge (u, v) that is incident at S, S owns endpoint v of edge (u, v) that is incident at a child of S, and an edge is incident at two children of S.)

23.4 Prove that there must be a set in $\mathcal{L}$ that has degree at most 3, and thus some edge must have $x_e \geq 1/3$. The counting argument required for this is much simpler. Notice that this fact leads to a factor 3 algorithm. (The counting argument requires the use of Lemma 23.19.)

The next two exercises develop a factor $2H_k$ algorithm for the Steiner network problem using the primal–dual schema, where k is the largest connectivity requirement specified in the instance. For simplicity, assume that the upper bounds, u_e, are 1 for each edge e.

23.5 (Williamson, Goemans, Mihail, and Vazirani [266]) Say that a function $h : 2^V \rightarrow \{0, 1\}$ is *uncrossable* if $h(V) = 0$, and for any two sets $A, B \subset V$, if

$h(A) = h(B) = 1$ then $h(A-B) = h(B-A) = 1$ or $h(A \cap B) = h(A \cup B) = 1$. Exercise 22.7 asked for a factor 2 approximation algorithm for IP (22.1) for the case that f was a proper function. In this exercise, we will extend this further to the case that f is an uncrossable function. Now, we need to enhance the last step of Algorithm 22.3; the pruning step needs to be done using reverse delete. Again, F denotes the forest of edges picked by the algorithm. Let us say that a set $A \subset V$ is *unsatisfied* w.r.t. the picked edges F if $h(A) = 1$ and $\delta_F(A) = \emptyset$. A minimal unsatisfied set will be said to be *active*. The algorithm is as follows.

Algorithm 23.22 (Uncrossable function)

1. **(Initialization)** $F \leftarrow \emptyset$; for each $S \subseteq V$, $y_S \leftarrow 0$.
2. **(Edge augmentation)** while there exists an unsatisfied set do:
   simultaneously raise y_S for each active set S, until some edge e goes tight;
   $F \leftarrow F \cup \{e\}$.
3. Let $e_1, e_2, \ldots, e_l$ be the ordered list of edges in F.
4. **(Reverse delete)** For $j = l$ downto 1 do:
   If $F - \{e_j\}$ satisfies h, then $F \leftarrow F - \{e_j\}$.
5. Return F.

Show that in each iteration, active sets must be disjoint. Assuming that active sets can be efficiently found, show that Algorithm 23.22 finds a primal solution of cost at most twice the dual, i.e.,

$$\sum_{e \in F} c_e \leq 2 \sum_S y_S.$$

Hint: Corresponding to each edge $e \in F$, there must be a set $A \subset V$ such that $h(A) = 1$ and $\delta_F(A) = \{e\}$. Call such a set a *witness* for e. A family $\mathcal{C}$ consisting of a witness for each $e \in F$ is called a *witness family*. Include V in this family. Show, by uncrossing, that $\mathcal{C}$ can be assumed to be laminar and therefore can be viewed as a tree. Use this to prove that in each iteration, the average degree of active sets is at most two, as in Lemma 22.8.

23.6 Give an example to show that if reverse delete is replaced by a forward delete, then the approximation factor for Algorithm 23.22 can be unbounded for some uncrossable function.

23.7 (Goemans, Goldberg, Plotkin, Shmoys, Tardos, and Williamson [108]) We will solve the Steiner network problem in k phases, numbered $0, 1, \ldots, k-1$. In each phase, we will pick a forest from the remaining graph. The solution will be the union of the k forests. Let F_{p-1} denote the set of edges picked in phases numbered $0, 1, \ldots, p-1$. At the beginning of the pth phase, define the *deficiency* of set $S \subset V$ to be $\max\{f(S) - |\delta_{F_{p-1}}(A)|, 0\}$. The first $p-1$ phases ensure that every set has deficiency at most $k-p$. In the pth phase, define function h as

$$h(S) = \begin{cases} 1 \text{ if deficiency}(S) = k-p \\ 0 \text{ otherwise} \end{cases}$$

Show that h is an uncrossable function. Show that Algorithm 23.22 can be implemented in polynomial time for this uncrossable function, i.e., active sets can be found in polynomial time. Let F be the set of edges picked by Algorithm 23.22 from $E - F_{p-1}$, and $\boldsymbol{y}$ be the dual solution constructed when run with function h. Construct the dual program to LP (23.2), and show that there is a feasible solution, say $\boldsymbol{d}$, to this program such that

$$\sum_{e \in F} c_e \le 2 \sum_S y_S \le \frac{2}{k-p} g(\boldsymbol{d}),$$

where $g(\boldsymbol{d})$ is the objective function value of dual solution $\boldsymbol{d}$. Adding over all k phases leads to the required factor.
Hint: Use a max-flow algorithm for finding active sets. The dual program will have a variable z_e for each edge e. For edges $e \in F_{p-1}$, set $z_e = \sum_{S:\ e \in \delta(S)} y_S$, for constructing a dual feasible solution.

23.8 Give an infinite family of graphs to show that the performance guarantee of the algorithm in Exercise 23.7 is tight within constant factors.

The following definitions will be useful for the next three exercises. These notions are connected to the theme of this chapter, i.e., small subgraphs with specified numbers of disjoint paths, via Menger's theorem (see Exercise 12.6). An undirected graph is said to be *k-vertex (k-edge) connected* if it has at least $k+1$ vertices, and the removal of any set of at most $k-1$ vertices (edges) from it leaves a connected graph. A directed graph is said to be *k-vertex (k-edge) connected* if it has at least $k+1$ vertices, and the deletion of any set of at most $k-1$ vertices (edges) leaves a strongly connected graph.

23.9 (Cheriyan and Thurimella [44]) This exercise develops a $1+2/k$ factor algorithm for the following problem.

Problem 23.23 (Minimum k-vertex connected subgraph) Given a nonnegative integer k and an undirected graph $G = (V, E)$ that is k-vertex connected, find a minimum cardinality set $E' \subset E$ such that the subgraph $G' = (V, E')$ is k-vertex connected.

Let $G = (V, E)$ be k-vertex connected. We will say that edge $e \in E$ is *critical* if its removal leaves a graph that is not k-vertex connected. A simple cycle C in G is *critical* if every edge on C is critical. A theorem of Mader, which states that a critical cycle in G must have a vertex of degree exactly k, is central to the algorithm.

Algorithm 23.24 (k-vertex connected subgraph)

1. Find a minimum cardinality set $M \subset E$ such that $\forall v \in V : \deg_M(v) \geq k-1$.
2. Find a minimal set F such that $M \cup F$ is k-vertex connected.
3. Output $G' = (V, M \cup F)$.

1. Give a polynomial time algorithm for Step 1 of Algorithm 23.24. Observe that $|M| \leq \text{OPT}$.
   Hint: Use a b-matching algorithm on the complement of G. Given an undirected graph $G = (V, E)$ and a function $b : V \rightarrow \mathbf{Z}^+$ specifying an upper bound for each vertex, the b-matching problem asks for a maximum cardinality set $M \subseteq E$ such that $\forall v \in V, \deg_M(v) \leq b(v)$. This problem is in **P**.
2. Give an efficient implementation for Step 2 of Algorithm 23.24.
3. Use Mader's theorem to show that F must be acyclic, and hence $|F| \leq |V| - 1$. Use this to show that Algorithm 23.24 achieves an approximation factor of $1 + 2/k$.
   Hint: Use the fact that $k|V|/2$ is a lower bound on OPT.

23.10 (Cheriyan and Thurimella [44]) Consider the problem of finding a minimum k-vertex connected subgraph of a directed graph. Give an algorithm similar to that in Exercise 23.9 for achieving factor $1 + 2/k$ for this problem. Use the following two facts.

1. In a directed graph, an *alternating cycle, C,* is an even length sequence of distinct edges $(v_0, v_1)(v_2, v_1)(v_2, v_3)(v_4, v_3) \ldots (v_{m-1}, v_m)(v_0, v_m)$, where vertices are allowed to repeat. Notice that alternate vertices on C have two out-edges (two in-edges). Vertices having two out-edges (two in-edges) will be called *C-out* (*C-in*) vertices. Mader showed that if G is a k-vertex connected directed graph containing an alternating cycle C, each of whose edges is critical, then C contains either a C-out vertex having out-degree exactly k or a C-in vertex having in-degree exactly k.
2. Given a directed graph $G = (V, E)$, define its *associated bipartite graph* H to be the following. Corresponding to each vertex $v \in V$, H has two vertices, v_- and v_+, and corresponding to each edge $(u, v) \in E$, H has the edge (u_+, v_-). There is an alternating cycle in G iff its associated bipartite graph contains a cycle.

23.11 (Khuller and Vishkin [179], using Edmonds [78]) This exercise develops a factor 2 algorithm for the following problem.

Problem 23.25 (Minimum k-edge connected subgraph) Given an undirected graph $G = (V, E)$, a function $w : E \rightarrow \mathbf{Q}^+$, and an integer k, find a minimum weight subgraph of G that is k-edge connected.

1. Let $r \in V$ be any vertex of G. Consider the problem of finding a minimum weight subgraph G' of G such that for each vertex $v \in V$, there are k edge-disjoint paths from r to v in G'. Show that this problem is the same as Problem 23.25, i.e., any solution to one is also a solution to the other.
2. Let $G = (V, E)$ be an edge-weighted directed graph and $r \in V$ be one of its vertices. A set $E' \subseteq E$ is said to be an *r-arborescence* if every vertex, other than r, has in-degree 1. In effect, an r-arborescence is a spanning tree directed out of r. Define the *r-connectivity* of G to be

   $$\max\{k \mid \forall v \in V\ \exists k \text{ edge-disjoint paths from } r \text{ to } v \text{ in } G\}.$$

   Edmonds showed that the maximum number of edge-disjoint r-arborescences in G is equal to the r-connectivity of G. Use this to show that the problem of finding a minimum weight subgraph of G that has an r-connectivity of k is the same as the problem of finding a minimum weight subgraph of G that has k edge-disjoint r-arborescences.
3. Edmonds showed that the edges of a directed graph $G = (V, E)$ can be partitioned into k edge-disjoint r-arborescences iff, on ignoring directions, E can be partitioned into k spanning trees, and the in-degree of every vertex, other than r, is exactly k. Use this characterization to show that the problem of finding a minimum weight subgraph of G that has k edge-disjoint arborescences can be solved in polynomial time.
   Hint: This problem can be expressed as a matroid intersection problem, the two matroids being a partition matroid and the k-fold union of a graphic matroid (which is also a matroid).
4. Let $G = (V, E)$ be an edge-weighted undirected graph and $r \in V$ be one of its vertices. Let $\mathrm{OPT}(G)$ denote the weight of an optimal solution to Problem 23.25 on instance G. Obtain graph H by bidirecting G, i.e., by replacing each edge $(u, v) \in E$ with the two edges $(u \rightarrow v)$ and $(v \rightarrow u)$, each having the same weight as (u, v). Let $\mathrm{OPT}(H)$ denote the weight of a minimum weight subgraph of H that can be partitioned into k r-arborescences. Show that

   $$\mathrm{OPT}(G) \leq \mathrm{OPT}(H) \leq 2 \cdot \mathrm{OPT}(G).$$

   Use this to obtain a factor 2 approximation algorithm for Problem 23.25.

23.12 (Goemans and Bertsimas [107]) The metric Steiner network problem is the Steiner network problem with the restrictions that G is a complete graph, the cost function on edges satisfies the triangle inequality, and $u_e = \infty$ for each edge. It generalizes the metric Steiner tree problem to arbitrary connectivity requirements. For $D \subseteq V$, define $LP_S(D)$ to be LP-relaxation (23.2), together with a set of equality constraints for vertices in D, as follows.

$$\begin{aligned}
\text{minimize} \quad & \sum_{e \in E} c_e x_e && && (23.4)\\
\text{subject to} \quad & \sum_{e:\ e \in \delta(S)} x_e \geq f(S), && S \subseteq V \\
& \sum_{e:\ e \text{ incident at } v} x_e = f(\{v\}), && v \in D \\
& x_e \geq 0, && e \in E
\end{aligned}$$

It turns out that the equality constraints are redundant for the metric Steiner network problem. For any choice of $D \subseteq V$, an optimal solution to $LP_S(D)$ is also an optimal solution to $LP_S(\emptyset)$. This is called the *parsimonious property.* Let us say that a vertex v is *Steiner* if it has no connectivity requirements, i.e., if $\forall u \in V, r(u,v) = 0$. Use the parsimonious property to prove that there is a fractional optimal solution to the metric Steiner network problem which has no edges incident at Steiner vertices.

23.13 Consider the following integer program for the traveling salesman problem (Problem 3.5).

$$\begin{aligned}
\text{minimize} \quad & \sum_{e \in E} c_e x_e && && (23.5)\\
\text{subject to} \quad & \sum_{e:\ e \text{ incident at } v} x_e = 2, && v \in V \\
& \sum_{e:\ e \in \delta(S)} x_e \geq 2, && S \subset V \\
& x_e \in \{0,1\}, && e \in E
\end{aligned}$$

Show that optimal solutions to this integer program are optimal TSP tours. The linear relaxation of this program is called the *subtour elimination* LP-relaxation for TSP.

The rest of this exercise deals with the special case of metric TSP and develops a proof that the solution found by Christofides' algorithm, Algorithm 3.10, is within a factor of 3/2 of the optimal solution to this LP-relaxation.

1. Give an example that puts a lower bound of (essentially) 4/3 on the integrality gap of this relaxation.
   Hint: Use the following graph.

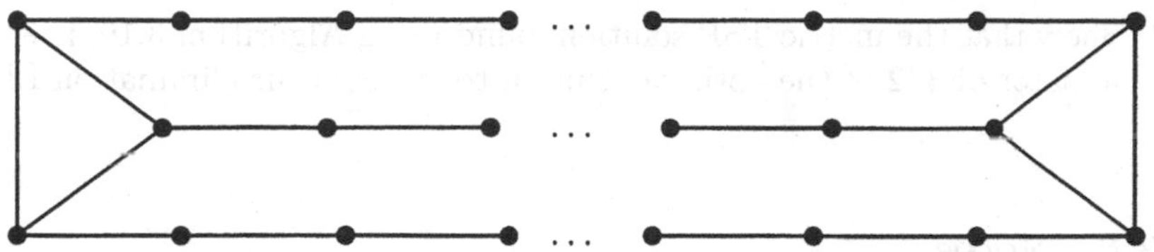

2. Let v_1 be an arbitrary vertex in the given graph $G = (V, E)$. Define a *1-tree* in G to be a spanning tree on the vertices $V - \{v_1\}$, together with two distinct edges incident at vertex v_1. Clearly, the cost of an optimal 1-tree is a lower bound on the cost of an optimal TSP tour. LP (12.12) stated in Exercise 12.10 was an exact relaxation for the MST problem. Use it to obtain an exact LP-relaxation for the minimum 1-tree problem.
3. (Held and Karp [130]) Show that the cost of a minimum 1-tree is a lower bound on the cost of an optimal solution to the subtour elimination LP.
Hint: Compare the LP obtained above for minimum 1-tree with the following equivalent formulation of the subtour elimination LP. (By $e : e \in S$ we mean edges e that have both endpoints in S.)

$$\begin{aligned}
\text{minimize} \quad & \sum_{e \in E} c_e x_e & & (23.6)\\
\text{subject to} \quad & \sum_{e:\ e \text{ incident at } v} x_e = 2, & v \in V &\\
& \sum_{e:\ e \in S} x_e \leq |S| - 1, & S \subseteq V &\\
& x_e \geq 0, & e \in E &
\end{aligned}$$

4. Use the parsimonious property, introduced in Exercise 23.12, to show that the equality constraints, on vertices, in the subtour elimination LP are redundant. (Observe that the LP obtained on removing these constraints is also an LP-relaxation for the problem of finding a minimum cost spanning two-edge connected subgraph of G.)
5. For $D \subseteq V$, let $LP_T(D)$ denote the subtour elimination LP for G_D, the subgraph of G induced on D. Let $\mathrm{OPT}_f(LP_T(D))$ denote the cost of an optimal solution to $LP_T(D)$. Show the following monotonicity property

$$\mathrm{OPT}_f(LP_T(D)) \leq \mathrm{OPT}_f(LP_T(V)).$$

Hint: Use the relaxation without equality constraints.
6. Let $D \subseteq V$ be of even cardinality. Show that the cost of a minimum cost perfect matching in the subgraph of G induced on D is $\leq \frac{1}{2}\mathrm{OPT}_f(LP_T(D))$.
Hint: Use LP (12.9), introduced in Exercise 12.9, for matching, and LP (23.6) for TSP.

7. Show that the metric TSP solution found using Algorithm 3.10, is within a factor of 3/2 of the optimal solution to the subtour elimination LP.

23.6 Notes

The first approximation algorithm for this problem, achieving a guarantee of $2k$, where k is the largest connectivity requirement specified in the instance, was given by Williamson, Goemans, Mihail, and Vazirani [266]. This algorithm was based on the primal-dual schema. Extending this approach, Goemans, Goldberg, Plotkin, Shmoys, Tardos, and Williamson [108] improved the approximation guarantee to $2H_k$. The result of this chapter is due to Jain [144]. Cheriyan and Thurimella [44] contains further results on finding small subgraphs of a given graph with a specified connectivity, as well as references to Mader's theorems. The subtour elimination LP-relaxation for TSP was given in Dantzig, Ford, and Fulkerson [62]. The result of Exercise 23.13 was first established by Wolsey [269]. The proof developed here is from Shmoys and Williamson [248].

24 Facility Location

The facility location problem has occupied a central place in operations research since the early 1960's. It models design situations such as deciding placements of factories, warehouses, schools, and hospitals. Modern day applications include placement of proxy servers on the web.

In this chapter, we will present a primal–dual schema based factor 3 approximation algorithm for the special case when connection costs satisfy the triangle inequality. The algorithm differs in two respects from previous primal–dual algorithms. First, the primal and dual pair of LPs have negative coefficients and do not form a covering-packing pair. Second, we will relax primal complementary slackness conditions rather than the dual ones. Also, the idea of synchronization, introduced in the primal–dual schema in Chapter 22, is developed further, with an explicit timing of events playing a role.

Problem 24.1 (Metric uncapacitated facility location) Let G be a bipartite graph with bipartition (F, C), where F is the set of *facilities* and C is the set of *cities*. Let f_i be the cost of opening facility i, and c_{ij} be the cost of connecting city j to (opened) facility i. The connection costs satisfy the triangle inequality. The problem is to find a subset $I \subseteq F$ of facilities that should be opened, and a function $\phi : C \rightarrow I$ assigning cities to open facilities in such a way that the total cost of opening facilities and connecting cities to open facilities is minimized.

Consider the following integer program for this problem. In this program, y_i is an indicator variable denoting whether facility i is open, and x_{ij} is an indicator variable denoting whether city j is connected to the facility i. The first set of constraints ensures that each city is connected to at least one facility, and the second ensures that this facility must be open.

$$
\begin{aligned}
\text{minimize} \quad & \sum_{i \in F,\, j \in C} c_{ij} x_{ij} + \sum_{i \in F} f_i y_i && && (24.1)\\
\text{subject to} \quad & \sum_{i \in F} x_{ij} \geq 1, && j \in C \\
& y_i - x_{ij} \geq 0, && i \in F,\ j \in C \\
& x_{ij} \in \{0, 1\}, && i \in F,\ j \in C \\
& y_i \in \{0, 1\}, && i \in F
\end{aligned}
$$

The LP-relaxation of this program is:

$$\begin{array}{lll} \text{minimize} & \sum_{i\in F,\ j\in C} c_{ij}x_{ij} + \sum_{i\in F} f_i y_i & \\ \text{subject to} & \sum_{i\in F} x_{ij} \geq 1, & j \in C \\ & y_i - x_{ij} \geq 0, & i \in F,\ j \in C \\ & x_{ij} \geq 0, & i \in F,\ j \in C \\ & y_i \geq 0, & i \in F \end{array} \tag{24.2}$$

The dual program is:

$$\begin{array}{lll} \text{maximize} & \sum_{j\in C} \alpha_j & \\ \text{subject to} & \alpha_j - \beta_{ij} \leq c_{ij}, & i \in F,\ j \in C \\ & \sum_{j\in C} \beta_{ij} \leq f_i, & i \in F \\ & \alpha_j \geq 0, & j \in C \\ & \beta_{ij} \geq 0, & i \in F,\ j \in C \end{array} \tag{24.3}$$

24.1 An intuitive understanding of the dual

Let us first give the reader some feel for how the dual variables "pay" for a primal solution by considering the following simple setting. Suppose LP (24.2) has an optimal solution that is integral, say $I \subseteq F$ and $\phi : C \to I$. Thus, under this solution, $y_i = 1$ iff $i \in I$, and $x_{ij} = 1$ iff $i = \phi(j)$. Let $(\boldsymbol{\alpha}, \boldsymbol{\beta})$ denote an optimal dual solution.

The primal and dual complementary slackness conditions are:

$$\begin{array}{ll} \text{(i)} & \forall i \in F, j \in C : x_{ij} > 0 \Rightarrow \alpha_j - \beta_{ij} = c_{ij} \\ \text{(ii)} & \forall i \in F : y_i > 0 \Rightarrow \sum_{j\in C} \beta_{ij} = f_i \\ \text{(iii)} & \forall j \in C : \alpha_j > 0 \Rightarrow \sum_{i\in F} x_{ij} = 1 \\ \text{(iv)} & \forall i \in F, j \in C : \beta_{ij} > 0 \Rightarrow y_i = x_{ij} \end{array}$$

By condition (ii), each open facility must be fully paid for, i.e., if $i \in I$, then

$$\sum_{j:\ \phi(j)=i} \beta_{ij} = f_i.$$

Consider condition (iv). Now, if facility i is open, but $\phi(j) \neq i$, then $y_i \neq x_{ij}$, and so $\beta_{ij} = 0$, i.e., city j does not contribute to opening any facility besides the one it is connected to.

By condition (i), if $\phi(j) = i$, then $\alpha_j - \beta_{ij} = c_{ij}$. Thus, we can think of α_j as the total price paid by city j; of this, c_{ij} goes towards the use of edge (i, j), and β_{ij} is the contribution of j towards opening facility i.

24.2 Relaxing primal complementary slackness conditions

Suppose the primal complementary slackness conditions were relaxed as follows, while maintaining the dual conditions:

$$\forall j \in C: \quad (1/3)c_{\phi(j)j} \leq \alpha_j - \beta_{\phi(j)j} \leq c_{\phi(j)j},$$

and

$$\forall i \in I: \quad (1/3)f_i \leq \sum_{j:\ \phi(j)=i} \beta_{ij} \leq f_i.$$

Then, the cost of the (integral) solution found would be within thrice the dual found, thus leading to a factor 3 approximation algorithm. However, we would like to obtain the stronger inequality stated in Theorem 24.7. Now, the dual pays at least one-third the connection cost, but must pay completely for opening facilities. This stronger inequality will be needed in order to use this algorithm to solve the k-median problem in Chapter 25.

For this reason, we will relax the primal conditions as follows. The cities are partitioned into two sets, *directly connected* and *indirectly connected*. Only directly connected cities will pay for opening facilities, i.e., β_{ij} can be nonzero only if j is a directly connected city and $i = \phi(j)$. For an indirectly connected city j, the primal condition is relaxed as follows:

$$(1/3)c_{\phi(j)j} \leq \alpha_j \leq c_{\phi(j)j}.$$

All other primal conditions are maintained, i.e., for a directly connected city j,

$$\alpha_j - \beta_{\phi(j)j} = c_{\phi(j)j},$$

and for each open facility i,

$$\sum_{j:\ \phi(j)=i} \beta_{ij} = f_i.$$

24.3 Primal–dual schema based algorithm

The algorithm consists of two phases. In Phase 1, the algorithm operates in a primal–dual fashion. It finds a dual feasible solution and also determines a set of tight edges and temporarily open facilities, F_t. Phase 2 consists of choosing a subset I of F_t to open, and finding a mapping, ϕ, from cities to I.

Algorithm 24.2

Phase 1
We would like to find as large a dual solution as possible. This motivates the following underlying process for dealing with the non-covering-packing pair of LPs. Each city j raises its dual variable, α_j, until it gets connected to an open facility. All other primal and dual variables simply respond to this change, trying to maintain feasibility or satisfying complementary slackness conditions.

A notion of *time* is defined in this phase, so that each event can be associated with the time at which it happened; the phase starts at time 0. Initially, each city is defined to be *unconnected.* Throughout this phase, the algorithm raises the dual variable α_j for each unconnected city j uniformly at unit rate, i.e., α_j will grow by 1 in unit time. When $\alpha_j = c_{ij}$ for some edge (i, j), the algorithm will declare this edge to be *tight.* Henceforth, dual variable β_{ij} will be raised uniformly, thus ensuring that the first constraint in LP (24.3) is not violated. β_{ij} goes towards paying for facility i. Each edge (i, j) such that $\beta_{ij} > 0$ is declared *special.*

Facility i is said to be *paid for* if $\sum_j \beta_{ij} = f_i$. If so, the algorithm declares this facility *temporarily open.* Furthermore, all unconnected cities having tight edges to this facility are declared *connected* and facility i is declared the *connecting witness* for each of these cities. (Notice that the dual variables α_j of these cities are not raised anymore.) In the future, as soon as an unconnected city j gets a tight edge to i, j will also be declared connected and i will be declared the connecting witness for j (notice that $\beta_{ij} = 0$ and thus edge (i, j) is not special). When all cities are connected, the first phase terminates. If several events happen simultaneously, the algorithm executes them in arbitrary order.

Remark 24.3 At the end of Phase 1, a city may have paid towards temporarily opening several facilities. However, we want to ensure that a city pays only for the facility that it is eventually connected to. This is ensured in Phase 2, which chooses a subset of temporarily open facilities for opening permanently.

Phase 2
Let F_t denote the set of temporarily open facilities and T denote the subgraph of G consisting of all special edges. Let T^2 denote the graph that has edge (u, v) iff there is a path of length at most 2 between u and v in T, and let H

be the subgraph of T^2 induced on F_t. Find any maximal independent set in H, say I. All facilities in the set I are declared *open*.

For city j, define $\mathcal{F}_j = \{i \in F_t \mid (i,j) \text{ is special}\}$. Since I is an independent set, at most one of the facilities in $\mathcal{F}_j$ is opened. If there is a facility $i \in \mathcal{F}_j$ that is opened, then set $\phi(j) = i$ and declare city j *directly connected.* Otherwise, consider tight edge (i',j) such that i' was the connecting witness for j. If $i' \in I$, again set $\phi(j) = i'$ and declare city j *directly connected* (notice that in this case $\beta_{i'j} = 0$). In the remaining case that $i' \notin I$, let i be any neighbor of i' in graph H such that $i \in I$. Set $\phi(j) = i$ and declare city j *indirectly connected.*

I and ϕ define a primal integral solution: $x_{ij} = 1$ iff $\phi(j) = i$ and $y_i = 1$ iff $i \in I$. The values of α_j and β_{ij} obtained at the end of Phase 1 form a dual feasible solution.

24.4 Analysis

We will show how the dual variables α_j's pay for the primal costs of opening facilities and connecting cities to facilities. Denote by α_j^f and α_j^e the contributions of city j to these two costs respectively; $\alpha_j = \alpha_j^f + \alpha_j^e$. If j is indirectly connected, then $\alpha_j^f = 0$ and $\alpha_j^e = \alpha_j$. If j is directly connected, then the following must hold:

$$\alpha_j = c_{ij} + \beta_{ij},$$

where $i = \phi(j)$. Now, let $\alpha_j^f = \beta_{ij}$ and $\alpha_j^e = c_{ij}$.

Lemma 24.4 *Let $i \in I$. Then,*

$$\sum_{j:\ \phi(j)=i} \alpha_j^f = f_i.$$

Proof: Since i is temporarily open at the end of Phase 1, it is completely paid for, i.e.,

$$\sum_{j:\ (i,j) \text{ is special}} \beta_{ij} = f_i.$$

The critical observation is that each city j that has contributed to f_i must be directly connected to i. For each such city, $\alpha_j^f = \beta_{ij}$. Any other city j' that is connected to facility i must satisfy $\alpha_{j'}^f = 0$. The lemma follows. □

Corollary 24.5 $\sum_{i \in I} f_i = \sum_{j \in C} \alpha_j^f$.

Recall that α_j^f was defined to be 0 for indirectly connected cities. Thus, only the directly connected cities pay for the cost of opening facilities.

Lemma 24.6 *For an indirectly connected city j, $c_{ij} \leq 3\alpha_j^e$, where $i = \phi(j)$.*

Proof: Let i' be the connecting witness for city j. Since j is indirectly connected to i, (i, i') must be an edge in H. In turn, there must be a city, say j', such that (i, j') and (i', j') are both special edges. Let t_1 and t_2 be the times at which i and i' were declared temporarily open during Phase 1.

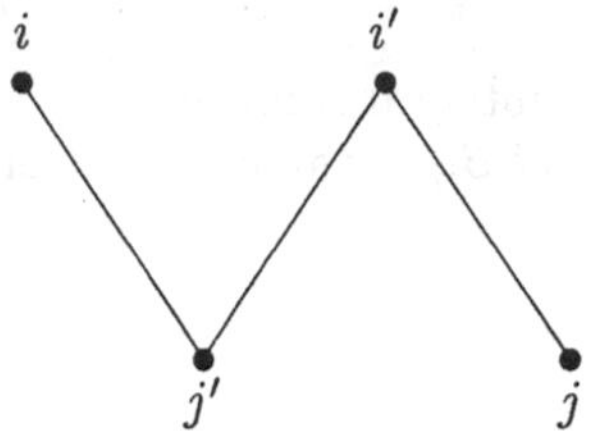

Since edge (i', j) is tight, $\alpha_j \geq c_{i'j}$. We will show that $\alpha_j \geq c_{ij'}$ and $\alpha_j \geq c_{i'j'}$. Then, the lemma will follow by using the triangle inequality.

Since edges (i', j') and (i, j') are tight, $\alpha_{j'} \geq c_{ij'}$ and $\alpha_{j'} \geq c_{i'j'}$. Since both these edges are special, they must both have gone tight before either i or i' is declared temporarily open. Consider the time $\min(t_1, t_2)$. Clearly, $\alpha_{j'}$ cannot be growing beyond this time. Therefore, $\alpha_{j'} \leq \min(t_1, t_2)$. Finally, since i' is the connecting witness for j, $\alpha_j \geq t_2$. Therefore, $\alpha_j \geq \alpha_{j'}$, and the required inequalities follow. □

Theorem 24.7 *The primal and dual solutions constructed by the algorithm satisfy:*

$$\sum_{i \in F,\, j \in C} c_{ij} x_{ij} + 3 \sum_{i \in F} f_i y_i \leq 3 \sum_{j \in C} \alpha_j.$$

Proof: For a directly connected city j, $c_{ij} = \alpha_j^e \leq 3\alpha_j^e$, where $\phi(j) = i$. Combining with Lemma 24.6 we get

$$\sum_{i \in F, j \in C} c_{ij} x_{ij} \leq 3 \sum_{j \in C} \alpha_j^e.$$

Adding to this the equality stated in Corollary 24.5 multiplied by 3 gives the theorem. □

24.4.1 Running time

A special feature of the primal–dual schema is that it yields algorithms with good running times. Since this is especially so for the current algorithm, we will provide some implementation details. We will adopt the following notation: $n_c = |C|$ and $n_f = |F|$. The total number of vertices $n_c + n_f = n$, and the total number of edges $n_c \times n_f = m$.

Sort all the edges by increasing cost – this gives the order and the times at which edges go tight. For each facility, i, we maintain the number of cities that are currently contributing towards it, and the *anticipated time*, t_i, at which it would be completely paid for if no other event happens on the way. Initially all t_i's are infinite, and each facility has 0 cities contributing to it. The t_i's are maintained in a binary heap so we can update each one and find the current minimum in $O(\log n_f)$ time. Two types of events happen, and they lead to the following updates.

- An edge (i, j) goes tight.
  - If facility i is not temporarily open, then it gets one more city contributing towards its cost. The amount contributed towards its cost at the current time can be easily computed. Therefore, the anticipated time for facility i to be paid for can be recomputed in constant time. The heap can be updated in $O(\log n_f)$ time.
  - If facility i is already temporarily open, city j is declared connected, and α_j is not raised anymore. For each facility i' that was counting j as a contributor, we need to decrease the number of contributors by 1 and recompute the anticipated time at which it gets paid for.
- Facility i is completely paid for. In this event, i will be declared temporarily open, and all cities contributing to i will be declared connected. For each of these cities, we will execute the second case of the previous event, i.e., update facilities that they were contributing towards.

The next theorem follows by observing that each edge (i, j) will be considered at most twice. First, when it goes tight. Second, when city j is declared connected. For each consideration of this edge, we will do $O(\log n_f)$ work.

Theorem 24.8 *Algorithm 24.2 achieves an approximation factor of 3 for the facility location problem and has a running time of* $O(m \log m)$.

24.4.2 Tight example

The following infinite family of examples shows that the analysis of our algorithm is tight: The graph has n cities, $c_1, c_2, \ldots, c_n$ and two facilities f_1 and f_2. Each city is at a distance of 1 from f_2. City c_1 is at a distance of 1 from f_1, and $c_2, \ldots, c_n$ are at a distance of 3 from f_1. The opening cost of f_1 and f_2 are ε and $(n+1)\varepsilon$, respectively, for a small number ε.

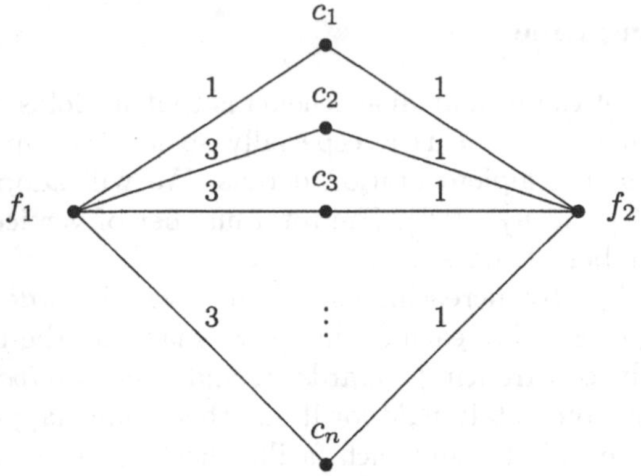

The optimal solution is to open f_2 and connect all cities to it, at a total cost of $(n+1)\varepsilon+n$. Algorithm 24.2 will however open facility f_1 and connect all cities to it, at a total cost of $\varepsilon+1+3(n-1)$.

24.5 Exercises

24.1 Consider the general uncapacitated facility location problem in which the connection costs are not required to satisfy the triangle inequality. Give a reduction from the set cover problem to show that approximating this problem is as hard as approximating set cover and therefore cannot be done better than $O(\log n)$ factor unless $\mathbf{NP} \subseteq \mathbf{DTIME}(n^{O(\log\log n)})$ (see Chapter 29). Also, give an $O(\log n)$ factor algorithm for this problem.

24.2 In Phase 2, instead of picking all special edges in T, pick all tight edges. Show that now Lemma 24.6 does not hold. Give a suitable modification to the algorithm that restores Lemma 24.6.
Hint: Order facilities in H in the order in which they were temporarily opened, and pick I to be the lexicographically first maximal independent set.

24.3 Give a factor 3 tight example for Algorithm 24.2 in which the set of cities and facilities is the same, i.e., $C = F$.

24.4 Consider the proof of Lemma 24.6. Give an example in which $\alpha_j > t_2$.

24.5 The vector $\boldsymbol{\alpha}$ found by Algorithm 24.2 is maximal in the sense that if we increase any α_j in this vector, then there is no way of setting the β_{ij}'s to get a feasible dual solution. Is every maximal solution $\boldsymbol{\alpha}$ within 3 times the optimal solution to dual program for facility location?

Hint: It is easy to construct a maximal solution that is $2/n$ times the optimal. Consider n facilities with an opening cost of 1 each and n cities connected to distinct facilities by edges of cost ε each. In addition, there is another city that is connected to each facility with an edge of cost 1.

24.6 Consider the following modification to the metric uncapacitated facility location problem. Define the cost of connecting city j to facility i to be c_{ij}^2. The c_{ij}'s still satisfy the triangle inequality (but the new connection costs, of c_{ij}^2, do not). Show that Algorithm 24.2 achieves an approximation guarantee of factor 9 for this case.

24.7 Consider the following generalization to arbitrary demands. For each city j, a nonnegative demand d_j is specified, and any open facility can serve this demand. The cost of serving this demand via facility i is $c_{ij}d_j$. Give an IP and LP-relaxation for this problem, and extend Algorithm 24.2 to get a factor 3 algorithm.
Hint: Raise α_j at rate d_j.

24.8 In the *capacitated* facility location problem, we are given a number u_i for each facility i, and facility i can serve at most u_i cities. Show that the modification of LP (24.2) to this problem has an unbounded integrality gap.

24.9 Consider the variant of the capacitated metric facility location problem in which each facility can be opened an unbounded number of times. If facility i is opened y_i times, it can serve at most $u_i y_i$ cities. Give an IP and LP-relaxation for this problem, and extend Algorithm 24.2 to obtain a constant factor algorithm.

24.10 (Charikar, Khuller, Mount, and Narshimhan [41]) Consider the *prize-collecting* variant of the facility location problem, in which there is a specified penalty for not connecting a city to an open facility. The objective is to minimize the sum of the connection costs, facility opening costs, and penalties. Give a factor 3 approximation algorithm for this problem.

24.11 (Jain and Vazirani [147]) Consider the *fault tolerant* variant of the facility location problem, in which the additional input is a connection requirement r_j for each city j. In the solution, city j needs to be connected to r_j distinct open facilities. The objective, as before, is to minimize the sum of the connection costs and the facility opening costs.

Decompose the problem into k phases, numbered k down to 1, as in Exercise 23.7. In phase p, all cities having a residual requirement of p are provided one more connection to an open facility. In phase p, the facility location algorithm of this chapter is run on the following modified graph, G_p. The cost of each facility that is opened in an earlier phase is set to 0. If city j is connected to facility i in an earlier phase, then c_{ij} is set to ∞.

1. Show that even though G_p violates the triangle inequality at some places, the algorithm gives a solution within factor 3 of the optimal solution for this graph.
   Hint: Every time short-cutting is needed; the triangle inequality holds.
2. Show that the solution found in phase p is of cost at most $3 \cdot \text{OPT}/p$, where OPT is the cost of the solution to the entire problem.
   Hint: Remove ∞ cost edges of G_p from the optimal solution and divide the rest by p. Show that this is a feasible fractional solution for phase p.
3. Show that this algorithm achieves an approximation factor of $3 \cdot H_k$ for the fault tolerant facility location problem.

24.12 (Mahdian, Markakis, Saberi, and Vazirani [208]) This exercise develops a factor 3 greedy algorithm for the metric uncapacitated facility location problem, together with an analysis using the method of dual fitting.

Consider the following modification to Algorithm 24.2. As before, dual variables, α_j, of all unconnected cities, j, are raised uniformly. If edge (i, j) is tight, β_{ij} is raised. As soon as a facility, say i, is paid for, it is declared open. Let S be the set of unconnected cities having tight edges to i. Each city $j \in S$ is declared connected and stops raising its α_j. So far, the new algorithm is the same as Algorithm 24.2. The main difference appears at this stage: Each city $j \in S$ withdraws its contribution from other facilities, i.e., for each facility $i' \neq i$, set $\beta_{i'j} = 0$. When all cities have been declared connected, the algorithm terminates. Observe that each city contributes towards the opening cost of at most one facility – the facility it gets connected to.

1. This algorithm actually has a simpler description as a greedy algorithm. Provide this description.
   Hint: Use the notion of cost–effectiveness defined for the greedy set cover algorithm.
2. The next 3 parts use the method of dual fitting to analyze this algorithm. First observe that the primal solution found is fully paid for by the dual computed.
3. Let i be an open facility and let $\{1, \ldots, k\}$ be the set of cities that contributed to opening i at some point in the algorithm. Assume w.l.o.g. that $\alpha_1 \leq \alpha_j$ for $j \leq k$. Show that for $j \leq k$, $\alpha_j - c_{ij} \leq 2\alpha_1$. Also, show that

$$\sum_{j=1}^{k} \alpha_j \leq 3 \sum_{j=1}^{k} c_{ij} + f_i.$$

   Hint: Use the triangle inequality and the following inequality which is a consequence of the fact that at any point, the total amount contributed for opening facility i is at most f_i:

$$\sum_{j:\ c_{ij} \leq \alpha_1} \alpha_1 - c_{ij} \leq f_i.$$

4. Hence show that $\boldsymbol{\alpha}/3$ is a dual feasible solution.
5. How can the analysis be improved – a factor 1.86 analysis is known for this algorithm.
6. Give a time efficient implementation of this algorithm, matching the running time of Algorithm 24.2
7. Do you see room for improving the algorithm?
   Hint: Suppose city j is connected to open facility i at some point in the algorithm. Later, facility i' is opened, and suppose that $c_{ij} > c_{i'j}$. Then, connecting j to i' will reduce the cost of the solution.

24.13 (Mahdian, Markakis, Saberi, and Vazirani [208]) Consider the following variant of the metric uncapacitated facility location problem. Instead of f_i, the opening cost for each facility $i \in F$, we are provided a startup cost s_i and an incremental cost t_i. Define the new opening cost for connecting $k > 0$ cities to facility i to be $s_i + kt_i$. Connection costs are specified by a metric, as before. The object again is to connect each city to an open facility so as to minimize the sum of connection costs and opening costs. Give an approximation factor preserving reduction from this problem to the metric uncapacitated facility location problem.
Hint: Modify the metric appropriately.

24.6 Notes

The first approximation algorithm for the metric uncapacitated facility location problem, due to Hochbaum [131], achieved an approximation guarantee of $O(\log n)$. The first constant factor approximation algorithm, achieving a guarantee of 3.16, was due to Shmoys, Tardos, and Aardal [247]. It was based on LP-rounding. The result of this chapter is due to Jain and Vazirani [148]. The current best factor is 1.52, due to Mahdian, Ye and Zhang [209]. On the other hand, Guha and Khuller [124] have used the hardness result for set cover (see Chapter 29) to show that achieving a factor better than 1.463 is not possible, unless $\mathbf{NP} \subseteq \mathbf{DTIME}(n^{O(\log\log n)})$.

25 k-Median

The k-median problem differs from the facility location problem in two respects – there is no cost for opening facilities and there is an upper bound, k, on the number of facilities that can be opened. It models the problem of finding a minimum cost clustering, and therefore has numerous applications.

The primal–dual schema works by making judicious local improvements and is not suitable for ensuring a global constraint, such as the constraint in the k-median problem that at most k facilities be opened. We will get around this difficulty by borrowing the powerful technique of Lagrangian relaxation from combinatorial optimization.

Problem 25.1 (Metric k-median) Let G be a bipartite graph with bipartition (F, C), where F is the set of *facilities* and C is the set of *cities*, and let k be a positive integer specifying the number of facilities that are allowed to be opened. Let c_{ij} be the cost of connecting city j to facility i. The connection costs satisfy the triangle inequality. The problem is to find a subset $I \subseteq F, |I| \leq k$, of facilities that should be opened and a function $\phi : \ C \to I$ assigning cities to open facilities in such a way that the total connecting cost is minimized.

25.1 LP-relaxation and dual

The following is an integer program for the k-median problem. The indicator variables y_i and x_{ij} play the same role as in (24.1).

$$
\begin{array}{lll}
\text{minimize} & \displaystyle\sum_{i \in F,\, j \in C} c_{ij} x_{ij} & \\
\text{subject to} & \displaystyle\sum_{i \in F} x_{ij} \geq 1, & j \in C \\
& y_i - x_{ij} \geq 0, & i \in F,\ j \in C \\
& \displaystyle\sum_{i \in F} -y_i \geq -k & \\
& x_{ij} \in \{0, 1\}, & i \in F,\ j \in C \\
& y_i \in \{0, 1\}, & i \in F
\end{array}
\tag{25.1}
$$

The LP-relaxation of this program is:

$$
\begin{aligned}
\text{minimize} \quad & \sum_{i \in F,\, j \in C} c_{ij} x_{ij} & & (25.2)\\
\text{subject to} \quad & \sum_{i \in F} x_{ij} \geq 1, & j \in C & \\
& y_i - x_{ij} \geq 0, & i \in F,\ j \in C & \\
& \sum_{i \in F} -y_i \geq -k & & \\
& x_{ij} \geq 0, & i \in F,\ j \in C & \\
& y_i \geq 0, & i \in F &
\end{aligned}
$$

The dual program is:

$$
\begin{aligned}
\text{maximize} \quad & \sum_{j \in C} \alpha_j - zk & & (25.3)\\
\text{subject to} \quad & \alpha_j - \beta_{ij} \leq c_{ij}, & i \in F,\ j \in C & \\
& \sum_{j \in C} \beta_{ij} \leq z, & i \in F & \\
& \alpha_j \geq 0, & j \in C & \\
& \beta_{ij} \geq 0, & i \in F,\ j \in C & \\
& z \geq 0 & &
\end{aligned}
$$

25.2 The high-level idea

The similarity between the two problems, facility location and k-median, leads to a similarity in their linear programs, which will be exploited as follows. Take an instance of the k-median problem, assign a cost of z for opening each facility, and find optimal solutions to LP (24.2) and LP (24.3), say $(\boldsymbol{x}, \boldsymbol{y})$ and $(\boldsymbol{\alpha}, \boldsymbol{\beta})$, respectively. By the strong duality theorem,

$$\sum_{i \in F,\, j \in C} c_{ij} x_{ij} + \sum_{i \in F} z y_i = \sum_{j \in C} \alpha_j .$$

Now, suppose that the primal solution $(\boldsymbol{x}, \boldsymbol{y})$ happens to open exactly k facilities (fractionally), i.e., $\sum_i y_i = k$. Then, we claim that $(\boldsymbol{x}, \boldsymbol{y})$ and $(\boldsymbol{\alpha}, \boldsymbol{\beta}, z)$ are optimal solutions to LP (25.2) and LP (25.3), respectively. Feasibility is easy to check. Optimality follows by substituting $\sum_i y_i = k$ in the above equality and rearranging terms to show that the primal and dual solutions achieve the same objective function value:

$$\sum_{i\in F,\, j\in C} c_{ij}x_{ij} = \sum_{j\in C} \alpha_j - zk.$$

Let's use this idea, together with Algorithm 24.2 and Theorem 24.7, to obtain a "good" integral solution to LP (25.2). Suppose with a cost of z for opening each facility, Algorithm 24.2, happens to find solutions $(\boldsymbol{x}, \boldsymbol{y})$ and $(\boldsymbol{\alpha}, \boldsymbol{\beta})$, where the primal solution opens exactly k facilities. By Theorem 24.7,

$$\sum_{i\in F,\, j\in C} c_{ij}x_{ij} + 3zk \le 3\sum_{j\in C} \alpha_j.$$

Now, observe that $(\boldsymbol{x}, \boldsymbol{y})$ and $(\boldsymbol{\alpha}, \boldsymbol{\beta}, z)$ are primal (integral) and dual feasible solutions to the k-median problem satisfying

$$\sum_{i\in F,\, j\in C} c_{ij}x_{ij} \le 3(\sum_{j\in C} \alpha_j - zk).$$

Therefore, $(\boldsymbol{x}, \boldsymbol{y})$ is a solution to the k-median problem within thrice the optimal.

Notice that the factor 3 proof given above would not work if less than k facilities were opened; if more than k facilities are opened, the solution is infeasible for the k-median problem. The remaining problem is to find a value of z so that *exactly* k facilities are opened. Several ideas are required for this. The first is the following principle from economics. Taxation is an effective way of controlling the amount of goods coming across a border – raising tariffs will reduce inflow and vice versa. In a similar manner, raising z should reduce the number of facilities opened and vice versa.

It is natural now to seek a modification to Algorithm 24.2 that can find a value of z so that exactly k facilities are opened. This would lead to a factor 3 approximation algorithm. Such a modification is not known. Instead, we present the following strategy which leads to a factor 6 algorithm. For the rest of the discussion, assume that we never encountered a run of the algorithm which resulted in exactly k facilities being opened.

Clearly, when $z = 0$ the algorithm will open all facilities, and when z is very large it will open only one facility. The latter value of z can be picked to be $nc_{\max}$, where $c_{\max}$ is the length of the longest edge. We will conduct a binary search on the interval $[0, nc_{\max}]$ to find z_2 and z_1 for which the algorithm opens $k_2 > k$ and $k_1 < k$ facilities, respectively, and, furthermore, $z_1 - z_2 \le c_{\min}/(12n_c^2)$, where $c_{\min}$ is the length of the shortest nonzero edge. As before, we will adopt the following notation: $n_c = |C|$ and $n_f = |F|$. The total number of vertices $n_c + n_f = n$ and the total number of edges $n_c \times n_f = m$. Let $(\boldsymbol{x}^s, \boldsymbol{y}^s)$ and $(\boldsymbol{x}^l, \boldsymbol{y}^l)$ be the two primal solutions found, with $\sum_{i\in F} y_i^s = k_1$ and $\sum_{i\in F} y_i^l = k_2$ (the superscripts s and l denote "small" and "large," respectively). Further, let $(\boldsymbol{\alpha}^s, \boldsymbol{\beta}^s)$ and $(\boldsymbol{\alpha}^l, \boldsymbol{\beta}^l)$ be the corresponding dual solutions found.

Let $(\boldsymbol{x}, \boldsymbol{y}) = a(\boldsymbol{x}^s, \boldsymbol{y}^s) + b(\boldsymbol{x}^l, \boldsymbol{y}^l)$ be a convex combination of these two solutions, with $ak_1 + bk_2 = k$. Under these conditions, $a = (k_2 - k)/(k_2 - k_1)$ and $b = (k - k_1)/(k_2 - k_1)$. Since $(\boldsymbol{x}, \boldsymbol{y})$ is a feasible (fractional) solution to the facility location problem that opens exactly k facilities, it is also a feasible (fractional) solution to the k-median problem. In this solution each city is connected to at most two facilities.

Lemma 25.2 *The cost of $(\boldsymbol{x}, \boldsymbol{y})$ is within a factor of $(3 + 1/n_c)$ of the cost of an optimal fractional solution to the k-median problem.*

Proof: By Theorem 24.7 we have

$$\sum_{i \in F,\, j \in C} c_{ij} x^s_{ij} \leq 3\Big(\sum_{j \in C} \alpha^s_j - z_1 k_1\Big),$$

and

$$\sum_{i \in F,\, j \in C} c_{ij} x^l_{ij} \leq 3\Big(\sum_{j \in C} \alpha^l_j - z_2 k_2\Big).$$

Since $z_1 > z_2$, $(\boldsymbol{\alpha}^l, \boldsymbol{\beta}^l)$ is a feasible dual solution to the facility location problem even if the cost of facilities is z_1. We would like to replace z_2 with z_1 in the second inequality, at the expense of the increased factor. This is achieved using the upper bound on $z_1 - z_2$ and the fact that $\sum_{i \in F,\, j \in C} c_{ij} x^l_{ij} \geq c_{\min}$. We get

$$\sum_{i \in F,\, j \in C} c_{ij} x^l_{ij} \leq \left(3 + \frac{1}{n_c}\right) \left(\sum_{j \in C} \alpha^l_j - z_1 k_2\right).$$

Adding this inequality multiplied by b with the first inequality multiplied by a gives

$$\sum_{i \in F,\, j \in C} c_{ij} x_{ij} \leq \left(3 + \frac{1}{n_c}\right) \left(\sum_{j \in C} \alpha_j - z_1 k\right),$$

where $\boldsymbol{\alpha} = a\boldsymbol{\alpha}^s + b\boldsymbol{\alpha}^l$. Let $\boldsymbol{\beta} = a\boldsymbol{\beta}^s + b\boldsymbol{\beta}^l$. Observe that $(\boldsymbol{\alpha}, \boldsymbol{\beta}, z_1)$ is a feasible solution to the dual of the k-median problem. The lemma follows. □

In Section 25.3 we give a randomized rounding procedure that obtains an integral solution to the k-median problem from $(\boldsymbol{x}, \boldsymbol{y})$, with a small increase in cost. In Section 25.3.1 we derandomize this procedure.

25.3 Randomized rounding

We give a randomized rounding procedure that produces an integral solution to the k-median problem from $(\boldsymbol{x}, \boldsymbol{y})$. In the process, it increases the cost by a multiplicative factor of $1 + \max(a, b)$.

Let A and B be the sets of facilities opened in the two solutions, $|A| = k_1$ and $|B| = k_2$. For each facility in A, find the closest facility in B – these facilities are not required to be distinct. Let $B' \subset B$ be these facilities. If $|B'| < k_1$, arbitrarily include additional facilities from $B - B'$ into B' until $|B'| = k_1$.

With probability a, open all facilities in A, and with probability $b = 1-a$, open all facilities in B'. In addition, a set of cardinality $k - k_1$ is picked randomly from $B - B'$ and facilities in this set are opened. Notice that each facility in $B - B'$ has a probability of b of being opened. Let I be the set of facilities opened, $|I| = k$.

The function $\phi : C \to I$ is defined as follows. Consider city j and suppose that it is connected to $i_1 \in A$ and $i_2 \in B$ in the two solutions. If $i_2 \in B'$, then one of i_1 and i_2 is opened by the procedure given above, i_1 with probability a and i_2 with probability b. City j is connected to the open facility.

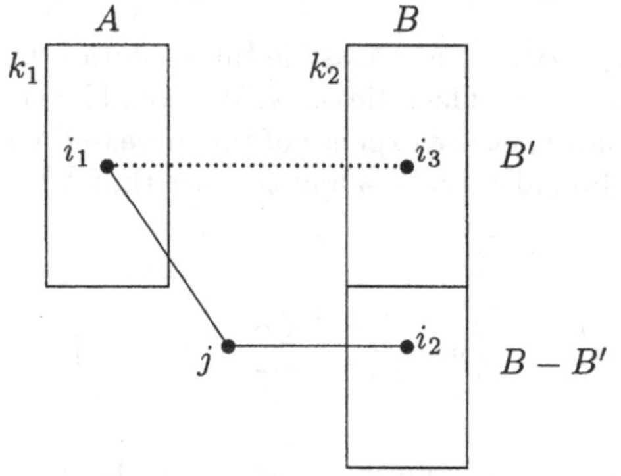

If $i_2 \in B - B'$, let $i_3 \in B'$ be the facility in B that is closest to i_1. City j is connected to i_2 if it is open. Otherwise, it is connected to i_1 if it is open. If neither i_2 or i_1 is open, then j is connected to i_3.

Denote by $\text{cost}(j)$ the connection cost for city j in the fractional solution $(\boldsymbol{x}, \boldsymbol{y})$; $\text{cost}(j) = ac_{i_1 j} + bc_{i_2 j}$.

Lemma 25.3 *The expected connection cost for city j in the integral solution, $\mathbf{E}[c_{\phi(j)j}]$, is $\leq (1 + \max(a, b))\text{cost}(j)$. Moreover, $\mathbf{E}[c_{\phi(j)j}]$ can be efficiently computed.*

Proof: If $i_2 \in B'$, $\mathbf{E}[c_{\phi(j)j}] = ac_{i_1 j} + bc_{i_2 j} = \text{cost}(j)$. Consider the second case, that $i_2 \notin B'$. Now, i_2 is open with probability b. The probability that i_2 is not open and i_1 is open is $(1 - b)a = a^2$, and the probability that both i_2 and i_1 are not open is $(1 - b)(1 - a) = ab$. This gives

$$\mathbf{E}[c_{\phi(j)j}] \leq bc_{i_2j} + a^2c_{i_1j} + abc_{i_3j}.$$

Since i_3 is the facility in B that is closest to i_1, $c_{i_1i_3} \leq c_{i_1i_2} \leq c_{i_1j} + c_{i_2j}$, where the second inequality follows from the triangle inequality. Again, by the triangle inequality, $c_{i_3j} \leq c_{i_1j} + c_{i_1i_3} \leq 2c_{i_1j} + c_{i_2j}$. Therefore,

$$\mathbf{E}[c_{\phi(j)j}] \leq bc_{i_2j} + a^2c_{i_1j} + ab(2c_{i_1j} + c_{i_2j}).$$

Now, $a^2c_{i_1j} + abc_{i_1j} = ac_{i_1j}$. Therefore,

$$\begin{aligned}\mathbf{E}[c_{\phi(j)j}] &\leq (ac_{i_1j} + bc_{i_2j}) + ab(c_{i_1j} + c_{i_2j}) \\ &\leq (ac_{i_1j} + bc_{i_2j})(1 + \max(a, b)).\end{aligned}$$

Clearly, in both cases, $\mathbf{E}[c_{\phi(j)j}]$ is easy to compute. □

Let $(\boldsymbol{x}^k, \boldsymbol{y}^k)$ denote the integral solution obtained to the k-median problem by this randomized rounding procedure. Then,

Lemma 25.4 $$\mathbf{E}\left[\sum_{i\in F,\ j\in C} c_{ij}x_{ij}^k\right] \leq (1 + \max(a, b))\left(\sum_{i\in F,\ j\in C} c_{ij}x_{ij}\right)$$

and, moreover, the expected cost of the solution found can be computed efficiently.

25.3.1 Derandomization

Derandomization follows in a straightforward manner using the method of conditional expectation. First, the algorithm opens the set A with probability a and the set B' with probability $b = 1-a$. Pick A, and compute the expected value if $k - k_1$ facilities are randomly chosen from $B - B'$. Next, do the same by picking B' instead of A. Choose to open the set that gives the smaller expectation.

Second, the algorithm opens a random subset of $k - k_1$ facilities from $B - B'$. For a choice $D \subset B - B'$, $|D| \leq k - k_1$, denote by $\mathbf{E}[D, B - (B' \cup D)]$ the expected cost of the solution if all facilities in D and additionally $k - k_1 - |D|$ facilities are randomly opened from $B - (B' \cup D)$. Since each facility of $B - (B' \cup D)$ is equally likely to be opened, we get

$$\mathbf{E}[D, B - (B' \cup D)] = \frac{1}{|B - (B' \cup D)|} \sum_{i\in B-(B'\cup D)} \mathbf{E}[D \cup \{i\}, B - (B' \cup D \cup \{i\})].$$

This implies that there is an i such that

$$\mathbf{E}[D \cup \{i\}, B - (B' \cup D \cup \{i\})] \leq \mathbf{E}[D, B - (B' \cup D)].$$

Choose such an i and replace D with $D \cup \{i\}$. Notice that the computation of $\mathbf{E}[D \cup \{i\}, B - (B' \cup D \cup \{i\})]$ can be done as in Lemma 25.4.

25.3.2 Running time

It is easy to see that $a \leq 1 - 1/n_c$ (this happens for $k_1 = k - 1$ and $k_2 = n_c$) and $b \leq 1 - 1/k$ (this happens for $k_1 = 1$ and $k_2 = k + 1$). Therefore, $1 + \max(a, b) \leq 2 - 1/n_c$. Altogether, the approximation guarantee is $(2 - 1/n_c)(3 + 1/n_c) < 6$. This procedure can be derandomized using the method of conditional probabilities, as in Section 25.3.1. The binary search will make $O(\log_2(n^3 c_{\max}/c_{\min})) = O(L + \log n)$ probes, where $L = \log(c_{\max}/c_{\min})$. The running time for each probe is dominated by the time taken to run Algorithm 24.2; randomized rounding takes $O(n)$ time and derandomization takes $O(m)$ time. Hence we get

Theorem 25.5 *The algorithm given above achieves an approximation factor of 6 for the k-median problem, and has a running time of $O(m \log m(L + \log(n)))$.*

25.3.3 Tight example

A tight example for the factor 6 k-median algorithm is not known. However, below we give an infinite family of instances which show that the analysis of the randomized rounding procedure cannot be improved.

The two solutions $(\boldsymbol{x}^s, \boldsymbol{y}^s)$ and $(\boldsymbol{x}^l, \boldsymbol{y}^l)$ open one facility, f_0, and $k + 1$ facilities, $f_1, \ldots, f_{k+1}$, respectively. The distance between f_0 and any other f_i is 1, and that between two facilities in the second set is 2. All n cities are at a distance of 1 from f_0, and at a distance of ε from f_{k+1}. The rest of the distances are given by the triangle inequality. The convex combination is constructed with $a = 1/k$ and $b = 1 - 1/k$.

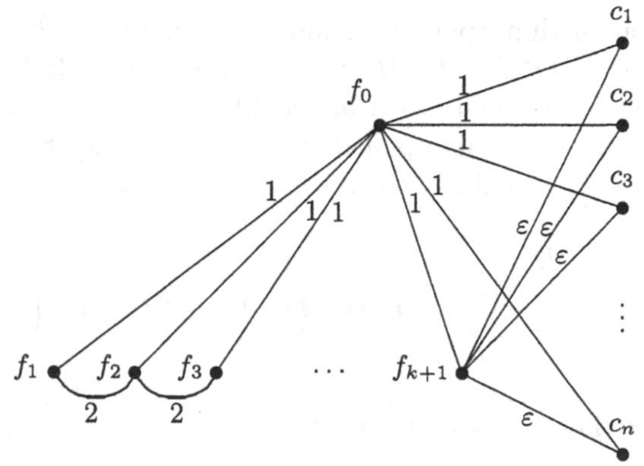

Now, the cost of the convex combination is $an + b\varepsilon n$. Suppose the algorithm picks f_1 as the closest neighbor of f_0. The expected cost of the solutions produced by the randomized rounding procedure is then $n(b\varepsilon + a^2 + ab(2+\varepsilon))$. Letting ε tend to 0, the cost of the convex combination is essentially na and that of the rounded solution is $na(1+b)$.

25.3.4 Integrality gap

The algorithm given above places an upper bound of 6 on the integrality gap of relaxation (25.2). The following example places a lower bound of essentially 2. The graph is a star with $n+1$ vertices and unit cost edges. F consists of all $n+1$ vertices, C consists of all but the center vertex and $k = n-1$. An optimal integral solution is to open facilities at $n-1$ vertices of C and has a cost of 2. Consider the following fractional solution. Open a facility to the extent of $1/(n-1)$ on the center vertex and $(n-2)/(n-1)$ on each vertex of C. This has a cost of $n/(n-1)$, giving a ratio of $2(n-1)/n$.

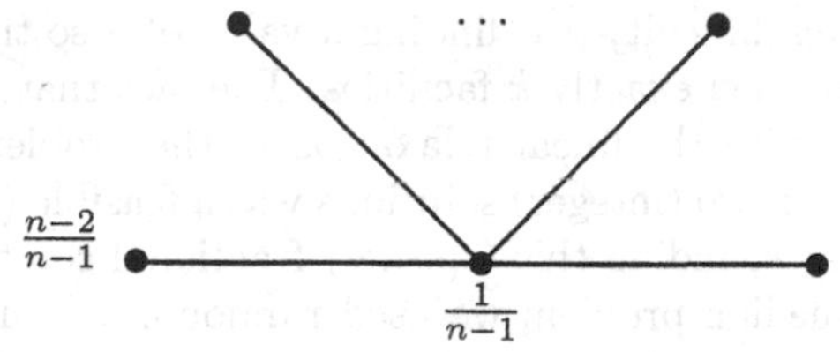

25.4 A Lagrangian relaxation technique for approximation algorithms

In this section we will abstract away the ideas developed above so they may be more widely applicable. First, let us recall the fundamental technique of Lagrangian relaxation from combinatorial optimization. This technique consists of relaxing a constraint by moving it into the objective function, together with an associated Lagrangian multiplier.

Let us apply this relaxation to the constraint, in the k-median IP (25.1), that at most k facilities be opened. Let λ be the Lagrangian multiplier.

$$\text{minimize} \quad \sum_{i\in F,\, j\in C} c_{ij}x_{ij} + \lambda\left(\sum_{i\in F} y_i - k\right) \tag{25.4}$$

$$\text{subject to} \quad \sum_{i\in F} x_{ij} \geq 1, \qquad j \in C$$

$$y_i - x_{ij} \geq 0, \qquad i \in F,\ j \in C$$
$$x_{ij} \in \{0,1\}, \qquad i \in F,\ j \in C$$
$$y_i \in \{0,1\}, \qquad i \in F$$

This is precisely the facility location IP, with the restriction that the cost of each facility is the same, i.e., λ. It contains an additional constant term of $-\lambda k$ in the objective function. We may assume w.l.o.g. that an optimal solution, $(\boldsymbol{x}, \boldsymbol{y})$, to IP (25.1) opens exactly k facilities. Now, $(\boldsymbol{x}, \boldsymbol{y})$ is a feasible solution to IP (25.4) as well, with the same objective function value. Hence, for each value of λ, IP (25.4) is a lower bound on IP (25.1).

We have shown that a Lagrangian relaxation of the k-median problem is the facility location problem. In doing so, the global constraint that at most k facilities be opened has been replaced with a penalty for opening facilities, the penalty being the Lagrangian multiplier. (See Exercise 25.4 for another application of this idea.)

The next important observation was to notice that in the facility location approximation algorithm, Theorem 24.7, the duals pay one-for-one for the cost of opening facilities, i.e., with approximation factor 1. (See Exercise 22.9 for another such algorithm.)

The remaining difficulty was finding a value of λ so that the facility location algorithm opened exactly k facilities. The fact that the facility location algorithm works with the linear relaxation of the problem helped. The convex combination of two (integer) solutions was a feasible (fractional) solution. The last step was rounding this (special) fractional solution into an integral one. For the k-median problem we used randomized rounding (see Exercise 25.4 for a different rounding procedure).

25.5 Exercises

25.1 (Lin and Vitter [195]) Consider the general k-median problem in which the connection costs are not required to satisfy the triangle inequality. Give a reduction from the set cover problem to show that approximating this problem is as hard as approximating set cover, and therefore cannot be done with a factor better than $O(\log n)$, unless $\mathbf{NP} \subseteq \mathbf{DTIME}(n^{O(\log\log n)})$ (see Chapter 29).

25.2 Obtain the dual of LP-relaxation to (25.4). (The constant term in the objective function will simply carry over.) How does it relate with the dual of the k-median LP?

25.3 Use the Lagrangian relaxation technique to give a constant factor approximation algorithm for the following common generalization of the facility location and k-median problems. Consider the uncapacitated facility location problem with the additional constraint that at most k facilities can be opened. This is a common generalization of the two problems solved in this paper: if k is made n_f, we get the first problem, and if the facility costs are set to zero, we get the second problem.

25.4 (Garg [100] and Chudak, Roughgarden, and Williamson [49]) Consider the following variant of the metric Steiner tree problem.

Problem 25.6 (Metric k-MST) We are given a complete undirected graph $G = (V, E)$, a special vertex $r \in V$, a positive integer k, and a function cost : $E \to \mathbf{Q}^+$ satisfying the triangle inequality. The problem is to find a minimum cost tree containing exactly k vertices, including r.

We will develop a factor 5 algorithm for this problem.

1. Observe that a Lagrangian relaxation of this problem is the prize-collecting Steiner tree problem, Problem 22.12, stated in Exercise 22.9.
2. Observe that the approximation algorithm for the latter problem, given in Exercise 22.9, pays for the penalties one-for-one with the dual, i.e., with an approximation factor of 1.
3. Use the prize-collecting algorithm as a subroutine to obtain two trees, T_1 and T_2, for very close values of the penalty, containing k_1 and k_2 vertices, with $k_1 < k < k_2$. Obtain a convex combination of these solutions, with multipliers α_1 and α_2.
4. We may assume that every vertex in G is at a distance of $\leq$ OPT from r. (Use the idea behind parametric pruning, introduced in Chapter 5. The parameter t is the length of the longest edge used by the optimal solution, which is clearly a lower bound on OPT. For each value of t, instance $G(t)$ is obtained by restricting G to vertices that are within a distance of t of r. The algorithm is run on each graph of this family, and the best tree is output.) Consider the following procedure for rounding the convex combination. If $\alpha_2 \geq 1/2$, then $\text{cost}(T_2) \leq 4 \cdot \text{OPT}$; remove $k_2 - k$ vertices from T_2. Otherwise, double every edge of T_2, find an Euler tour, and shortcut the tour to a cycle containing only those vertices that are in T_2 and not in T_1 (i.e., at most $k_2 - k_1$ vertices). Pick the cheapest path of length $k - k_1 - 1$ from this cycle, and connect it by means of an edge to vertex r in T_1. The resulting tree has exactly k vertices. Show that the cost of this tree is $\leq 5 \cdot \text{OPT}$.
   Hint: Use the fact that $\alpha_2 = (k - k_1)/(k_2 - k_1)$.

25.5 Let us apply the Lagrangian relaxation technique to the following linear program; for simplicity, we have not imposed nonnegativity explicitly.

$$\begin{array}{ll} \text{minimize} & \boldsymbol{c}^T \boldsymbol{x} \\ \text{subject to} & \boldsymbol{A}\boldsymbol{x} = \boldsymbol{b} \end{array} \qquad (25.5)$$

where $\boldsymbol{A}$ is an $m \times n$ matrix. Suppose this LP attains its optimum at $\boldsymbol{x} = \boldsymbol{a}$, and let OPT denote the optimal objective function value.

We will define $\boldsymbol{y}$ to be an m-dimensional Lagrange multiplier and will move all the constraints into the objective function:

$$\min_{\boldsymbol{x}} \left(\boldsymbol{c}^T \boldsymbol{x} - \boldsymbol{y}^T (\boldsymbol{A}\boldsymbol{x} - \boldsymbol{b})\right).$$

For any $\boldsymbol{y}$, the above expression is a lower bound on OPT. To see this, simply substitute $\boldsymbol{x} = \boldsymbol{a}$. Hence, the following is also a lower bound on OPT:

$$\max_{\boldsymbol{y}} \min_{\boldsymbol{x}} \left(\boldsymbol{c}^T \boldsymbol{x} - \boldsymbol{y}^T (\boldsymbol{A}\boldsymbol{x} - \boldsymbol{b})\right) = \max_{\boldsymbol{y}} \left(\left(\min_{\boldsymbol{x}} (\boldsymbol{c}^T - \boldsymbol{y}^T \boldsymbol{A})\boldsymbol{x}\right) + \boldsymbol{y}^T \boldsymbol{b}\right).$$

If $\boldsymbol{y}$ does not satisfy $\boldsymbol{A}^T \boldsymbol{y} = \boldsymbol{c}$, then by a suitable choice of $\boldsymbol{x}$, the lower bound given by this expression can be made as small as desired and therefore meaningless. Meaningful lower bounds arise only if we insist that $\boldsymbol{A}^T \boldsymbol{y} = \boldsymbol{c}$. But then we get the following LP:

$$\begin{aligned} &\text{maximize} && \boldsymbol{y}^T \boldsymbol{b} \\ &\text{subject to} && \boldsymbol{A}^T \boldsymbol{y} = \boldsymbol{c} \end{aligned} \tag{25.6}$$

Notice that this is the dual of LP (25.5)! Hence, the Lagrangian relaxation of a linear program is simply its dual and is therefore tight.

Obtain the Lagrangian relaxation of the following LP:

$$\begin{aligned} &\text{minimize} && \boldsymbol{c}^T \boldsymbol{x} \\ &\text{subject to} && \boldsymbol{A}\boldsymbol{x} \geq \boldsymbol{b} \\ &&& \boldsymbol{x} \geq 0 \end{aligned} \tag{25.7}$$

25.6 (Jain and Vazirani [148]) Consider the ℓ_2^2 *clustering* problem. Given a set of n points $S = \{v_1, \ldots, v_n\}$ in $\mathbf{R}^d$ and a positive integer k, the problem is to find a minimum cost k-clustering, i.e., to find k points, called *centers*, $f_1, \ldots, f_k \in \mathbf{R}^d$, so as to minimize the sum of squares of distances from each point v_i to its closest center. This naturally defines a partitioning of the n points into k clusters. Give a constant factor approximation algorithm for this problem.
Hint: First show that restricting the centers to be a subset S increases the cost of the optimal solution by a factor of at most 2. Apply the solution of Exercise 24.6 to this modified problem.

25.7 (Korupolu, Plaxton, and Rajaraman [183] and Arya et al. [16]) For a set S of k facilities, define cost(S) to be the total cost of connecting each city to its closest facility in S. Define a *swap* to be the process of replacing one facility in S by a facility from $\overline{S}$. A natural algorithm for metric k-median, based on local search, is: Start with an arbitrary set S of k facilities. In each iteration, check if there is a swap that leads to a lower cost solution. If so, execute any such swap and go to the next iteration. If not, halt. The terminating solution is said to be *locally optimal.*

Let $G = \{o_1, \ldots, o_k\}$ be an optimal solution and $L = \{s_1, \ldots, s_k\}$ be a locally optimal solution. This exercise develops a proof showing $\text{cost}(L) \leq 5 \cdot \text{cost}(G)$, as well as a constant factor approximation algorithm.

1. For $o \in G$, let $N_G(o)$ denote the set of cities connected to facility o in the optimal solution. Similarly, for $s \in L$, let $N_L(s)$ denote the set of cities connected to facility s in the locally optimal solution. Say that $s \in L$ *captures* $o \in G$ if $|N_G(o) \cap N_L(s)| > |N_G(o)|/2$. Clearly, each $o \in G$ is captured by at most one facility in L. In this part let us make the simplifying assumption that each facility $s \in L$ captures a unique facility in G. Assume that the facilities are numbered so that s_i captures o_i, for $1 \leq i \leq k$. Use the fact that for $1 \leq i \leq k$, $\text{cost}(L + o_i - s_i) \geq \text{cost}(L)$ to show that $\text{cost}(L) \leq 3 \cdot \text{cost}(G)$.
   Hint: $\text{cost}(L + o_i - s_i)$ is bounded by the cost of the following solution: The cities in $\overline{N_L(s_i) \cup N_G(o_i)}$ are connected as in the locally optimal solution. Those in $N_G(o_i)$ are connected to facility o_i. Cities in $N_L(s_i) - N_G(o_i)$ are connected to facilities in $L - s_i$ using "3 hops" in such a way that each connecting edge of G and each connecting edge of L is used at most once in the union of all these hops.
2. Show that without the simplifying assumption of the previous part, $\text{cost}(L) \leq 5 \cdot \text{cost}(G)$.
   Hint: Consider k appropriately chosen swaps so that each facility $o \in G$ is swapped in exactly once and each facility $s \in L$ is swapped out at most twice.
3. Strengthen the condition for swapping so as to obtain, for any $\varepsilon > 0$ a factor $5 + \varepsilon$ algorithm running in time polynomial in $1/\varepsilon$ and the size of the instance.

25.6 Notes

The first approximation algorithm, achieving a factor of $O(\log n \log \log n)$, was given by Bartal [22]. The first constant factor approximation algorithm for the k-median problem, achieving a guarantee of $6\frac{2}{3}$, was given by Charikar, Guha, Tardos, and Shmoys [40], using ideas from Lin and Vitter [196]. This algorithm used LP-rounding. The results of this chapter are due to Jain and Vazirani [148]. The current best factor is $3 + 2/p$, with a running time of $O(n^p)$, due to Arya et al. [16]. This is a local search algorithm that swaps p facilities at a time (see Exercise 25.7 for the algorithm for $p = 1$).

The example of Section 25.3.4 is due to Jain, Mahdian, and Saberi [145]. The best upper bound on the integrality gap of relaxation (25.2) is 4, due to Charikar and Guha [39]. For a factor 2 approximation algorithm for the ℓ_2^2 clustering problem (Exercise 25.6), see Drineas, Kannan, Frieze, Vempala, and Vinay [66].

26 Semidefinite Programming

In the previous chapters of Part II of this book we have shown how linear programs provide a systematic way of placing a good lower bound on OPT (assuming a minimization problem), for numerous **NP**-hard problems. As stated earlier, this is a key step in the design of an approximation algorithm for an **NP**-hard problem. It is natural, then, to ask if there are other widely applicable ways of doing this.

In this chapter we provide another class of relaxations, called *vector programs.* These serve as relaxations for several **NP**-hard problems, in particular, for problems that can be expressed as *strict quadratic programs* (see Section 26.1 for a definition). Vector programs are equivalent to a powerful and well-studied generalization of linear programs, called semidefinite programs. Semidefinite programs, and consequently vector programs, can be solved within an additive error of ε, for any $\varepsilon > 0$, in time polynomial in n and $\log(1/\varepsilon)$, using the ellipsoid algorithm (see Section 26.3).

We will illustrate the use of vector programs by deriving a 0.87856 factor algorithm for the following problem (see Exercises 2.1 and 16.6 for a factor 1/2 algorithm).

Problem 26.1 (Maximum cut (MAX-CUT)) Given an undirected graph $G = (V, E)$, with edge weights $w : E \rightarrow \mathbf{Q}^+$, find a partition $(S, \overline{S})$ of V so as to maximize the total weight of edges in this cut, i.e., edges that have one endpoint in S and one endpoint in $\overline{S}$.

26.1 Strict quadratic programs and vector programs

A *quadratic program* is the problem of optimizing (minimizing or maximizing) a quadratic function of integer valued variables, subject to quadratic constraints on these variables. If each monomial in the objective function, as well as in each of the constraints, is of degree 0 (i.e., is a constant) or 2, then we will say that this is a *strict quadratic program.*

Let us give a strict quadratic program for MAX-CUT. Let y_i be an indicator variable for vertex v_i which will be constrained to be either $+1$ or -1. The partition $(S, \overline{S})$ will be defined as follows. $S = \{v_i \mid y_i = 1\}$ and $\overline{S} = \{v_i \mid y_i = -1\}$. If v_i and v_j are on opposite sides of this partition, then $y_i y_j = -1$, and edge (v_i, v_j) contributes w_{ij} to the objective function. On the other hand, if they are on the same side, then $y_i y_j = 1$, and edge (v_i, v_j) makes no contribution. Hence, an optimal solution to this program is a maximum cut in G.

$$\begin{aligned} \text{maximize} \quad & \frac{1}{2} \sum_{1 \le i < j \le n} w_{ij}(1 - y_i y_j) && \\ \text{subject to} \quad & y_i^2 = 1, && v_i \in V \\ & y_i \in \mathbf{Z}, && v_i \in V \end{aligned} \tag{26.1}$$

We will relax this program to a vector program. A *vector program* is defined over n *vector variables* in $\mathbf{R}^n$, say $\boldsymbol{v}_1, \ldots, \boldsymbol{v}_n$, and is the problem of optimizing (minimizing or maximizing) a linear function of the inner products $\boldsymbol{v}_i \cdot \boldsymbol{v}_j, 1 \le i \le j \le n$, subject to linear constraints on these inner products. Thus, a vector program can be thought of as being obtained from a linear program by replacing each variable with an inner product of a pair of these vectors.

A strict quadratic program over n integer variables defines a vector program over n vector variables in $\mathbf{R}^n$ as follows. Establish a correspondence between the n integer variables and the n vector variables, and replace each degree 2 term with the corresponding inner product. For instance, the term $y_i y_j$ in (26.1) is replaced with $\boldsymbol{v}_i \cdot \boldsymbol{v}_j$. In this manner, we obtain the following vector program for MAX-CUT.

$$\begin{aligned} \text{maximize} \quad & \frac{1}{2} \sum_{1 \le i < j \le n} w_{ij}(1 - \boldsymbol{v}_i \cdot \boldsymbol{v}_j) && \\ \text{subject to} \quad & \boldsymbol{v}_i \cdot \boldsymbol{v}_i = 1, && v_i \in V \\ & \boldsymbol{v}_i \in \mathbf{R}^n, && v_i \in V \end{aligned} \tag{26.2}$$

Because of the constraint $\boldsymbol{v}_i \cdot \boldsymbol{v}_i = 1$, the vectors $\boldsymbol{v}_1, \ldots, \boldsymbol{v}_n$ are constrained to lie on the n-dimensional sphere, S_{n-1}. Any feasible solution to (26.1) yields a solution to (26.2) having the same objective function value, by assigning the vector $(y_i, 0, \ldots, 0)$ to $\boldsymbol{v}_i$. (Notice that under this assignment, $\boldsymbol{v}_i \cdot \boldsymbol{v}_j$ is simply $y_i y_j$.) Therefore, the vector program (26.2) is a relaxation of the strict quadratic program (26.1). Clearly, this holds in general as well; the vector program corresponding to a strict quadratic program is a relaxation of the quadratic program.

Interestingly enough, vector programs are approximable to any desired degree of accuracy in polynomial time, and thus relaxation (26.2) provides an upper bound on OPT for MAX-CUT. To show this, we need to recall some interesting and powerful properties of positive semidefinite matrices.

Remark 26.2 Vector programs do not always come about as relaxations of strict quadratic programs. Exercise 26.13 gives an **NP**-hard problem that has vector program relaxation; however, we do not know of a strict quadratic program for it.

26.2 Properties of positive semidefinite matrices

Let $\boldsymbol{A}$ be a real, symmetric $n \times n$ matrix. Then $\boldsymbol{A}$ has real eigenvalues and has n linearly independent eigenvectors (even if the eigenvalues are not distinct). We will say that $\boldsymbol{A}$ is *positive semidefinite* if

$$\forall \boldsymbol{x} \in \mathbf{R}^n, \ \boldsymbol{x}^T \boldsymbol{A} \boldsymbol{x} \geq 0.$$

We will use the following two equivalent conditions crucially. We provide a proof sketch for completeness.

Theorem 26.3 *Let $\boldsymbol{A}$ be a real symmetric $n \times n$ matrix. Then, the following are equivalent:*

1. $\forall \boldsymbol{x} \in \mathbf{R}^n$, $\boldsymbol{x}^T \boldsymbol{A} \boldsymbol{x} \geq 0$.
2. *All eigenvalues of $\boldsymbol{A}$ are nonnegative.*
3. *There is an $n \times n$ real matrix $\boldsymbol{W}$, such that $\boldsymbol{A} = \boldsymbol{W}^T \boldsymbol{W}$.*

Proof: $(1 \Rightarrow 2)$: Let λ be an eigenvalue of $\boldsymbol{A}$, and let $\boldsymbol{v}$ be a corresponding eigenvector. Therefore, $\boldsymbol{A}\boldsymbol{v} = \lambda \boldsymbol{v}$. Pre-multiplying by $\boldsymbol{v}^T$ we get $\boldsymbol{v}^T \boldsymbol{A} \boldsymbol{v} = \lambda \boldsymbol{v}^T \boldsymbol{v}$. Now, by (1), $\boldsymbol{v}^T \boldsymbol{A} \boldsymbol{v} \geq 0$. Therefore, $\lambda \boldsymbol{v}^T \boldsymbol{v} \geq 0$. Since $\boldsymbol{v}^T \boldsymbol{v} > 0$, $\lambda \geq 0$.

$(2 \Rightarrow 3)$: Let $\lambda_1, \ldots, \lambda_n$ be the n eigenvalues of $\boldsymbol{A}$, and $\boldsymbol{v}_1, \ldots, \boldsymbol{v}_n$ be the corresponding complete collection of orthonormal eigenvectors. Let $\mathbf{Q}$ be the matrix whose columns are $\boldsymbol{v}_1, \ldots, \boldsymbol{v}_n$, and $\boldsymbol{\Lambda}$ be the diagonal matrix with entries $\lambda_1, \ldots, \lambda_n$. Since for each i, $\boldsymbol{A}\boldsymbol{v}_i = \lambda_i \boldsymbol{v}_i$, we have $\boldsymbol{A}\mathbf{Q} = \mathbf{Q}\boldsymbol{\Lambda}$. Since $\mathbf{Q}$ is orthogonal, i.e., $\mathbf{Q}\mathbf{Q}^T = I$, we get that $\mathbf{Q}^T = \mathbf{Q}^{-1}$. Therefore,

$$\boldsymbol{A} = \mathbf{Q}\boldsymbol{\Lambda}\mathbf{Q}^T.$$

Let $\boldsymbol{D}$ be the diagonal matrix whose diagonal entries are the positive square roots of $\lambda_1, \ldots, \lambda_n$ (by (2), $\lambda_1, \ldots, \lambda_n$ are nonnegative, and thus their square roots are real). Then, $\boldsymbol{\Lambda} = \boldsymbol{D}\boldsymbol{D}^T$. Substituting, we get

$$\boldsymbol{A} = \mathbf{Q}\boldsymbol{D}\boldsymbol{D}^T\mathbf{Q}^T = (\mathbf{Q}\boldsymbol{D})(\mathbf{Q}\boldsymbol{D})^T.$$

Now, (3) follows by letting $\boldsymbol{W} = (\mathbf{Q}\boldsymbol{D})^T$.

$(3 \Rightarrow 1)$: For any

$$\boldsymbol{x} \in \mathbf{R}^n, \ \boldsymbol{x}^T \boldsymbol{A} \boldsymbol{x} = \boldsymbol{x}^T \boldsymbol{W}^T \boldsymbol{W} \boldsymbol{x} = (\boldsymbol{W}\boldsymbol{x})^T (\boldsymbol{W}\boldsymbol{x}) \geq 0. \qquad \square$$

Using Cholesky decomposition (see Section 26.7), a real symmetric matrix can be decomposed, in polynomial time, as $\boldsymbol{A} = \boldsymbol{U}\boldsymbol{\Lambda}\boldsymbol{U}^T$, where $\boldsymbol{\Lambda}$ is a diagonal matrix whose diagonal entries are the eigenvalues of $\boldsymbol{A}$. Now $\boldsymbol{A}$ is positive semidefinite iff all the entries of $\boldsymbol{\Lambda}$ are nonnegative, thus giving a polynomial time test for positive semidefiniteness. The decomposition $\boldsymbol{W}\boldsymbol{W}^T$ is not polynomial time computable because in general it may contain irrational entries. However, it can be approximated to any desired degree by approximating the square roots of the entries of $\boldsymbol{\Lambda}$. In the rest of this chapter we will assume that we have an exact decomposition, since the inaccuracy resulting from an approximate decomposition can be absorbed into the approximation factor (see Exercise 26.6).

It is easy to see that the sum of two $n \times n$ positive semidefinite matrices is also positive semidefinite (e.g., using characterization (1) of Theorem 26.3). This is also true of any convex combination of such matrices.

26.3 The semidefinite programming problem

Let $\boldsymbol{Y}$ be an $n \times n$ matrix of real valued variables whose (i, j)th entry is y_{ij}. The problem of maximizing a linear function of the y_{ij}'s, subject to linear constraints on them, and the additional constraint that $\boldsymbol{Y}$ be symmetric and positive semidefinite, is called the *semidefinite programming problem.*

Let us introduce some notation to state this formally. Denote by $\mathbf{R}^{n\times n}$ the space of $n \times n$ real matrices. Recall that the *trace* of a matrix $\boldsymbol{A} \in \mathbf{R}^{n\times n}$ is the sum of its diagonal entries and is denoted by $\mathrm{tr}(\boldsymbol{A})$. The Frobenius inner product of matrices $\boldsymbol{A}, \boldsymbol{B} \in \mathbf{R}^{n\times n}$, denoted $\boldsymbol{A} \bullet \boldsymbol{B}$, is defined to be

$$\boldsymbol{A} \bullet \boldsymbol{B} = \mathrm{tr}(\boldsymbol{A}^T\boldsymbol{B}) = \sum_{i=1}^{n}\sum_{j=1}^{n} a_{ij}b_{ij},$$

where a_{ij} and b_{ij} are the (i, j)th entries of $\boldsymbol{A}$ and $\boldsymbol{B}$, respectively. Let M_n denote the cone of symmetric $n \times n$ real matrices. For $\boldsymbol{A} \in M_n$, $\boldsymbol{A} \succeq 0$ denotes the fact that matrix $\boldsymbol{A}$ is positive semidefinite.

Let $\boldsymbol{C}, \boldsymbol{D}_1, \ldots, \boldsymbol{D}_k \in M_n$ and $d_1, \ldots, d_k \in \mathbf{R}$. Following is a statement of the general semidefinite programming problem. Let us denote it by $\mathcal{S}$.

$$\begin{aligned} \text{maximize} \quad & \boldsymbol{C} \bullet \boldsymbol{Y} && (26.3)\\ \text{subject to} \quad & \boldsymbol{D}_i \bullet \boldsymbol{Y} = d_i, \quad 1 \le i \le k \\ & \boldsymbol{Y} \succeq 0, \\ & \boldsymbol{Y} \in M_n. \end{aligned}$$

Observe that if $\boldsymbol{C}, \boldsymbol{D}_1, \ldots, \boldsymbol{D}_k$ are all diagonal matrices, this is simply a linear programming problem. As in the case of linear programs, it is easy to see that allowing linear inequalities, in addition to equalities, does not make the problem more general.

Let us call a matrix in $\mathbf{R}^{n\times n}$ satisfying all the constraints of $\mathcal{S}$ a *feasible solution.* Since a convex combination of positive semidefinite matrices is positive semidefinite, it is easy to see that the set of feasible solutions is *convex*, i.e., if $\boldsymbol{A} \in \mathbf{R}^{n\times n}$ and $\boldsymbol{B} \in \mathbf{R}^{n\times n}$ are feasible solutions then so is any convex combination of these solutions.

Let $\boldsymbol{A} \in \mathbf{R}^{n\times n}$ be an infeasible point. Let $\boldsymbol{C} \in \mathbf{R}^{n\times n}$. A hyperplane $\boldsymbol{C} \bullet \boldsymbol{Y} \leq b$ is called a *separating hyperplane for* $\boldsymbol{A}$ if all feasible points satisfy it and point $\boldsymbol{A}$ does not satisfy it. In the next theorem we show how to find a separating hyperplane in polynomial time. As a consequence, for any $\varepsilon > 0$, semidefinite programs can be solved within an additive error of ε, in time polynomial in n and $\log(1/\varepsilon)$, using the ellipsoid algorithm (see Section 26.7 for more efficient methods).

Theorem 26.4 *Let $\mathcal{S}$ be a semidefinite programming problem, and $\boldsymbol{A}$ be a point in $\mathbf{R}^{n\times n}$. We can determine, in polynomial time, whether $\boldsymbol{A}$ is feasible for $\mathcal{S}$ and, if it is not, find a separating hyperplane.*

Proof: Testing for feasibility involves ensuring that $\boldsymbol{A}$ is symmetric and positive semidefinite and that it satisfies all the linear constraints. By remarks made in Section 26.2, this can be done in polynomial time. If $\boldsymbol{A}$ is infeasible, a separating hyperplane is obtained as follows.

- If $\boldsymbol{A}$ is not symmetric, $a_{ij} > a_{ji}$ for some i, j. Then $y_{ij} \leq y_{ji}$ is a separating hyperplane.
- If $\boldsymbol{A}$ is not positive semidefinite, then it has a negative eigenvalue, say λ. Let $\boldsymbol{v}$ be the corresponding eigenvector. Now $(\boldsymbol{v}\boldsymbol{v}^T) \bullet \boldsymbol{Y} = \boldsymbol{v}^T\boldsymbol{Y}\boldsymbol{v} \geq 0$ is a separating hyperplane.
- If any of the linear constraints is violated, it directly yields a separating hyperplane.

□

Next, let us show that vector programs are equivalent to semidefinite programs, thereby showing that the former can be solved efficiently to any desired degree of accuracy. Let $\mathcal{V}$ be a vector program on n n-dimensional vector variables $\boldsymbol{v}_1, \ldots, \boldsymbol{v}_n$. Define the corresponding semidefinite program, $\mathcal{S}$, over n^2 variables $y_{ij}, 1 \leq i, j \leq n$, as follows. Replace each inner product $\boldsymbol{v}_i \cdot \boldsymbol{v}_j$ occurring in $\mathcal{V}$ by the variable y_{ij}. The objective function and constraints are now linear in the y_{ij}'s. Additionally, require that matrix $\boldsymbol{Y}$, whose (i, j)th entry is y_{ij}, be symmetric and positive semidefinite.

Lemma 26.5 *Vector program $\mathcal{V}$ is equivalent to semidefinite program $\mathcal{S}$.*

Proof: We will show that corresponding to each feasible solution to $\mathcal{V}$, there is a feasible solution to $\mathcal{S}$ of the same objective function value, and vice versa. Let $\boldsymbol{a}_1, \ldots, \boldsymbol{a}_n$ be a feasible solution to $\mathcal{V}$. Let W be the matrix whose columns are $\boldsymbol{a}_1, \ldots, \boldsymbol{a}_n$. Then, it is easy to see that $\boldsymbol{A} = \boldsymbol{W}^T\boldsymbol{W}$ is a feasible solution to $\mathcal{S}$ having the same objective function value.

For the other direction, let $\boldsymbol{A}$ be a feasible solution to $\mathcal{S}$. By Theorem 26.3, there is an $n \times n$ matrix $\boldsymbol{W}$ such that $\boldsymbol{A} = \boldsymbol{W}^T\boldsymbol{W}$. Let $\boldsymbol{a}_1, \ldots, \boldsymbol{a}_n$ be the columns of $\boldsymbol{W}$. Then, it is easy to see that $\boldsymbol{a}_1, \ldots, \boldsymbol{a}_n$ is a feasible solution to $\mathcal{V}$ having the same objective function value. □

Finally, we give the semidefinite programming relaxation to MAX-CUT that is equivalent to vector program (26.2).

$$\begin{aligned} \text{maximize} \quad & \frac{1}{2} \sum_{1 \le i < j \le n} w_{ij}(1 - y_i y_j) && (26.4) \\ \text{subject to} \quad & y_i^2 = 1, \qquad v_i \in V \\ & \boldsymbol{Y} \succeq 0, \\ & \boldsymbol{Y} \in M_n. \end{aligned}$$

26.4 Randomized rounding algorithm

We now present the algorithm for MAX-CUT. For convenience, let us assume that we have an optimal solution to the vector program (26.2). The slight inaccuracy in solving it can be absorbed into the approximation factor (see Exercise 26.6). Let $\boldsymbol{a}_1, \ldots, \boldsymbol{a}_n$ be an optimal solution, and let OPT_v denote its objective function value. These vectors lie on the n-dimensional unit sphere S_{n-1}. We need to obtain a cut $(S, \overline{S})$ whose weight is a large fraction of OPT_v.

Let θ_{ij} denote the angle between vectors $\boldsymbol{a}_i$ and $\boldsymbol{a}_j$. The contribution of this pair of vectors to OPT_v is

$$\frac{w_{ij}}{2}(1 - \cos\theta_{ij}).$$

Clearly, the closer θ_{ij} is to π, the larger this contribution will be. In turn, we would like vertices v_i and v_j to be separated if θ_{ij} is large. The following method accomplishes precisely this. Pick $\boldsymbol{r}$ to be a uniformly distributed vector on the unit sphere S_{n-1}, and let $S = \{v_i \mid \boldsymbol{a}_i \cdot \boldsymbol{r} \ge 0\}$.

Lemma 26.6 $\quad \mathbf{Pr}[v_i \text{ and } v_j \text{ are separated}] = \dfrac{\theta_{ij}}{\pi}.$

Proof: Project $\boldsymbol{r}$ onto the plane containing $\boldsymbol{a}_i$ and $\boldsymbol{a}_j$. Now, vertices v_i and v_j will be separated iff the projection lies in one of the two arcs of angle θ_{ij} shown below.

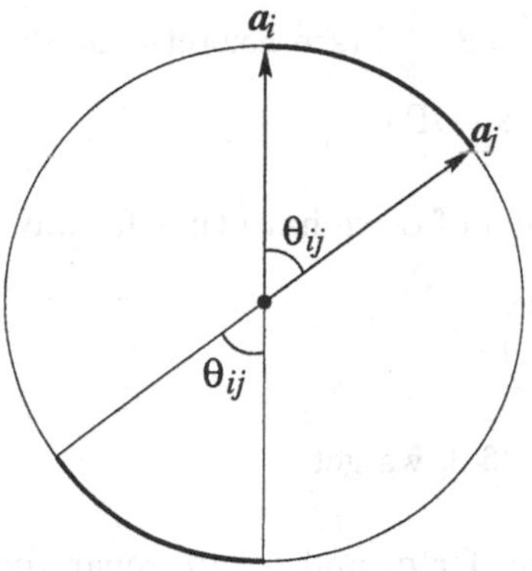

Since r has been picked from a spherically symmetric distribution, its projection will be a random direction on this plane. The lemma follows. □

The next lemma shows how to generate vectors that are uniformly distributed on the unit sphere S_{n-1}.

Lemma 26.7 *Let $x_1, \ldots, x_n$ be picked independently from the normal distribution with mean 0 and unit standard deviation. Let $d = (x_1^2 + \ldots + x_n^2)^{1/2}$. Then, $(x_1/d, \ldots, x_n/d)$ is a random vector on the unit sphere S_{n-1}.*

Proof: Consider the vector $r = (x_1, \ldots, x_n)$. The distribution function for r has density

$$f(y_1, \ldots, y_n) = \prod_{i=1}^{n} \frac{1}{\sqrt{2\pi}} e^{-y_i^2/2} = \frac{1}{(2\pi)^{n/2}} e^{-\frac{1}{2}\sum_i y_i^2}.$$

Notice that the density function depends only on the distance of the point from the origin. Therefore, the distribution of r is spherically symmetric. Hence, dividing by the length of r, i.e., d, we get a random vector on S_{n-1}. □

The algorithm is summarized below.

Algorithm 26.8 (MAX-CUT)

1. Solve vector program (26.2). Let $a_1, \ldots, a_n$ be an optimal solution.
2. Pick r to be a uniformly distributed vector on the unit sphere S_{n-1}.
3. Let $S = \{v_i \mid a_i \cdot r \geq 0\}$.

Let W be the random variable denoting the weight of edges in the cut picked by Algorithm 26.8, and let

$$\alpha = \frac{2}{\pi} \min_{0 \leq \theta \leq \pi} \frac{\theta}{1 - \cos\theta}.$$

One can show that $\alpha > 0.87856$ (see Exercise 26.3).

Lemma 26.9 $\mathbf{E}[W] \geq \alpha \cdot \mathrm{OPT}_v$.

Proof: By the definition of α we have that for any θ, $0 \leq \theta \leq \pi$,

$$\frac{\theta}{\pi} \geq \alpha \left(\frac{1 - \cos \theta}{2} \right). \tag{26.5}$$

Using this and Lemma 26.6, we get

$$\mathbf{E}[W] = \sum_{1 \leq i < j \leq n} w_{ij} \mathbf{Pr}[\boldsymbol{a}_i \text{ and } \boldsymbol{a}_j \text{ are separated}]$$

$$= \sum_{1 \leq i < j \leq n} w_{ij} \frac{\theta_{ij}}{\pi} \geq \alpha \cdot \sum_{1 \leq i < j \leq n} \frac{1}{2} w_{ij} (1 - \cos \theta_{ij}) = \alpha \cdot \mathrm{OPT}_v.$$

□

Let us define the *integrality gap* for relaxation (26.2) to be

$$\inf_I \frac{\mathrm{OPT}(I)}{\mathrm{OPT}_v(I)},$$

where the infimum is over all instances I of MAX-CUT.

Corollary 26.10 *The integrality gap for relaxation (26.2) is at least* $\alpha > 0.87856$.

Theorem 26.11 *There is a randomized approximation algorithm for MAX-CUT achieving an approximation factor of* 0.87856.

Proof: Let us first obtain a "high probability" statement using the bound on expectation established in Lemma 26.9. Let T denote the sum of weights of all edges in G, and define a so that $\mathbf{E}[W] = aT$. Let

$$p = \mathbf{Pr}[W < (1 - \varepsilon) aT],$$

where $\varepsilon > 0$ is a constant. Since the random variable W is always bounded by T, we get

$$aT \leq p(1 - \varepsilon) aT + (1 - p) T.$$

Therefore,

$$p \leq \frac{1 - a}{1 - a + a\varepsilon}.$$

Now,

$$T \geq \mathbf{E}[W] = aT \geq \alpha \cdot \mathrm{OPT}_v \geq \alpha \cdot \mathrm{OPT} \geq \frac{\alpha T}{2},$$

where the last inequality follows from the fact that $\mathrm{OPT} \geq T/2$ (see Exercise 2.1). Therefore, $1 \geq a \geq \alpha/2$. Using this upper and lower bound on a, we get

$$p \leq 1 - \frac{\varepsilon\alpha/2}{1 + \varepsilon\alpha/2 - \alpha/2} \leq 1 - c,$$

where

$$c = \frac{\varepsilon\alpha/2}{1 + \varepsilon\alpha/2 - \alpha/2}.$$

Run Algorithm 26.8 $1/c$ times, and output the heaviest cut found in these runs. Let W' be the weight of this cut. Then,

$$\mathbf{Pr}[W' \geq (1-\varepsilon)aT] \geq 1 - (1-c)^{1/c} \geq 1 - \frac{1}{e}.$$

Since $aT \geq \alpha \cdot \mathrm{OPT} > 0.87856$ OPT, we can pick a value of $\varepsilon > 0$ so that $(1-\varepsilon)aT \geq 0.87856$ OPT. □

Example 26.12 The following example shows that the bound on the integrality gap of relaxation (26.2) given in Corollary 26.10 is almost tight. Consider a graph which is a 5-cycle $v_1, v_2, v_3, v_4, v_5, v_1$. Then, an optimal solution to relaxation (26.2) is to place the five vectors in a 2-dimensional subspace within which they are given by $\boldsymbol{a}_i = (\cos(\frac{4i\pi}{5}), \sin(\frac{4i\pi}{5}))$, for $1 \leq i \leq 5$ (see Exercise 26.5). The cost of this solution is $\mathrm{OPT}_v = \frac{5}{2}(1 + \cos\frac{\pi}{5}) = \frac{25+5\sqrt{5}}{8}$. Since $\mathrm{OPT} = 4$ for this graph, the integrality gap for this example is $\frac{32}{25+5\sqrt{5}} = 0.88445....$ □

26.5 Improving the guarantee for MAX-2SAT

MAX-2SAT is the restriction of MAX-SAT (Problem 16.1) to formulae in which each clause contains at most two literals. In Chapter 16 we obtained a factor 3/4 algorithm for this problem using randomization, followed by the method of conditional expectation. We will give an improved algorithm using semidefinite programming.

The key new idea needed is a way of converting the obvious quadratic program (see Exercise 26.8) for this problem into a strict quadratic program. We will accomplish this as follows. Corresponding to each Boolean variable

x_i, introduce variable y_i which is constrained to be either $+1$ or -1, for $1 \le i \le n$. In addition, introduce another variable, say y_0, which is also constrained to be $+1$ or -1. Let us impose the convention that Boolean variable x_i is true if $y_i = y_0$ and false otherwise. Under this convention we can write the value of a clause in terms of the y_i's, where the *value*, $v(C)$, of clause C is defined to be 1 if C is satisfied and 0 otherwise. Thus, for clauses containing only one literal,

$$v(x_i) = \frac{1 + y_0 y_i}{2} \text{ and } v(\overline{x_i}) = \frac{1 - y_0 y_i}{2}.$$

Consider a clause containing 2 literals, e.g., $(x_i \vee x_j)$. Its value is

$$\begin{aligned} v(x_i \vee x_j) &= 1 - v(\overline{x_i})v(\overline{x_j}) = 1 - \frac{1 - y_0 y_i}{2} \frac{1 - y_0 y_j}{2} \\ &= \frac{1}{4}\left(3 + y_0 y_i + y_0 y_j - y_0^2 y_i y_j\right) \\ &= \frac{1 + y_0 y_i}{4} + \frac{1 + y_0 y_j}{4} + \frac{1 - y_i y_j}{4}. \end{aligned}$$

Observe that in this derivation we have used the fact that $y_0^2 = 1$. In all the remaining cases as well, it is easy to check that the value of a 2 literal clause consists of a linear combination of terms of the form $(1 + y_i y_j)$ or $(1 - y_i y_j)$. Therefore, a MAX-2SAT instance can be written as the following strict quadratic program, where the a_{ij}'s and b_{ij}'s are computed by collecting terms appropriately.

$$\begin{aligned} \text{maximize} \quad & \sum_{0 \le i < j \le n} a_{ij}(1 + y_i y_j) + b_{ij}(1 - y_i y_j) && (26.6) \\ \text{subject to} \quad & y_i^2 = 1, && 0 \le i \le n \\ & y_i \in \mathbf{Z}, && 0 \le i \le n \end{aligned}$$

Following is the vector program relaxation for (26.6), where vector variable $\boldsymbol{v}_i$ corresponds to y_i.

$$\begin{aligned} \text{maximize} \quad & \sum_{0 \le i < j \le n} a_{ij}(1 + \boldsymbol{v}_i \cdot \boldsymbol{v}_j) + b_{ij}(1 - \boldsymbol{v}_i \cdot \boldsymbol{v}_j) && (26.7) \\ \text{subject to} \quad & \boldsymbol{v}_i \cdot \boldsymbol{v}_i = 1, && 0 \le i \le n \\ & \boldsymbol{v}_i \in \mathbf{R}^{n+1}, && 0 \le i \le n \end{aligned}$$

The algorithm is similar to that for MAX-CUT. We solve vector program (26.7). Let $\boldsymbol{a}_0, \ldots, \boldsymbol{a}_n$ be an optimal solution. Pick a vector $\boldsymbol{r}$ uniformly distributed on the unit sphere in $(n+1)$ dimensions, S_n, and let $y_i = 1$ iff

$\boldsymbol{r} \cdot \boldsymbol{a}_i \geq 0$, for $0 \leq i \leq n$. This gives a truth assignment for the Boolean variables. Let W be the random variable denoting the weight of this truth assignment.

Lemma 26.13 $\mathbf{E}[W] \geq \alpha \cdot \text{OPT}_v$.

Proof:

$$\mathbf{E}[W] = 2 \sum_{0 \leq i < j \leq n} a_{ij} \mathbf{Pr}[y_i = y_j] + b_{ij} \mathbf{Pr}[y_i \neq y_j].$$

Let θ_{ij} denote the angle between $\boldsymbol{a}_i$ and $\boldsymbol{a}_j$. By inequality (26.5),

$$\mathbf{Pr}[y_i \neq y_j] = \frac{\theta_{ij}}{\pi} \geq \frac{\alpha}{2}(1 - \cos\theta_{ij}).$$

By Exercise 26.4,

$$\mathbf{Pr}[y_i = y_j] = 1 - \frac{\theta_{ij}}{\pi} \geq \frac{\alpha}{2}(1 + \cos\theta_{ij}).$$

Therefore,

$$\mathbf{E}[W] \geq \alpha \cdot \sum_{0 \leq i < j \leq n} a_{ij}(1 + \cos\theta_{ij}) + b_{ij}(1 - \cos\theta_{ij}) = \alpha \cdot \text{OPT}_v.$$

□

26.6 Exercises

26.1 Is matrix $\boldsymbol{W}$ in Theorem 26.3 unique (up to multiplication by -1)?
Hint: Consider the matrix $\boldsymbol{QDQ}^T$.

26.2 Let $\boldsymbol{B}$ be obtained from matrix $\boldsymbol{A}$ by throwing away a set of columns *and* the corresponding set of rows. We will say that $\boldsymbol{B}$ is a *principal submatrix* of $\boldsymbol{A}$. Show that the following is another equivalent condition for a real symmetric matrix to be positive semidefinite: that all of its principal submatrices have nonnegative determinants. (See Theorem 26.3 for other conditions.)

26.3 Show, using elementary calculus, that $\alpha > 0.87856$.

26.4 Show that for any ϕ, $0 \leq \phi \leq \pi$,

$$1 - \frac{\phi}{\pi} \geq \frac{\alpha}{2}(1 + \cos\phi).$$

Hint: Substitute $\theta = \pi - \phi$ in inequality (26.5).

26.5 Show that for a 5-cycle, the solution given in Example 26.12 is indeed an optimal solution to the vector program relaxation for MAX-CUT.

26.6 Show that the inaccuracies resulting from the fact we do not have an optimal solution to the vector program (26.2) and that matrix $\boldsymbol{A}$ is not exactly decomposed as $\boldsymbol{W}\boldsymbol{W}^T$ (see end of Section 26.2) can be absorbed into the approximation factor for MAX-CUT.
Hint: Use the idea behind the proof of Theorem 26.11 and the fact that the solution to program (26.2) lies in the range $[T/2, T]$, where T is the sum of weights of all edges in G.

26.7 Theorem 26.11 shows how to obtain a "high probability" statement from Lemma 26.9. Obtain a similar statement for MAX-2SAT, using Lemma 26.13, thereby obtaining a 0.87856 factor algorithm for MAX-2SAT.

26.8 Give a quadratic program for MAX-2SAT.

26.9 (Linial, London, and Rabinovich [197]) Let G be the complete undirected graph on n vertices, V, and let w be a function assigning nonnegative weights to the edges of G. The object is to find an optimal distortion ℓ_2^2-embedding of the vertices of G. Let vertex i be mapped to $\boldsymbol{v}_i \in \mathbf{R}^n$ by such an embedding. The embedding should satisfy:

1. no edge is overstretched, i.e., for $1 \le i < j \le n$, $||\boldsymbol{v}_i - \boldsymbol{v}_j||^2 \le w_{ij}$, and
2. the maximum shrinkage is minimized, i.e.,

$$\text{maximize} \quad \min_{(i,j):w_{ij} \neq 0} (||\boldsymbol{v}_i - \boldsymbol{v}_j||^2 / w_{ij}).$$

Give a vector program for finding such an optimal embedding and give the equivalent semidefinite program.
Hint: The vector program is:

$$\begin{array}{lll} \text{minimize} & c & \qquad (26.8) \\ \text{subject to} & \boldsymbol{v}_i \cdot \boldsymbol{v}_i + \boldsymbol{v}_j \cdot \boldsymbol{v}_j - 2\boldsymbol{v}_i \cdot \boldsymbol{v}_j \le w_{ij}, & 1 \le i < j \le n \\ & \boldsymbol{v}_i \cdot \boldsymbol{v}_i + \boldsymbol{v}_j \cdot \boldsymbol{v}_j - 2\boldsymbol{v}_i \cdot \boldsymbol{v}_j \ge c w_{ij}, & 1 \le i < j \le n \\ & \boldsymbol{v}_i \in \mathbf{R}^n, & 1 \le i \le n \end{array}$$

26.10 (Knuth [181]) Give an efficient algorithm for sampling from the normal distribution with mean 0 and unit standard deviation, given a source of unbiased random bits.

26.11 Give a strict quadratic program for the MAX k-CUT and maximum directed cut problems, Problems 2.14 and 2.15 stated in Exercises 2.3 and 2.4. Give a vector program relaxation and an equivalent semidefinite program as well.

26.12 (Goemans and Williamson [112]) Consider MAX-CUT with the additional constraint that specified pairs of vertices be on the same/opposite sides of the cut. Formally, we are specified two sets of pairs of vertices, S_1 and S_2. The pairs in S_1 need to be separated, and those in S_2 need to be on the same side of the cut sought. Under these constraints, the problem is to find a maximum weight cut. Assume that the constraints provided by S_1 and S_2 are not inconsistent. Give a strict quadratic program and vector program relaxation for this problem. Show how Algorithm 26.8 can be adapted to this problem so as to maintain the same approximation factor.

26.13 (Karger, Motwani, and Sudan [165]) Let $G = (V, E)$ be an undirected graph. Consider a vector program with n n-dimensional vectors corresponding to the vertices of G, and constraints that the vectors lie on the unit sphere, S_{n-1}, and that for each edge $(i, j) \in G$,

$$v_i \cdot v_j \leq -\frac{1}{k-1}.$$

Show that this vector program is a relaxation of the k-coloring problem, i.e., if G is k-colorable, then this vector program has a feasible solution.
Hint: Consider the following k vectors in $\mathbf{R}^n$. Each vector has 0 in the last $n-k$ positions. Vector i has $-\sqrt{\frac{k-1}{k}}$ in the ith position and $1/\sqrt{k(k-1)}$ in the remaining positions.

26.14 (Chor and Sudan [45]) Consider the following problem:

Problem 26.14 (Betweenness) We are given a set $S = \{x_1, x_2, \ldots, x_n\}$ of n items and a set T of m triplets $T \subseteq S \times S \times S$. Each triplet consists of three distinct items. A total ordering (permutation) of S, $x_{\pi_1} < x_{\pi_2} < \ldots < x_{\pi_n}$ *satisfies* a triplet $(x_i, x_j, x_k) \in T$ if x_j occurs between x_i and x_k in the ordering, i.e., if either $x_i < x_j < x_k$ holds or $x_k < x_j < x_i$ holds. The problem is to find a total ordering that maximizes the number of satisfied triplets.

1. Show that a random ordering (i.e., a permutation chosen uniformly at random among all possible permutations) will satisfy in expectation one third of all triplets in T.
2. Use the method of conditional expectation to derandomize the above algorithm, thereby obtaining a factor 1/3 approximation algorithm. What upper bound on OPT is this algorithm using? Give an example showing that with this upper bound a better algorithm is not possible.

3. The rest of the exercise develops an algorithm based on semidefinite programming. The ideas can be illustrated more simply by assuming that the instance is satisfiable, i.e., that all m triplets can be satisfied simultaneously. Note that checking for this condition is **NP**-hard, so the restriction of the betweenness problem to such instances is not an **NP**-optimization problem (see Exercise 1.9). Show that an instance is satisfiable iff the following strict quadratic program in variables $p_i \in \mathbf{R}$, $i = 1, \ldots, n$, has a solution:

$$\begin{aligned} (p_i - p_j)^2 &\geq 1 \quad \text{for all } i, j, \\ (p_i - p_j)(p_k - p_j) &\leq 0 \quad \text{for all } (x_i, x_j, x_k) \in T. \end{aligned}$$

4. Obtain the vector programming relaxation of this strict quadratic program as well as the equivalent semidefinite program.
5. Give an instance where the above semidefinite program is satisfiable but the instance itself is not satisfiable.
6. Let us assume that $n \times n$ matrix Y is a feasible solution to the above semidefinite program, and let $v_i \in \mathbf{R}^n$ for $i = 1, \cdots, n$ be vectors such that $Y_{ij} = v_i^T v_j$. Now select r uniformly at random on the unit sphere S_{n-1}. Consider the random ordering obtained by sorting $r^T v_i$. Show that, in expectation, this random ordering satisfies at least half of the constraints in T.
   Hint: What is the probability that a single triplet is satisfied? What is the angle between $v_i - v_j$ and $v_k - v_j$?

26.7 Notes

The results of this chapter are based on the seminal work of Goemans and Williamson [112] that introduced the use of semidefinite programs in approximation algorithms. Experimental results reported in their paper show that Algorithm 26.8 performs much better on typical instances than the worst case guarantee. Mahajan and Ramesh [207] give a derandomization of Algorithm 26.8, as well as the MAX-2SAT algorithm, using the method of conditional expectation. Karloff [168] provides a family of tight examples for Algorithm 26.8, for which the expected weight of the cut produced is arbitrarily close to $\alpha \cdot \mathrm{OPT}_v$. Feige and Schechtman [90] strengthen this to showing that there are graphs such that even the best hyperplane (rather than a random one, as prescribed in Algorithm 26.8) gives a cut of weight only $\alpha \cdot \mathrm{OPT}_v$. They also show that the integrality gap of the semidefinite relaxation (26.2) for MAX-CUT is α.

For efficient algorithms, using interior point methods, for approximating semidefinite programs, see Alizadeh [5], Nesterov and Nemirovskii [222] and Overton [223]. For a duality theory for semidefinite programs, see Wolkowitz [268] and Vandenberghe and Boyd [258].

Lovász and Schrijver [203] use semidefinite programming to provide an automatic way of strengthening any convex relaxation (having a convex feasible region) of a 0/1 integer program. They also show that if the original relaxation can be optimized in polynomial time, then so can the strengthened relaxation (however, in order to guarantee polynomial running time, this process can be applied only a constant number of times).

Feige and Goemans [87] improve the approximation factor for MAX-2SAT to 0.931. They also give a 0.859 factor for the maximum directed cut problem (see Exercise 26.11). For semidefinite-programming-based algorithms for the MAX k-CUT problem see Frieze and Jerrum [96]. Karger, Motwani, and Sudan [165] use the relaxation in Exercise 26.13 to obtain an $O(n^{1-3/(k+1)} \log^{1/2} n)$ coloring for k-colorable graphs.

Part III

Other Topics

27 Shortest Vector

The shortest vector problem is a central computational problem in the classical area of geometry of numbers. The approximation algorithm presented below has many applications in computational number theory and cryptography. Two of its most prominent applications are the derivation of polynomial time algorithms for factoring polynomials over the rationals and for simultaneous diophantine approximation.

Problem 27.1 (Shortest vector) Given n linearly independent vectors $\boldsymbol{a}_1, \ldots, \boldsymbol{a}_n \in \mathbf{Q}^n$, find the shortest vector, in Euclidean norm, in the lattice generated by these vectors. The lattice $\mathcal{L}$ generated by $\boldsymbol{a}_1, \ldots, \boldsymbol{a}_n$ is the set of all integer linear combinations of these vectors, i.e., $\mathcal{L} = \{\lambda_1 \boldsymbol{a}_1 + \ldots + \lambda_n \boldsymbol{a}_n \mid \lambda_i \in \mathbf{Z}\}$.

Remark 27.2 We will consider only full rank lattices, i.e., lattices that span the entire space on which they are defined. Most of the results apply to the general case as well, although the details can get more involved.

We will present an exponential (in n) factor algorithm for this problem. The problem of finding a polynomial factor algorithm for shortest vector has remained open for over two decades. It is worth pointing out that the exponential factor algorithm does perform well in practice and is widely used.

The shortest vector in the 1-dimensional lattice generated by two integers is simply the greatest common divisor (gcd) of these integers, which is polynomial time computable by Euclid's algorithm. When restricted to two dimensions, the shortest vector problem is polynomial time solvable. This follows from an algorithm of Gauss that was originally formulated in the language of quadratic forms. It will be instructive to first study these two algorithms, since Gauss' algorithm can be viewed as a generalization of Euclid's algorithm, and the n-dimensional algorithm can be seen as a generalization of Gauss' algorithm.

Several peculiarities set the shortest vector problem apart from other problems studied in this book. Unlike other **NP**-hard problems, we do not know if there is an instance of shortest vector that has exponentially many solutions, i.e., a lattice that has exponentially many shortest vectors. Indeed, this problem is not known to be **NP**-hard in the usual sense; it is only known to be **NP**-hard under randomized reductions. Another point of difference is that the lower bounding scheme used for obtaining the exponential factor algorithm can, in principle, be used to obtain a polynomial factor algorithm – by an existential argument we will prove that this scheme gives rise to a factor n approximate No certificate for the shortest vector problem. In contrast, for all other problems studied in this book, the best known lower bound is actually polynomial time computable.

27.1 Bases, determinants, and orthogonality defect

In this chapter all vectors will be assumed to be row vectors, unless otherwise stated. Let $\boldsymbol{A}$ denote the $n \times n$ matrix whose rows are the given vectors, $\boldsymbol{a}_1, \ldots, \boldsymbol{a}_n$. Let vectors $\boldsymbol{b}_1, \ldots, \boldsymbol{b}_n \in \mathcal{L}$, and let $\boldsymbol{B}$ denote the $n \times n$ matrix whose rows are $\boldsymbol{b}_1, \ldots, \boldsymbol{b}_n$. Since $\boldsymbol{a}_1, \ldots, \boldsymbol{a}_n$ can generate $\boldsymbol{b}_1, \ldots, \boldsymbol{b}_n$, $\boldsymbol{B} = \boldsymbol{\Lambda A}$, where $\boldsymbol{\Lambda}$ is an $n \times n$ integer matrix. Therefore $\det(\boldsymbol{B})$ is an integer multiple of $\det(\boldsymbol{A})$. We will say that $\boldsymbol{b}_1, \ldots, \boldsymbol{b}_n$ form a *basis* for lattice $\mathcal{L}$ if the lattice generated by these vectors is precisely $\mathcal{L}$. A square matrix with integer entries whose determinant is ± 1 will be called *unimodular*. Observe that its inverse is also unimodular.

Theorem 27.3 *Let vectors $\boldsymbol{b}_1, \ldots, \boldsymbol{b}_n \in \mathcal{L}$. The following conditions are equivalent:*

1. *$\boldsymbol{b}_1, \ldots, \boldsymbol{b}_n$ form a basis for lattice $\mathcal{L}$*
2. *$|\det(\boldsymbol{B})| = |\det(\boldsymbol{A})|$*
3. *there is an $n \times n$ unimodular matrix $\boldsymbol{U}$ such that $\boldsymbol{B} = \boldsymbol{UA}$*

Proof: Since $\boldsymbol{a}_1, \ldots, \boldsymbol{a}_n$ can generate $\boldsymbol{b}_1, \ldots, \boldsymbol{b}_n$, $\boldsymbol{B} = \boldsymbol{\Lambda A}$, where $\boldsymbol{\Lambda}$ is an $n \times n$ integer matrix.

$1 \Rightarrow 2$: If $\boldsymbol{b}_1, \ldots, \boldsymbol{b}_n$ form a basis for $\mathcal{L}$, they can generate $\boldsymbol{a}_1, \ldots, \boldsymbol{a}_n$. Therefore, $\boldsymbol{A} = \boldsymbol{\Lambda}'\boldsymbol{B}$, where $\boldsymbol{\Lambda}'$ is an $n \times n$ integer matrix. Hence, $\det(\boldsymbol{\Lambda})\det(\boldsymbol{\Lambda}') = 1$. Since both $\boldsymbol{\Lambda}$ and $\boldsymbol{\Lambda}'$ have integer determinants, $\det(\boldsymbol{\Lambda}) = \pm 1$, and so $|\det(\boldsymbol{B})| = |\det(\boldsymbol{A})|$.

$2 \Rightarrow 3$: Since $|\det(\boldsymbol{B})| = |\det(\boldsymbol{A})|$, $\det(\boldsymbol{\Lambda}) = \pm 1$, and hence $\boldsymbol{\Lambda}$ is unimodular.

$3 \Rightarrow 1$: Since $\boldsymbol{U}$ is unimodular, $\boldsymbol{U}^{-1}$ is also unimodular. Now, $\boldsymbol{A} = \boldsymbol{U}^{-1}\boldsymbol{B}$, and so each of $\boldsymbol{a}_1, \ldots, \boldsymbol{a}_n$ can be written as an integer linear combination of $\boldsymbol{b}_1, \ldots, \boldsymbol{b}_n$. Therefore, $\boldsymbol{b}_1, \ldots, \boldsymbol{b}_n$ form a basis for $\mathcal{L}$.

□

By Theorem 27.3, the determinant of a basis for $\mathcal{L}$ is invariant, up to sign. We will call $|\det(\boldsymbol{A})|$ the *determinant of lattice* $\mathcal{L}$ and will denote it as $\det \mathcal{L}$. Observe that $\det \mathcal{L}$ is the volume of the parallelohedron defined by the basis vectors. Theorem 27.3 also tells us that we can move from basis to basis by applying unimodular transforms. We will use this fact in our algorithms.

The most desirable basis to obtain is an orthogonal basis, since it must contain a shortest vector of $\mathcal{L}$ (see Exercise 27.1). However, not every lattice admits such a basis. For instance, the following 2-dimensional lattice has no orthogonal basis. (The two shaded parallelohedra have volume det $\mathcal{L}$ each.)

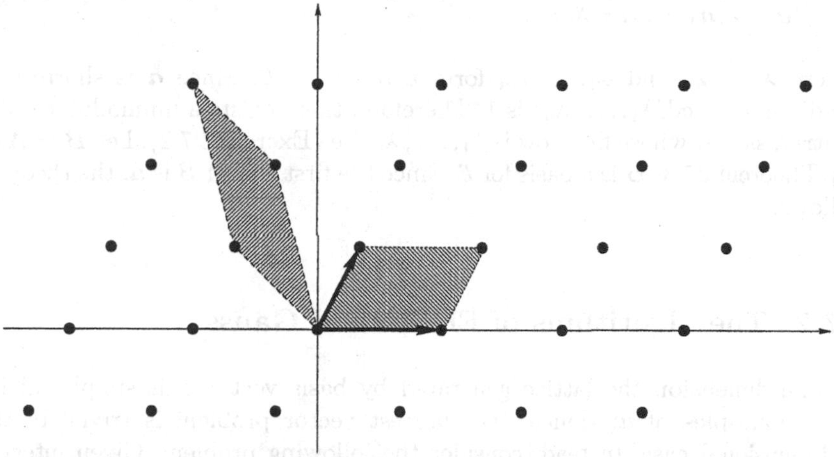

We will denote the Euclidean norm of vector $\boldsymbol{a}$ by $\|\boldsymbol{a}\|$. Recall Hadamard's inequality, which states that for an $n \times n$ real matrix $\boldsymbol{A}$,

$$|\det(\boldsymbol{A})| \leq \|\boldsymbol{a}_1\| \ldots \|\boldsymbol{a}_n\|.$$

This inequality holds with equality iff one of the rows of $\boldsymbol{A}$ is the zero vector or else the rows are mutually orthogonal. Applying this inequality we get that any basis $\boldsymbol{b}_1, \ldots, \boldsymbol{b}_n$ satisfies

$$\det \mathcal{L} \leq \|\boldsymbol{b}_1\| \ldots \|\boldsymbol{b}_n\|.$$

Since none of the basis vectors is the zero vector, the inequality holds with equality iff the basis is orthogonal. Let us define the *orthogonality defect* of basis $\boldsymbol{b}_1, \ldots, \boldsymbol{b}_n$ to be

$$\frac{\|\boldsymbol{b}_1\| \ldots \|\boldsymbol{b}_n\|}{\det \mathcal{L}}.$$

Since det $\mathcal{L}$ is invariant, the smaller the orthogonality defect of a basis, the shorter its vectors must be.

Let us say that linearly independent vectors $\boldsymbol{b}_1, \ldots, \boldsymbol{b}_k \in \mathcal{L}$ are *primitive* if they can be extended to a basis of $\mathcal{L}$. For a single vector, primitivity is easy to characterize. Let us say that $\boldsymbol{a} \in \mathcal{L}$ is *shortest in its direction* if $x\boldsymbol{a}$ is not a vector in $\mathcal{L}$ for $0 < |x| < 1$.

Theorem 27.4 *Vector $\boldsymbol{a} \in \mathcal{L}$ is primitive iff $\boldsymbol{a}$ is shortest in its direction.*

Proof: Suppose there is a basis $\boldsymbol{B}$ of $\mathcal{L}$ containing $\boldsymbol{a}$. Any vectors of the form $x\boldsymbol{a}$, where x is a scalar, that can be generated using $\boldsymbol{B}$ must have $x \in \mathbf{Z}$. Therefore, $\boldsymbol{a}$ is shortest in its direction.

Suppose $\boldsymbol{a}$ is shortest its direction. Since $\boldsymbol{a} \in \mathcal{L}$, we can write

$$\boldsymbol{a} = \lambda_1 \boldsymbol{a}_1 + \ldots + \lambda_n \boldsymbol{a}_n,$$

where $\lambda_i \in \mathbf{Z}$, and $\boldsymbol{a}_1, \ldots, \boldsymbol{a}_n$ form a basis for $\mathcal{L}$. Since $\boldsymbol{a}$ is shortest in its direction, $\gcd(\lambda_1, \ldots, \lambda_n)$ is 1. Therefore, there exists a unimodular $n \times n$ matrix, say $\boldsymbol{\Lambda}$, whose first row is $\lambda_1, \ldots, \lambda_n$ (see Exercise 27.2). Let $\boldsymbol{B} = \boldsymbol{\Lambda A}$. By Theorem 27.3, $\boldsymbol{B}$ is a basis for $\mathcal{L}$. Since the first row of B is $\boldsymbol{a}$, the theorem follows. □

27.2 The algorithms of Euclid and Gauss

In one dimension the lattice generated by basis vector $\boldsymbol{a}$ is simply all integer multiples of $\boldsymbol{a}$. Hence, the shortest vector problem is trivial in the 1-dimensional case. Instead, consider the following problem. Given integers a and b, consider all integer linear combinations of a and b, and find the smallest number in this lattice. This will be the gcd of a and b, denoted by (a, b).

Assume w.l.o.g. that $a \geq b \geq 0$. The idea behind Euclid's algorithm is that since $(a, b) = (a - b, b)$, we can replace the original problem by a smaller one. Continuing in this manner we are left with finding the smallest number, in absolute value, in the sequence $\{|a - mb|, \text{ for } m \in \mathbf{Z}\}$. Let this be c. If $c = 0$, then $(a, b) = b$, and we are done. Otherwise, $(a, b) = (b, c)$, and we proceed with the pair b, c. Observe that $c \leq b/2$. Hence this process terminates in $\log_2 b$ iterations.

Next, let us present Gauss' algorithm in 2 dimensions. For this case, a weaker condition than orthogonality is sufficient for ensuring that a basis contains a shortest vector. Let θ denote the angle between basis vectors $\boldsymbol{b}_1$ and $\boldsymbol{b}_2$, $0° < \theta < 180°$. Thus, $\det \mathcal{L} = \|\boldsymbol{b}_1\| \cdot \|\boldsymbol{b}_2\| \sin \theta$. Number the vectors so that $\|\boldsymbol{b}_1\| \leq \|\boldsymbol{b}_2\|$.

Theorem 27.5 *If $60° \leq \theta \leq 120°$, then $\boldsymbol{b}_1$ is a shortest vector in lattice $\mathcal{L}$.*

Proof: Assume for the purpose of contradiction that there is a vector $\boldsymbol{b} \in \mathcal{L}$ that is shorter than $\boldsymbol{b}_1$. Since $\boldsymbol{b}_1$ and $\boldsymbol{b}_2$ are both primitive, $\boldsymbol{b}$ cannot be linearly dependent on $\boldsymbol{b}_1$ or $\boldsymbol{b}_2$. By a simple case analysis it is easy to see that $\boldsymbol{b}$ must form an angle of at most $60°$ with one of $\boldsymbol{b}_1$, $\boldsymbol{b}_2$, $-\boldsymbol{b}_1$, or $-\boldsymbol{b}_2$.

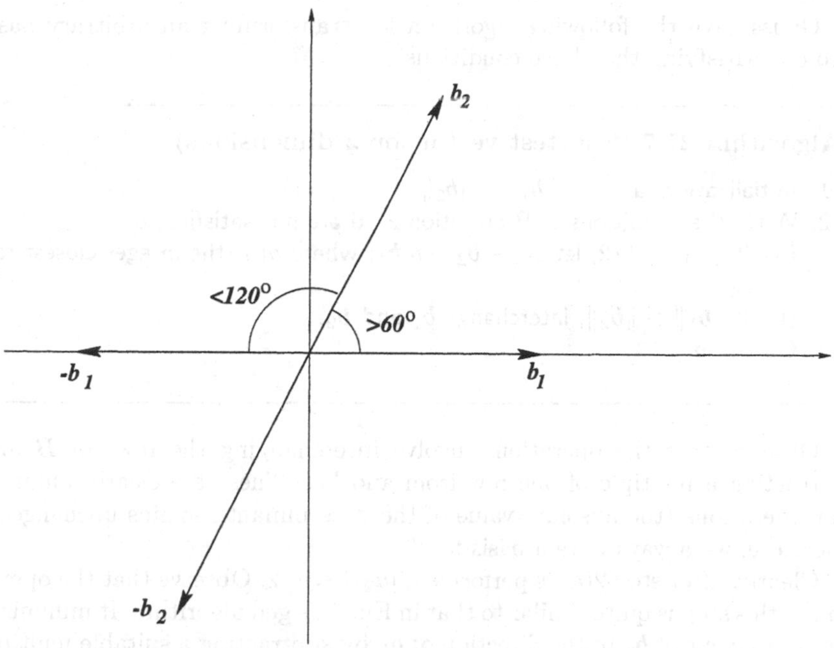

Let $\boldsymbol{D}$ be the 2×2 matrix whose rows consist of this vector and $\boldsymbol{b}$. Observe that $|\det(\boldsymbol{D})|$ is nonzero and is strictly smaller than $\det \mathcal{L} = \|\boldsymbol{b}_1\| \cdot \|\boldsymbol{b}_2\| \sin \theta$. This contradicts the fact that $\det(\boldsymbol{D})$ is an integer multiple of $\det \mathcal{L}$. Hence $\boldsymbol{b}_1$ is a shortest vector in $\mathcal{L}$. □

Define

$$\mu_{21} = \frac{\boldsymbol{b}_2 \cdot \boldsymbol{b}_1}{\|\boldsymbol{b}_1\|^2},$$

where $\cdot$ represents inner product. Observe that $\mu_{21}\boldsymbol{b}_1$ is the component of $\boldsymbol{b}_2$ in the direction of $\boldsymbol{b}_1$. The following proposition suggests an algorithm for finding a basis satisfying the condition of Theorem 27.5.

Proposition 27.6 *If basis* $(\boldsymbol{b}_1, \boldsymbol{b}_2)$ *satisfies*

- $\|\boldsymbol{b}_1\| \leq \|\boldsymbol{b}_2\|$ *and*
- $|\mu_{21}| \leq 1/2$,

then $60° \leq \theta \leq 120°$.

Proof: Note that,

$$\cos \theta = \frac{\boldsymbol{b}_1 \cdot \boldsymbol{b}_2}{\|\boldsymbol{b}_1\| \cdot \|\boldsymbol{b}_2\|} = \frac{\mu_{21}\|\boldsymbol{b}_1\|}{\|\boldsymbol{b}_2\|}.$$

Therefore, by the two conditions, $|\cos \theta| \leq 1/2$. Hence, $60° \leq \theta \leq 120°$. □

Gauss gave the following algorithm for transforming an arbitrary basis into one satisfying the above conditions.

Algorithm 27.7 (Shortest vector for 2 dimensions)

1. Initialization: assume $\|\boldsymbol{b}_1\| \leq \|\boldsymbol{b}_2\|$.
2. While the conditions of Proposition 27.6 are not satisfied, do:
   (a) If $|\mu_{21}| > 1/2$, let $\boldsymbol{b}_2 \leftarrow \boldsymbol{b}_2 - m\boldsymbol{b}_1$, where m is the integer closest to μ_{21}.
   (b) If $\|\boldsymbol{b}_1\| > \|\boldsymbol{b}_2\|$, interchange $\boldsymbol{b}_1$ and $\boldsymbol{b}_2$.
3. Output $\boldsymbol{b}_1$.

Observe that the operations involve interchanging the rows of $\boldsymbol{B}$ and subtracting a multiple of one row from another. These are clearly unimodular operations (the absolute value of the determinant remains unchanged). Therefore, we always have a basis for $\mathcal{L}$.

Clearly, after step 2(a) is performed, $|\mu_{21}| \leq 1/2$. Observe that the operation in this step is quite similar to that in Euclid's gcd algorithm. It minimizes the component of $\boldsymbol{b}_2$ in the direction of $\boldsymbol{b}_1$ by subtracting a suitable multiple of $\boldsymbol{b}_1$. Since $\|\boldsymbol{b}_1\| \cdot \|\boldsymbol{b}_2\|$ decreases after each iteration, the algorithm must terminate (there are only a bounded number of vectors of $\mathcal{L}$ within a given radius). For polynomial time termination of Algorithm 27.7, see Exercises 27.3, 27.4, and 27.5, and the comments in Section 27.7.

27.3 Lower bounding OPT using Gram–Schmidt orthogonalization

In this section we will present the Gram–Schmidt lower bound on OPT, the length of a shortest vector in lattice $\mathcal{L}$.

Intuitively, the *Gram–Schmidt orthogonalization* of basis $\boldsymbol{b}_1, \ldots, \boldsymbol{b}_n$ gives the n "altitudes" of the parallelohedron defined by this basis. Formally, it is a set of mutually orthogonal vectors $\boldsymbol{b}_1^*, \ldots, \boldsymbol{b}_n^*$, where $\boldsymbol{b}_1^* = \boldsymbol{b}_1$ and $\boldsymbol{b}_i^*$ is the component of $\boldsymbol{b}_i$ orthogonal to $\boldsymbol{b}_1^*, \ldots, \boldsymbol{b}_{i-1}^*$, for $2 \leq i \leq n$. $\boldsymbol{b}_i^*$ can be obtained from $\boldsymbol{b}_i$ by subtracting from it components in the directions of $\boldsymbol{b}_1^*, \ldots, \boldsymbol{b}_{i-1}^*$, as given by the following recurrence:

$$\boldsymbol{b}_1^* = \boldsymbol{b}_1$$

$$\boldsymbol{b}_i^* = \boldsymbol{b}_i - \sum_{j=1}^{i-1} \frac{\boldsymbol{b}_i \cdot \boldsymbol{b}_j^*}{\|\boldsymbol{b}_j^*\|^2} \boldsymbol{b}_j^*, \quad i = 2, \ldots, n. \tag{27.1}$$

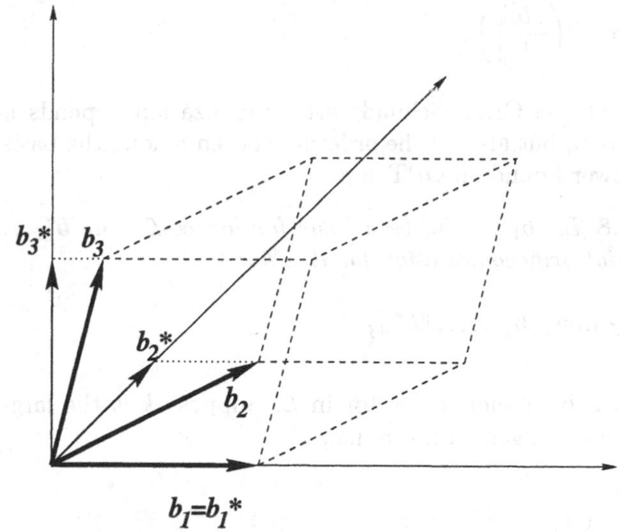

For $1 \leq j < i \leq n$, define

$$\mu_{ij} = \frac{\boldsymbol{b}_i \cdot \boldsymbol{b}_j^*}{\|\boldsymbol{b}_j^*\|^2},$$

and define $\mu_{ii} = 1$. Then

$$\boldsymbol{b}_i = \sum_{j=1}^{i} \mu_{ij} \boldsymbol{b}_j^*, \quad i = 1, \ldots, n. \tag{27.2}$$

For $j \leq i$, define $\boldsymbol{b}_i(j)$ to be the component of $\boldsymbol{b}_i$ orthogonal to $\boldsymbol{b}_1, \ldots, \boldsymbol{b}_{j-1}$, i.e.,

$$\boldsymbol{b}_i(j) = \mu_{ij} \boldsymbol{b}_j^* + \mu_{i,j+1} \boldsymbol{b}_{j+1}^* + \cdots + \boldsymbol{b}_i^*.$$

It is easy to see that

$$\det \mathcal{L} = \|\boldsymbol{b}_1^*\| \ldots \|\boldsymbol{b}_n^*\|.$$

Therefore, the orthogonality defect of this basis can be written as

$$\frac{\|\boldsymbol{b}_1\| \ldots \|\boldsymbol{b}_n\|}{\|\boldsymbol{b}_1^*\| \ldots \|\boldsymbol{b}_n^*\|} = \frac{1}{\sin \theta_2 \ldots \sin \theta_n},$$

where θ_i is the angle between $\boldsymbol{b}_i$ and the vector space spanned by $\boldsymbol{b}_1, \ldots, \boldsymbol{b}_{i-1}$, for $2 \leq i \leq n$. This angle is defined as follows. Let $\boldsymbol{b}_i'$ be the projection of $\boldsymbol{b}_i$ onto the space spanned by $\boldsymbol{b}_1, \ldots, \boldsymbol{b}_{i-1}$. Then

$$\theta_i = \cos^{-1}\left(\frac{\|\boldsymbol{b}'_i\|}{\|\boldsymbol{b}_i\|}\right).$$

Notice that the Gram–Schmidt orthogonalization depends not only on the basis chosen, but also on the ordering chosen among the basis vectors. It provides a lower bound on OPT as follows.

Lemma 27.8 *Let $\boldsymbol{b}_1, \ldots, \boldsymbol{b}_n$ be a basis for lattice $\mathcal{L}$, and $\boldsymbol{b}_1^*, \ldots, \boldsymbol{b}_n^*$ be the Gram–Schmidt orthogonalization for it. Then*

$$\text{OPT} \geq \min\{\|\boldsymbol{b}_1^*\|, \ldots, \|\boldsymbol{b}_n^*\|\}.$$

Proof: Let $\boldsymbol{v}$ be a shortest vector in $\mathcal{L}$. Suppose k is the largest indexed basis vector used in generating $\boldsymbol{v}$, i.e.,

$$\boldsymbol{v} = \sum_{i=1}^{k} \lambda_i \boldsymbol{b}_i,$$

where $\lambda_k \neq 0$. Using (27.2), $\boldsymbol{v}$ can also be expressed as a linear combination of the vectors $\boldsymbol{b}_1^*, \ldots, \boldsymbol{b}_k^*$. In this combination, the coefficient of $\boldsymbol{b}_k^*$ is λ_k, since $\mu_{k,k} = 1$. Since the vectors $\boldsymbol{b}_1^*, \ldots, \boldsymbol{b}_n^*$ are orthogonal,

$$\|\boldsymbol{v}\|^2 \geq \lambda_k^2 \|\boldsymbol{b}_k^*\|^2 \geq \|\boldsymbol{b}_k^*\|^2.$$

Hence

$$\text{OPT} = \|\boldsymbol{v}\| \geq \|\boldsymbol{b}_k^*\| \geq \min\{\|\boldsymbol{b}_1^*\|, \ldots, \|\boldsymbol{b}_n^*\|\}.$$

□

27.4 Extension to n dimensions

In this section we will extend Gauss' 2 dimensions algorithm to n dimensions. As usual, our main effort will be on improving the lower bound on OPT. Recall that for any basis, $\boldsymbol{b}_1, \ldots, \boldsymbol{b}_n$,

$$\|\boldsymbol{b}_1^*\| \ldots \|\boldsymbol{b}_n^*\| = \det \mathcal{L}$$

is an invariant, and $\min\{\|\boldsymbol{b}_1^*\|, \ldots, \|\boldsymbol{b}_n^*\|\}$ is a lower bound on OPT. In a typical basis, the minimum will be attained towards the end of the sequence $\|\boldsymbol{b}_1^*\|, \ldots, \|\boldsymbol{b}_n^*\|$. We will try to make this sequence lexicographically small, thereby ensuring that the last entries, and hence the minimum, are not too small. The conditions in Proposition 27.6 suggest a way of accomplishing this. By the first condition we have

$$\|b_1\|^2 \leq \|b_2\|^2, \quad \text{i.e.,} \quad \|b_1^*\|^2 \leq \mu_{21}^2\|b_1^*\|^2 + \|b_2^*\|^2.$$

Substituting for μ_{21} from the second condition, we get

$$\|b_1^*\|^2 \leq \frac{1}{4}\|b_1^*\|^2 + \|b_2^*\|^2$$

Therefore,

$$\|b_1^*\| \leq \frac{2}{\sqrt{3}}\|b_2^*\|.$$

Thus b_2^* cannot be smaller than b_1^* by more than a factor of $2/\sqrt{3}$. Let us impose similar conditions on each consecutive pair of basis vectors, so that for each i, b_{i+1}^* is not smaller than b_i^* by more than a factor of $2/\sqrt{3}$. Then the first vector, b_1, will be within an exponential factor of the lower bound. Algorithm 27.7 extends in a natural way to iteratively ensuring all these conditions. However, we do not know how to establish termination for this extended algorithm (see Exercise 27.7 for details).

To ensure termination in a polynomial number of iterations, we will use an idea similar to that in Exercise 27.3. Let us say that basis $b_1, \ldots, b_n$ is *Gauss reduced* if, for $1 \leq i \leq n-1$,

- $\|b_i(i)\| \leq \frac{2}{\sqrt{3}} \|b_{i+1}(i)\|$
- $|\mu_{i+1,i}| \leq 1/2$.

The algorithm for obtaining a Gauss reduced basis is a simple generalization of Algorithm 27.7.

Algorithm 27.9 (Shortest vector algorithm)

1. While basis $b_1, \ldots, b_n$ is not Gauss reduced, do:
   (a) For each $i, 1 \leq i \leq n-1$, ensure that $|\mu_{i+1,i}| \leq 1/2$.
   If $|\mu_{i+1,i}| > 1/2$ then $b_{i+1} \leftarrow b_{i+1} - mb_i$, where m is the integer closest to $\mu_{i+1,i}$.
   (b) Pick any i such that $\|b_i\| > \frac{2}{\sqrt{3}} \|b_{i+1}(i)\|$, and interchange b_i and b_{i+1}.
2. Output b_1.

Theorem 27.10 *Algorithm 27.9 terminates in a polynomial number of iterations and achieves an approximation guarantee of* $2^{(n-1)/2}$.

Proof: For $1 \le i \le n-1$ we have

$$\|\boldsymbol{b}_i(i)\|^2 = \|\boldsymbol{b}_i^*\|^2 \le \frac{4}{3} \|\boldsymbol{b}_{i+1}(i)\|^2 = \frac{4}{3}(\mu_{i+1,i}^2 \|\boldsymbol{b}_i^*\|^2 + \|\boldsymbol{b}_{i+1}^*\|^2)$$
$$\le \frac{4}{3} (\frac{1}{4} \|\boldsymbol{b}_i^*\|^2 + \|\boldsymbol{b}_{i+1}^*\|^2).$$

Therefore,

$$\|\boldsymbol{b}_i^*\|^2 \le 2 \|\boldsymbol{b}_{i+1}^*\|^2. \tag{27.3}$$

Hence,

$$\|\boldsymbol{b}_1\| \le 2^{(n-1)/2} \min_i\{\|\boldsymbol{b}_i^*\|\} \le 2^{(n-1)/2} \text{ OPT}.$$

In order to show termination in a polynomial number of iterations, we will use the following potential function.

$$\Phi = \prod_{i=1}^{n} \|\boldsymbol{b}_i^*\|^{(n-i)}.$$

Notice that step 1(a) in Algorithm 27.9 does not change the Gram–Schmidt orthogonalization of the basis and can be executed in $O(n)$ arithmetic operations (each addition, subtraction, multiplication, and division counts as one operation). In an iteration, at most one interchange of vectors is performed in Step 1(b). If this happens, Φ drops by a factor of at least $2/\sqrt{3}$. Indeed this was the reason to introduce the factor of $2/\sqrt{3}$ in the definition of a Gauss reduced basis. Let us assume w.l.o.g. that the initial basis has integer components. Then, the initial value of Φ is at most

$$(\max_i \|\boldsymbol{b}_i\|)^{n(n-1)/2}.$$

We will show that $\Phi \ge 1$ throughout the execution of the algorithm. Consequently, if the number of iterations is m, then

$$\left(\frac{2}{\sqrt{3}}\right)^m \le (\max_i \|\boldsymbol{b}_i\|)^{n(n-1)/2}.$$

Therefore,

$$m \le \frac{n(n-1)\max_i \log \|\boldsymbol{b}_i\|}{2\log(2/\sqrt{3})}.$$

Express basis vectors $\boldsymbol{b}_1, \ldots \boldsymbol{b}_n$ in the orthogonal basis $\boldsymbol{b}_1^*, \ldots, \boldsymbol{b}_n^*$. For $1 \le k \le n-1$, let $\boldsymbol{B}_k$ be the $k \times k$ lower-triangular matrix whose rows are $\boldsymbol{b}_1, \ldots, \boldsymbol{b}_k$. Clearly,

$$\det(\boldsymbol{B}_k) = \|\boldsymbol{b}_1^*\| \ldots \|\boldsymbol{b}_k^*\|.$$

Now, the (i,j)th entry of $\boldsymbol{B}_k \boldsymbol{B}_k^T$ is $\boldsymbol{b}_i \cdot \boldsymbol{b}_j$, which is an integer. Therefore,

$$\det(\boldsymbol{B}_k \boldsymbol{B}_k^T) = \|\boldsymbol{b}_1^*\|^2 \ldots \|\boldsymbol{b}_k^*\|^2 \geq 1.$$

Hence,

$$\Phi = \prod_{k=1}^{(n-1)} \det(\boldsymbol{B}_k \boldsymbol{B}_k^T) \geq 1.$$

□

We have shown that Algorithm 27.9 terminates in a polynomial number of arithmetic operations. Strictly speaking, we need to upper bound the number of bit operations, i.e., we need to show that the numbers being handled can be written using a polynomial number of bits. Below we define the stronger notion of a Lovász reduced basis. One can show that a simple extension of Algorithm 27.9 that finds a Lovász reduced basis executes only a polynomial number of bit operations. However, the proof is tedious and is omitted.

Another motivation for defining the stronger notion of a Lovász reduced basis is that its orthogonality defect can also be bounded. Let us say that basis $\boldsymbol{b}_1, \ldots, \boldsymbol{b}_n$ is *weakly reduced* if for $1 \leq j \leq i \leq n$, $|\mu_{ij}| \leq 1/2$. The basis is said to be *Lovász reduced* if it is Gauss reduced and weakly reduced. One can obtain a weakly reduced basis from a given basis without changing the Gram–Schmidt orthogonalization, using at most $n(n-1)/2$ operations (see Exercise 27.9). Substituting step 1(a) in Algorithm 27.9 by this procedure, we obtain an algorithm for finding a Lovász reduced basis. The bound on the number of iterations established in Theorem 27.10 applies to this algorithm as well.

Theorem 27.11 *The orthogonality defect of a Lovász reduced basis is bounded by* $2^{n(n-1)/4}$.

Proof: We will use inequality (27.3) established in Theorem 27.10.

$$\|\boldsymbol{b}_i\|^2 = \sum_{j=1}^{i} \mu_{ij}^2 \|\boldsymbol{b}_j^*\|^2 \leq \sum_{j=1}^{i-1} \frac{1}{4} \|\boldsymbol{b}_j^*\|^2 + \|\boldsymbol{b}_i^*\|^2$$

$$\leq (1 + \frac{1}{4}(2 + \cdots + 2^{i-1}))\|\boldsymbol{b}_i^*\|^2 \leq 2^{i-1}\|\boldsymbol{b}_i^*\|^2.$$

Therefore,

$$\prod_{i=1}^{n} \|b_i\|^2 \leq 2^{n(n-1)/2} \prod_{i=1}^{n} \|b_i^*\|^2 = 2^{n(n-1)/2} (\det \mathcal{L})^2.$$

The theorem follows. □

Example 27.12 Consider the following basis, together with its Gram–Schmidt orthogonalization.

$$\begin{pmatrix} 1 & 0 & 0 & \ldots & 0 & 0 \\ 1/2 & \rho & 0 & \ldots & 0 & 0 \\ 1/2 & \rho/2 & \rho^2 & \ldots & 0 & 0 \\ \vdots & & & & & \\ 1/2 & \rho/2 & \rho^2/2 & \ldots & \rho^{n-2}/2 & \rho^{n-1} \end{pmatrix}$$

Here $\rho = \sqrt{3}/2$. It is easy to verify that this basis is Lovász reduced. Each basis vector is of unit length. The Gram–Schmidt lower bound is $\rho^{(n-1)}$, i.e., exponentially smaller than any of the basis vectors. Subtracting the last two rows we obtain the vector

$$\left(0, \ldots, 0, \frac{\rho^{(n-2)}}{2}, -\rho^{(n-1)}\right).$$

This vector is exponentially smaller than any of the basis vectors. □

27.5 The dual lattice and its algorithmic use

As in the case of linear programs, one can define a structure that can be viewed as the dual of a given lattice. The power of duality in linear programs derives from the potent min–max relation connecting a linear program and its dual. In contrast, there is no such relation connecting a lattice with its dual. Yet, the dual lattice does seem to have algorithmic significance. In this section we will use it to show that the Gram–Schmidt lower bound is good by producing a basis for which this lower bound is at least OPT$/n$.

The *dual lattice*, $\mathcal{L}^*$, is defined by $\mathcal{L}^* = \{\boldsymbol{v} \in \mathbf{R}^n \mid \forall \boldsymbol{b} \in \mathcal{L}, \boldsymbol{b} \cdot \boldsymbol{v} \in \mathbf{Z}\}$. For an $n \times n$ matrix $\boldsymbol{B}$, we will denote $(\boldsymbol{B}^{-1})^T$ by $\boldsymbol{B}^{-T}$.

Theorem 27.13 *Let $\boldsymbol{b}_1, \ldots, \boldsymbol{b}_n$ be any basis for $\mathcal{L}$. Then, the rows of $\boldsymbol{B}^{-T}$ form a basis for the dual lattice $\mathcal{L}^*$. Furthermore,* $\det \mathcal{L}^* = 1/\det \mathcal{L}$.

Proof: Let us name the rows of $\boldsymbol{B}^{-T}$ as $\boldsymbol{v}_1, \ldots, \boldsymbol{v}_n$. Then,

$$\boldsymbol{b}_i \cdot \boldsymbol{v}_j = \begin{cases} 1 \text{ if } i = j \\ 0 \text{ otherwise.} \end{cases}$$

Therefore, any integer linear combination of $\boldsymbol{v}_1, \ldots, \boldsymbol{v}_n$ has an integer inner product with each of $\boldsymbol{b}_1, \ldots, \boldsymbol{b}_n$, and hence is in $\mathcal{L}^*$. Conversely, let $\boldsymbol{v} \in \mathcal{L}^*$. Let $\boldsymbol{v} \cdot \boldsymbol{b}_i = a_i \in \mathbf{Z}$, for $1 \leq i \leq n$. Let $\boldsymbol{a} = (a_1, \ldots, a_n)$. Thus, we have $\boldsymbol{B}\boldsymbol{v}^T = \boldsymbol{a}^T$. Therefore, $\boldsymbol{v} = \boldsymbol{a}\boldsymbol{B}^{-T}$, i.e., $\boldsymbol{v}$ is an integer linear combination of the vectors $\boldsymbol{v}_1, \ldots, \boldsymbol{v}_n$. Hence $\boldsymbol{v}_1, \ldots, \boldsymbol{v}_n$ form a basis for $\mathcal{L}^*$. Finally,

$$\det \mathcal{L}^* = \det(\boldsymbol{B}^{-T}) = \frac{1}{\det \mathcal{L}}.$$

□

Let $\boldsymbol{v} \in \mathbf{R}^n$ be a nonzero vector. Then, $\boldsymbol{v}^{\perp}$ will denote the $(n-1)$-dimensional vector space $\{\boldsymbol{b} \in \mathbf{R}^n \mid \boldsymbol{b} \cdot \boldsymbol{v} = 0\}$. A set $\mathcal{L}' \subset \mathcal{L}$ that is a lattice in its own right will be called a *sublattice* of $\mathcal{L}$. Its *dimension* is the dimension of the vector space spanned by $\mathcal{L}'$. Lemma 27.14 and Exercise 27.11 will establish that there is a one-to-one correspondence between $(n-1)$-dimensional sublattices of $\mathcal{L}$ and primitive vectors of $\mathcal{L}^*$.

Lemma 27.14 *Let $\boldsymbol{v} \in \mathcal{L}^*$ be primitive. Then*

- $\mathcal{L} \cap (\boldsymbol{v}^{\perp})$ *is an $(n-1)$-dimensional sublattice of $\mathcal{L}$.*
- *There is a vector $\boldsymbol{b} \in \mathcal{L}$ such that $\boldsymbol{v} \cdot \boldsymbol{b} = 1$.*

Proof: Since $\boldsymbol{v}$ is primitive, there is a basis of $\mathcal{L}^*$ containing $\boldsymbol{v}$. Let it be $\boldsymbol{v}, \boldsymbol{v}_2, \ldots, \boldsymbol{v}_n$, and let $\boldsymbol{V}$ be the $n \times n$ matrix whose rows are these basis vectors. Let $\boldsymbol{b}_1, \ldots, \boldsymbol{b}_n$ be the rows of $\boldsymbol{V}^{-T}$. Since $\boldsymbol{v} \cdot \boldsymbol{b}_i = 0$, for $2 \leq i \leq n$, the $(n-1)$-dimensional sublattice of $\mathcal{L}$ generated by $\boldsymbol{b}_2, \ldots, \boldsymbol{b}_n$ lies in $\boldsymbol{v}^{\perp}$. The second assertion follows from the fact that $\boldsymbol{v} \cdot \boldsymbol{b}_1 = 1$. □

Next we will present a key primitive for constructing a basis with a good Gram–Schmidt orthogonalization. Let $\boldsymbol{v} \in \mathcal{L}^*$ be primitive, $\mathcal{L}' = \mathcal{L} \cap (\boldsymbol{v}^{\perp})$, and $\boldsymbol{w} \in \mathcal{L}$ with $\boldsymbol{v} \cdot \boldsymbol{w} = 1$. Let $\boldsymbol{b}_1, \ldots, \boldsymbol{b}_{n-1}$ be a basis for $\mathcal{L}'$ with Gram–Schmidt orthogonalization $\boldsymbol{b}_1^*, \ldots, \boldsymbol{b}_{n-1}^*$.

Lemma 27.15 *$\boldsymbol{b}_1, \ldots, \boldsymbol{b}_{n-1}, \boldsymbol{w}$ is a basis for $\mathcal{L}$, with Gram–Schmidt orthogonalization*

$$\boldsymbol{b}_1^*, \ldots, \boldsymbol{b}_{n-1}^*, \frac{\boldsymbol{v}}{\|\boldsymbol{v}\|^2}.$$

Proof: Let $\boldsymbol{b} \in \mathcal{L}$ be an arbitrary vector. Let us first observe that $\boldsymbol{b} - (\boldsymbol{b} \cdot \boldsymbol{v})\boldsymbol{w} \in \mathcal{L}'$. This is so because $(\boldsymbol{b} - (\boldsymbol{b} \cdot \boldsymbol{v})\boldsymbol{w}) \cdot \boldsymbol{v} = 0$. Thus, we can write

$$\boldsymbol{b} - (\boldsymbol{b} \cdot \boldsymbol{v})\boldsymbol{w} = \sum_{i=1}^{n-1} \lambda_i \boldsymbol{b}_i,$$

where $\lambda_1, \ldots, \lambda_{n-1} \in \mathbf{Z}$. Therefore,

$$b = \sum_{i=1}^{n-1} \lambda_i b_i + (b \cdot v) w.$$

Hence, every vector of $\mathcal{L}$ can be written as an integer linear combination of $b_1, \ldots, b_{n-1}, w$, thus establishing the first part.

Let $\mu_i b_i^*$ be the component of w in the direction of b_i^*, for $1 \le i \le n-1$. Let

$$w^* = w - \sum_{i=1}^{n-1} \mu_i b_i^*.$$

Since w^* is orthogonal to each of $b_1, \ldots, b_{n-1}$, it is linearly dependent on v. Since $v \cdot w = 1$, $v \cdot w^* = 1$. Therefore,

$$\|w^*\| = \frac{1}{\|v\|}.$$

Hence,

$$w^* = \frac{v}{\|v\|^2}.$$

□

Let us use the primitive given above as follows. Let $\mathcal{L}_1 = \mathcal{L}$, and let $\mathcal{L}_1^*$ be its dual. Pick a primitive vector $v_1 \in \mathcal{L}_1^*$, and let sublattice $\mathcal{L}_2 = \mathcal{L}_1 \cap (v_1^\perp)$. Let $\mathcal{L}_2^*$ denote the dual of $\mathcal{L}_2$ (it lies in the vector space spanned by $\mathcal{L}_2$). Now we proceed by picking another primitive vector $v_2 \in \mathcal{L}_2^*$, and define $\mathcal{L}_3 = \mathcal{L}_2 \cap (v_2^\perp)$, and so on. In general, we pick a primitive vector $v_i \in \mathcal{L}_i^*$, and define sublattice $\mathcal{L}_{i+1} = \mathcal{L}_i \cap (v_i^\perp)$. Observe that although $\mathcal{L}_i$ is a sublattice of $\mathcal{L}$, $\mathcal{L}_i^*$ is not necessarily a sublattice of $\mathcal{L}^*$, for $1 \le i \le n$. Next, use the second part of Lemma 27.14 to pick $w_i \in \mathcal{L}_i$ such that $v_i \cdot w_i = 1$, for $1 \le i \le n$. We summarize this procedure below. (We call it a "procedure" rather than an "algorithm" because we do not know how to execute some of its steps in polynomial time.)

Procedure 27.16

1. Initialization: $\mathcal{L}_1 \leftarrow \mathcal{L}$, and $\mathcal{L}_1^* \leftarrow$ dual of $\mathcal{L}$.
2. For $i = 1$ to n do:
   Pick a primitive vector $v_i \in \mathcal{L}_i^*$.
   Pick $w_i \in \mathcal{L}_i$ such that $v_i \cdot w_i = 1$.
   $\mathcal{L}_{i+1} \leftarrow \mathcal{L}_i \cap (v_i^\perp)$.
   $\mathcal{L}_{i+1}^* \leftarrow$ dual of $\mathcal{L}_{i+1}$.
3. Output $w_n, w_{n-1}, \ldots, w_1$ and

$$\left(\frac{v_n}{\|v_n\|^2}, \ldots, \frac{v_1}{\|v_1\|^2}\right).$$

Using Lemma 27.15 and applying induction, we get

Lemma 27.17 *The output of Procedure 27.16 satisfies that $w_n, w_{n-1}, \ldots, w_1$ is a basis for $\mathcal{L}$ with Gram–Schmidt orthogonalization*

$$\left(\frac{v_n}{\|v_n\|^2}, \ldots, \frac{v_1}{\|v_1\|^2}\right).$$

We will need the following fundamental theorem of Minkowski.

Theorem 27.18 *There is a vector $b \in \mathcal{L}$ such that $\|b\| \le \sqrt{n}\sqrt[n]{\det \mathcal{L}}$.*

Proof: Let $\|a\|_\infty$ denote the ℓ_∞ norm of vector a. Now, $\|a\| \le \sqrt{n}\|a\|_\infty$. Therefore, it is sufficient to show that there is a vector $b \in \mathcal{L}$ such that $\|b\|_\infty \le \sqrt[n]{\det \mathcal{L}}$. Further, w.l.o.g. assume that $\det \mathcal{L} = 1$ (since we can scale). Define the *density* of a lattice, $\mathcal{L}'$, to be the number of lattice points per unit volume; clearly, the density of a lattice $\mathcal{L}'$ is $1/\det \mathcal{L}'$. Since $\det \mathcal{L} = 1$, its density is 1.

Let $C = \{v \in \mathbf{R}^n \mid \|v\|_\infty \le 1/2\}$ be the unit cube with center at the origin. Place a cube at each lattice point of $\mathcal{L}$, i.e., consider the cubes $C + b$, for each lattice point b. Now, we claim that two of these cubes must intersect. If not, each cube must contain exactly one lattice point and since the cubes are of unit volume, this implies that the density of lattice points is strictly smaller than 1, leading to a contradiction.

Let the intersecting cubes be centered at lattice points b_1 and b_2. Then, $\|b_1 - b_2\|_\infty \le 1$. Since $b = b_1 - b_2$ is also a lattice point, the theorem follows. □

Theorem 27.19 *There is a basis for $\mathcal{L}$ whose Gram–Schmidt lower bound is at least* OPT$/n$.

Proof: By Theorem 27.18, there are vectors $b \in \mathcal{L}$ and $v \in \mathcal{L}^*$ such that $\|b\|\|v\| \le n$. Refine Procedure 27.16 as follows. Let $v_i \in \mathcal{L}_i^*$ be a primitive vector, and $b_i \in \mathcal{L}_i$ such that $\|v_i\|\|b_i\| \le n+1-i \le n$. Vectors w_i are defined as above.

By Lemma 27.17, $w_n, \ldots, w_1$ is a basis for $\mathcal{L}$ with Gram–Schmidt orthogonalization

$$\left(\frac{v_n}{\|v_n\|^2}, \ldots, \frac{v_1}{\|v_1\|^2}\right).$$

Let b denote the shortest vector among $b_1, \ldots, b_n$. Then,

$$\|b\| = \min\{\|b_1\|, \ldots, \|b_n\|\} \le n \min\left\{\frac{1}{\|v_1\|}, \ldots, \frac{1}{\|v_n\|}\right\}.$$

Hence, the Gram–Schmidt lower bound,

$$\min\left\{\frac{1}{\|\boldsymbol{v}_1\|},\ldots,\frac{1}{\|\boldsymbol{v}_n\|}\right\} \geq \frac{\|\boldsymbol{b}\|}{n} \geq \frac{\mathrm{OPT}}{n}.$$

□

As a result of Theorem 27.19, there is a *factor n approximate No certificate* for the shortest vector problem (see Section 1.2 for definition). Thus, if the answer to the question "Is the shortest vector in lattice $\mathcal{L}$ of length at most α?" is "no", and in fact $\alpha < \mathrm{OPT}/n$, then there is a polynomial sized guess that enables us to verify in polynomial time that the answer is indeed "no". The guess is a basis whose existence is demonstrated in Theorem 27.19, and the verification simply involves confirming that α is less than the lower bound established by this basis.

27.6 Exercises

27.1 Show that if lattice $\mathcal{L}$ has a basis whose vectors are mutually orthogonal, then the shortest vector in the basis is a shortest vector in $\mathcal{L}$.
Hint: Write the length of a shortest vector $\boldsymbol{v} \in \mathcal{L}$ in terms of the basis vectors.

27.2 Recall that if $(a, b) = 1$, then Euclid's algorithm gives integers x, y such that $ax + by = 1$. Generalize this fact to the following: If $\gcd(a_1, \ldots, a_n) = 1$, then there is an $n \times n$ unimodular matrix $\boldsymbol{U}$ whose first row is $a_1, \ldots, a_n$.
Hint: Prove by induction on n.

27.3 For the purpose of showing polynomial time termination, let us relax the first condition of Gauss' algorithm in 2 dimensions to ensuring that vector $\boldsymbol{b}_1$ is not much bigger than $\boldsymbol{b}_2$, i.e., $\|\boldsymbol{b}_1\| \leq (1+\varepsilon)\|\boldsymbol{b}_2\|$. The second condition is as before. Show that the algorithm terminates after a polynomial – in the length of the initial vectors and $1/\varepsilon$ – number of iterations. What is the guarantee on the length of $\boldsymbol{b}_1$ at termination?
Hint: Use the Gram–Schmidt lower bound to establish the guarantee.

27.4 Polynomial time termination for Algorithm 27.7 is difficult to establish because in any one iteration, the progress may be minimal. Give an example to show this.
Hint: Consider for instance $\boldsymbol{b}_1 = (1, 0)$ and $\boldsymbol{b}_2 = (\frac{1}{2} + \varepsilon, \frac{\sqrt{3}}{2})$.

27.5 (Kaib and Schnorr [159]) To get around the difficulty mentioned in Exercise 27.4, consider the following small modification to Algorithm 27.7. Say that basis $(\boldsymbol{b}_1, \boldsymbol{b}_2)$ is *well ordered* if

$$\|b_1\| \leq \|b_1 - b_2\| < \|b_2\|.$$

1. Assume $\|b_1\| \leq \|b_2\|$. Show that one of the three bases (b_1, b_2), $(b_1, b_1 - b_2)$, and $(b_1 - b_2, b_1)$ is well ordered. Hence, modify the initialization step to start with a well ordered basis.
2. The main algorithm is as follows:
   (a) If $|\mu_{21}| > 1/2$, let $b_2 \leftarrow b_2 - mb_1$, where m is the integer closest to μ_{21}.
   (b) If $\|b_1 - b_2\| > \|b_1 + b_2\|$, then let $b_2 \leftarrow -b_2$.
   (c) If $\|b_1\| \leq \|b_2\|$, output (b_1, b_2) and halt.
   Otherwise, interchange b_1 and b_2 and go to step (a).

   Show that the basis obtained after step (b) is well ordered.
3. Suppose the modified algorithm, starting with well ordered basis (b_1, b_2), executes k iterations and ends with a reduced basis. Renumber the vectors encountered in reverse order as $(a_1, a_2, \ldots, a_k, a_{k+1})$, where $b_1 = a_k$, $b_2 = a_{k+1}$, and the terminating basis is (a_1, a_2). Prove that for $i > 1, \|a_i\| \leq (1/2)\|a_{i+1}\|$, thereby establishing a polynomial bound on the running time.

27.6 Give an example of a 3-dimensional lattice $\mathcal{L}$ and vectors $b_1, b_2 \in \mathcal{L}$ such that b_1 and b_2 are individually primitive, but b_1, b_2 is not.

27.7 Suppose basis $b_1, \ldots, b_n$ satisfies the following conditions for $1 \leq i \leq n-1$:

- $\|b_i(i)\| \leq \|b_{i+1}(i)\|$
- $|\mu_{i+1,i}| \leq 1/2$.

Show that

$$\|b_1\| \leq \left(\frac{2}{\sqrt{3}}\right)^i \|b^*_{i+1}\|,$$

for $1 \leq i \leq n-1$. Hence,

$$\|b_1\| \leq \left(\frac{2}{\sqrt{3}}\right)^{n-1} \min_i\{\|b^*_i\|\} \leq \left(\frac{2}{\sqrt{3}}\right)^{n-1} \text{OPT}.$$

Hint: By a calculation similar to the 2-dimensional case show that for $1 \leq i \leq n-1$,

$$\|b^*_i\| \leq \frac{2}{\sqrt{3}}\|b^*_{i+1}\|.$$

27.8 By modifying the definition of a Gauss reduced basis appropriately, obtain a factor $c^{(n-1)/2}$ algorithm, for any constant $c > 2/\sqrt{3}$, for the shortest vector problem.
Hint: Replace the factor of $2/\sqrt{3}$ in the definition of Gauss reduced basis by $1+\varepsilon$, for a suitable $\varepsilon > 0$.

27.9 Show that a basis can be made weakly reduced using at most $n(n-1)/2$ arithmetic operations.
Hint: In Algorithm 27.9, the operations of step 1(a) can be carried out in any order. In contrast, for obtaining a weakly reduced basis, order is important. Pick a pair (i,j) with $|\mu_{ij}| > 1/2$ such that j is the largest possible, and carry out the operation of step 1(a) in Algorithm 27.9.

27.10 Prove that the dual of lattice $\mathcal{L}^*$ is $\mathcal{L}$.

27.11 Prove the converse of Lemma 27.14.

- If $\mathcal{L}'$ is an $(n-1)$-dimensional sublattice of $\mathcal{L}$, then there is a vector $\boldsymbol{v} \in \mathcal{L}^*$ such that $\mathcal{L}' = \mathcal{L} \cap (\boldsymbol{v}^{\perp})$.
- If $\boldsymbol{v} \in \mathcal{L}^*$ and $\exists \boldsymbol{b} \in \mathcal{L}$ such that $\boldsymbol{v} \cdot \boldsymbol{b} = 1$, then $\boldsymbol{v}$ is primitive.

27.12 Let $\boldsymbol{v} \in \mathcal{L}^*$ be primitive and $\mathcal{L}'$ be an $(n-1)$-dimensional sublattice of $\mathcal{L}$ such that $\mathcal{L}' = \mathcal{L} \cap (\boldsymbol{v}^{\perp})$. Prove that $\det \mathcal{L}' = \|\boldsymbol{v}\| \cdot \det \mathcal{L}$.

27.13 This exercise develops a partial converse for Lemma 27.15. Let $\boldsymbol{v}$ and $\boldsymbol{b}_1, \ldots, \boldsymbol{b}_{n-1}$ be as defined in Lemma 27.15. Show that if $\boldsymbol{b}_1, \ldots, \boldsymbol{b}_{n-1}, \boldsymbol{w}$ form a basis for $\mathcal{L}$, then $\boldsymbol{v} \cdot \boldsymbol{w} = 1$.
Hint: By Lemma 27.14, there is a vector $\boldsymbol{w}' \in \mathcal{L}$ such that $\boldsymbol{v} \cdot \boldsymbol{w}' = 1$. Express $\boldsymbol{w}'$ in the basis $\boldsymbol{b}_1, \ldots, \boldsymbol{b}_{n-1}, \boldsymbol{w}$, and use this to argue that $\boldsymbol{v} \cdot \boldsymbol{w} = 1$.

27.14 Show that the basis whose existence is established in Theorem 27.19 can be found in polynomial time, given an oracle for the shortest vector problem.
Hint: A shortest vector satisfies the condition of Theorem 27.18.

27.15 (Korkhine and Zolotarav [182]) A basis $\boldsymbol{b}_1, \ldots, \boldsymbol{b}_n$ for lattice $\mathcal{L}$ is said to be *KZ reduced* if:

- $\boldsymbol{b}_1$ is a shortest vector in $\mathcal{L}$, $\boldsymbol{b}_2$ is such that $\boldsymbol{b}_2^*$ is shortest possible while ensuring the primitiveness of $\boldsymbol{b}_1, \boldsymbol{b}_2$, etc. In general, $\boldsymbol{b}_i$ is such that $\boldsymbol{b}_i^*$ is shortest possible while ensuring the primitiveness of $\boldsymbol{b}_1, \ldots, \boldsymbol{b}_i$, i.e., the component of $\boldsymbol{b}_i$ orthogonal to the vector space spanned by $\boldsymbol{b}_1, \ldots, \boldsymbol{b}_{i-1}$ is being minimized, while ensuring primitiveness.

- For $1 \leq j < i \leq n$, $|\mu_{ij}| \leq 1/2$.

Give a polynomial time algorithm for finding a KZ reduced basis in a lattice, given an oracle for the shortest vector problem.

27.16 (Lovász [200]) Let $p(n)$ be a polynomial function. Show that if there is a polynomial time algorithm that finds a vector $v \in \mathcal{L}$ such that

$$\|v\| \leq p(n) \sqrt[n]{\det \mathcal{L}},$$

then there is a factor $p^2(n)$ approximation algorithm for the shortest vector problem.
Hint: The main idea is present in Theorem 27.19.

27.17 (Designed by K. Jain and R. Venkatesan, based on Frieze [94]) Given a set of n nonnegative integers $S = \{a_1, \ldots, a_n\}$ and an integer b, the subset sum problem asks if there is a subset of S that adds up to b. This **NP**-hard problem lends itself naturally to the design of a one–way function, a basic cryptographic primitive, as follows. An n bit message $\boldsymbol{c} = c_1, \ldots, c_n$ can be encoded as

$$f_S(\boldsymbol{c}) = \sum_{i=1}^{n} a_i c_i = b, \text{ say}.$$

Clearly, f_S is easy to compute. The **NP**-hardness of the subset sum problem implies that f_S is hard to invert in general, i.e., given $a_1, \ldots, a_n$ and b, it is hard to compute $\boldsymbol{c}$.

This exercise shows how, using Algorithm 27.9, f_S can be inverted with high probability if $a_1, \ldots, a_n$ are picked uniformly and independently in the (large) range $[1, B]$, where $B \geq 2^{n^2}$. In practice, it is desirable to pick B to be a large number so that f_S turns out to be one-to-one.

Let $m = 2^{n/2}\sqrt{n+1}$ and $p = m+1$. Pick the following basis vectors for an $(n+1)$-dimensional lattice $\mathcal{L}$. $\boldsymbol{b}_0 = (pb, 0, ...0), \boldsymbol{b}_1 = (-pa_1, 1, 0...0), \ldots, \boldsymbol{b}_n = (-pa_n, 0, 0, ..., 1)$.

1. Let $u_1, \ldots, u_n$ be fixed integers. Show that if the a_i's are picked uniformly and independently in the range $[1, B]$, then

$$\mathbf{Pr}[\sum_{i=1}^{n} u_i a_i = 0] \leq \frac{1}{B},$$

where the probability is over the choice of a_i's.

2. Observe that $\mathcal{L}$ has a vector of length $\leq \sqrt{n}$. Show that if there is no other vector in $\mathcal{L}$ of length $\leq m$, then Algorithm 27.9 can be used to find $\boldsymbol{c}$ such that $f_S(\boldsymbol{c}) = b$.

3. Suppose there is another vector in $\mathcal{L}$ of length $\leq m$, say $\boldsymbol{v}$. Let $\boldsymbol{v} = \lambda \boldsymbol{b}_0 + \sum_{i=1}^{n} v_i \boldsymbol{b}_i$, say, with $\lambda, v_i \in \mathbf{Z}$. Clearly each $v_i \in [-m, m]$. Show that $\lambda \in [-2m, 2m]$.
Hint: We can assume w.l.o.g. that $b \geq (\frac{1}{2}) \sum_{i=1}^{n} a_i$ (if not, then decode $\sum_{i=1}^{n} a_i - b$). The first coordinate of $\boldsymbol{v}$ must be zero, which gives

$$|\lambda| b \leq \sum_{i=1}^{n} a_i |v_i| \leq (\sum_{i=1}^{n} a_i) \|\boldsymbol{v}\| \leq 2bm.$$

4. Show that if there is another vector in $\mathcal{L}$ of length $\leq m$ then there are integers $u_1, \ldots, u_n$, with each $u_i \in [-3m, 3m]$, such that $\sum_{i=1}^{n} u_i a_i = 0$.
Hint: $\sum_{i=1}^{n} a_i(\lambda c_i - v_i) = 0$.
5. Show that if the a_i's are picked uniformly and independently in the range $[1, B]$, then the probability that $\sum_{i=1}^{n} u_i a_i = 0$ for integers $u_1, \ldots, u_n$, with each $u_i \in [-3m, 3m]$, is very small.
Hint: Use the fact that the number of choices for $u_1, \ldots, u_n$ is $(6m+1)^n$.
6. Hence prove that f_S can be inverted with high probability.

27.18 (Goldreich, Micciancio, Safra, and Seifert [114]) Consider the *closest vector problem.* Given basis vectors $\boldsymbol{b}_1, \ldots, \boldsymbol{b}_n$ for lattice $\mathcal{L}$ and a target vector $\boldsymbol{b} \in \mathbf{R}^n$, find a lattice vector that is closest to $\boldsymbol{b}$ in Euclidean distance. Show that for any function $f(n)$, an $f(n)$ factor approximation algorithm for the closest vector problem can be used as a subroutine to give an $f(n)$ factor approximation algorithm for the shortest vector problem.
Hint: Given basis vectors $\boldsymbol{b}_1, \ldots, \boldsymbol{b}_n$ for a lattice $\mathcal{L}$, suppose $\boldsymbol{b} = \sum_i \lambda_i \boldsymbol{b}_i$ is a shortest vector in $\mathcal{L}$. Clearly all λ_i cannot be even. Suppose λ_i is odd. Now consider lattice $\mathcal{L}'$ whose basis vectors are $\boldsymbol{b}_1, \ldots, \boldsymbol{b}_{i-1}, 2\boldsymbol{b}_i, \boldsymbol{b}_{i+1}, \ldots, \boldsymbol{b}_n$. Observe that if the closest vector in lattice $\mathcal{L}'$ to target vector $\boldsymbol{b}_i$ is $\boldsymbol{v}$, then $\boldsymbol{b} = \boldsymbol{v} - \boldsymbol{b}_i$. Since i is not known, construct n such questions and pick the best answer.

27.7 Notes

Algorithm 27.9 and Theorem 27.11 are due to Lenstra, Lenstra, and Lovász [190]. Schnorr [243] modified this algorithm to achieve an approximation factor of $(1+\varepsilon)^n$, for any $\varepsilon > 0$. More precisely, Schnorr's algorithm runs in polynomial time for a suitable choice of $\varepsilon = o(1)$, which results in a slightly subexponential factor algorithm, i.e., an algorithm with an approximation factor of $2^{o(n)}$. This is currently the best approximation guarantee known for shortest vector.

Algorithm 27.7 appears in Gauss [105], though it is stated there in terms of quadratic forms. Using a complex potential function argument, Lagarias [187] showed that this algorithm has polynomial running time (in the number of bits needed to write the input basis). For a polynomial time algorithm for shortest vector in d-dimensions, for fixed constant d, see Kannan [161]. Example 27.12 is due to Kannan [160]. Ajtai [3] showed that the shortest vector problem is **NP**-hard under randomized reductions. Theorem 27.19 is due to Lagarias, Lenstra, and Schnorr [188].

28 Counting Problems

The techniques for approximately counting the number of solutions to **#P**-complete problems are quite different from those for obtaining approximation algorithms for **NP**-hard optimization problems. Much of the former theory is built around the Markov chain Monte Carlo method, see Section 28.4 for references. In this chapter, we will present combinatorial algorithms (not using Markov chains) for two fundamental problems, counting the number of satisfying truth assignments for a DNF formula, and estimating the failure probability of an undirected network.

Intuitively, the class **#P** captures the problems of counting the number of solutions to **NP** problems. Let us formalize this notion. Let L be a language in **NP**, M be its associated verifier, and polynomial p be the bound on the length of its Yes certificates (see Section A.1). For string $x \in \Sigma^*$, define $f(x)$ to be the number of strings y such that $|y| \leq p(|x|)$ and $M(x, y)$ accepts. Functions $f : \Sigma^* \rightarrow \mathbf{Z}^+$ that arise in this manner constitute the class **#P**.

Function $f \in$ **#P** is said to be **#P**-*complete* if every function $g \in$ **#P** can be reduced to f in the following sense. There is a polynomial time transducer $R : \Sigma^* \rightarrow \Sigma^*$, that, given an instance, x, of g, produces an instance, $R(x)$, of f. Furthermore, there is a polynomial time computable function $S : \Sigma^* \times \mathbf{Z}^+ \rightarrow \mathbf{Z}^+$ that given x and $f(R(x))$ computes $g(x)$, i.e.,

$$\forall x \in \Sigma^*, g(x) = S(x, f(R(x))).$$

In other words, an oracle for f can be used to compute g in polynomial time.

The solution counting versions of almost all known **NP**-complete problems are **#P**-complete[1] (see Section 28.4 for an exception). Interestingly enough, other than a handful of exceptions, this is true of problems in **P** as well. This raises the question of designing polynomial time algorithms for approximately counting the number of solutions to these latter problems (see Exercise 28.3 regarding the question of approximately counting the number of solutions to **NP**-complete problems). These problems admit only two interesting possibilities: they either allow approximability to any required degree,

[1] In fact, typically a polynomial time reduction from one **NP**-complete problem to another maps solutions of the given instance to solutions of the transformed instance, and so preserves the number of solutions; hence, the proof of **#P**-completeness follows directly from the proof of **NP**-completeness.

or essentially not at all (see Section 28.4). The former possibility is captured in the definition of a *fully polynomial randomized approximation scheme*, abbreviated FPRAS.

Consider a problem in **P** whose counting version, f, is #**P**-complete. An algorithm $\mathcal{A}$ is an FPRAS for this problem if for each instance $x \in \Sigma^*$, and error parameter $\varepsilon > 0$,

$$\mathbf{Pr}[|\mathcal{A}(x) - f(x)| \leq \varepsilon f(x)] \geq \frac{3}{4},$$

and the running time of $\mathcal{A}$ is polynomial in $|x|$ and $1/\varepsilon$. (See Exercise 28.1 for a method for reducing the error probability of an FPRAS.)

28.1 Counting DNF solutions

Problem 28.1 (Counting DNF solutions) Let $f = C_1 \vee C_2 \vee \ldots \vee C_m$ be a formula in disjunctive normal form on n Boolean variables $x_1, \ldots, x_n$. Each clause C_i is of the form $C_i = l_1 \wedge l_2 \wedge \ldots \wedge l_{r_i}$, where each l_j is a literal, i.e., it is either a Boolean variable or its negation. We may assume w.l.o.g. that each clause is satisfiable, i.e., does not contain a variable and its negation. The problem is to compute #f, the number of satisfying truth assignments of f.

The main idea is to define an efficiently samplable random variable X which is an unbiased estimator for #f, i.e., $\mathbf{E}[X] = $ #f. If in addition, the standard deviation of X is within a polynomial factor of $\mathbf{E}[X]$, then an FPRAS for #f can be obtained in a straightforward manner by sampling X a polynomial number of times (in n and $1/\varepsilon$) and outputting the mean.

Constructing an unbiased estimator for #f is easy. Let random variable Y have uniform distribution on all 2^n truth assignments, and let $Y(\tau)$ be 2^n if τ satisfies f, and 0 otherwise (see Exercise 28.4). However, this random variable can have a very large standard deviation, and does not yield an FPRAS. For instance, suppose f has only polynomially many satisfying truth assignments. Then, with high probability, a polynomial number of randomly picked truth assignments will all have $Y = 0$, giving a poor estimate for #f.

We will rectify this by defining a random variable that assigns nonzero probability to only the satisfying truth assignments of f. Let S_i denote the set of truth assignments to $x_1, \ldots, x_n$ that satisfy clause C_i. Clearly, $|S_i| = 2^{n-r_i}$, where r_i is the number of literals in clause C_i. Also, #f $= |\cup_{i=1}^m S_i|$. Let $c(\tau)$ denote the number of clauses that truth assignment τ satisfies. Let M denote the multiset union of the sets S_i, i.e., it contains each satisfying truth assignment, τ, $c(\tau)$ number of times. Notice that $|M| = \sum_i |S_i|$ is easy to compute.

Pick a satisfying truth assignment, τ, for f with probability $c(\tau)/|M|$, and define $X(\tau) = |M|/c(\tau)$. We will first show that X can be efficiently sampled, i.e., using a randomized polynomial time algorithm.

Lemma 28.2 *Random variable X can be efficiently sampled.*

Proof: Picking a random element from the multiset M ensures that each truth assignment is picked with the desired probability. The following two-step process will accomplish this. First pick a clause so that the probability of picking clause C_i is $|S_i|/|M|$. Next, among the truth assignments satisfying the picked clause, pick one at random.

Now, the probability with which truth assignment τ is picked is

$$\sum_{i:\tau \text{ satisfies } C_i} \frac{|S_i|}{|M|} \times \frac{1}{|S_i|} = \frac{c(\tau)}{|M|}.$$

□

Lemma 28.3 *X is an unbiased estimator for #f.*

Proof:

$$\mathbf{E}[X] = \sum_{\tau} \mathbf{Pr}[\tau \text{ is picked}] \cdot X(\tau) = \sum_{\tau \text{ satisfies } f} \frac{c(\tau)}{|M|} \times \frac{|M|}{c(\tau)} = \#\text{f}.$$

□

X takes values only in a "polynomial range", thereby ensuring that its standard deviation is not large compared to its expectation. This fact is proved in the next lemma, and leads to the FPRAS construction.

Lemma 28.4 *If m denotes the number of clauses in f, then*

$$\frac{\sigma(X)}{\mathbf{E}[X]} \leq m - 1.$$

Proof: Denote $|M|/m$ by α. Clearly, $\mathbf{E}[X] \geq \alpha$. For each satisfying truth assignment τ of f, $1 \leq c(\tau) \leq m$. Therefore, $X(\tau)$ lies in the range $[\alpha, m\alpha]$, and so the random variable deviates from its mean by at most $(m-1)\alpha$, i.e., $|X(\tau) - \mathbf{E}[X]| \leq (m-1)\alpha$. Therefore, the standard deviation of X is bounded by $(m-1)\alpha$. Using the lower bound on $\mathbf{E}[X]$ stated above, we get the lemma. □

Finally, we will show that sampling X polynomially many times (in n and $1/\varepsilon$) and simply outputting the mean leads to an FPRAS for #f. Let X_k denote the mean of k samples of X.

Lemma 28.5 *For any* $\varepsilon > 0$,

$$\mathbf{Pr}[|X_k - \#f\,| \le \varepsilon \#f\,] \ge 3/4,$$

where $k = 4(m-1)^2/\varepsilon^2$.

Proof: We will use Chebyshev's inequality (see Section B.2), with $a = \varepsilon \mathbf{E}[X_k]$. Using the value of k stated above we get

$$\mathbf{Pr}[|X_k - \mathbf{E}[X_k]| \ge \varepsilon \mathbf{E}[X_k]] \le \left(\frac{\sigma(X_k)}{\varepsilon \mathbf{E}[X_k]}\right)^2 = \left(\frac{\sigma(X)}{\varepsilon \sqrt{k}\mathbf{E}[X]}\right)^2 \le \frac{1}{4},$$

where the equality follows by noting that $\mathbf{E}[X_k] = \mathbf{E}[X]$ and $\sigma(X_k) = \sigma(X)/\sqrt{k}$, and the last inequality follows by applying Lemma 28.4. The lemma follows. □

Theorem 28.6 *There is an FPRAS for the problem of counting DNF solutions.*

28.2 Network reliability

Problem 28.7 (Network reliability) Given a connected, undirected graph $G = (V, E)$, with failure probability p_e specified for each edge e, compute the probability that the graph becomes disconnected.

Graph G will become disconnected if all edges in some cut $(C, \overline{C})$, $C \subset V$ fail. We will present an FPRAS for this problem.

Let us first handle the case that each edge has the same failure probability, denoted by p. However, we will allow G to have parallel edges between any two vertices. Denote by FAIL(p) the probability that G gets disconnected. If FAIL(p) is at least inverse polynomial, then it can be efficiently estimated by Monte Carlo sampling (see proof of Theorem 28.11 for details). Let us handle the difficult case that FAIL(p) is small. Assume that FAIL$(p) \le n^{-4}$. The reason for this choice will become clear below.

The probability that cut $(C, \overline{C})$ gets disconnected is simply p^c where c is the number of edges crossing this cut. Since the failure probability of a cut decreases exponentially with capacity, the most important cuts for the purpose of estimating FAIL(p) are cuts with "small" capacity. The algorithm is built around two ideas:

1. For any $\varepsilon > 0$, we will show that only polynomially many "small" cuts (in n and $1/\varepsilon$) are responsible for $1-\varepsilon$ fraction of the total failure probability FAIL(p). Moreover, these cuts, say $E_1, \ldots E_k$, $E_i \subseteq E$, can be enumerated in polynomial time.

2. We will construct a polynomial sized DNF formula f whose probability of being satisfied is precisely the probability that at least one of these cuts fails.

As a result of the first idea, it is sufficient to estimate the probability that one of the cuts $E_1, \ldots, E_k$ fails. However, because of correlations, this is nontrivial. The second idea reduces this problem to counting DNF solutions, for which we have an FPRAS.

Formula f has a Boolean variable x_e for each edge e. x_e is set to true with probability p, the failure probability of edge e. Suppose cut $E_i = \{e_1, \ldots, e_j\}$. Construct the clause $D_i = x_{e_1} \wedge \cdots \wedge x_{e_j}$, i.e., the conjunct of all variables corresponding to edges in this cut. The probability that this clause is satisfied is precisely the failure probability of cut E_i. Finally, $f = D_1 \vee \cdots \vee D_k$, i.e., the disjunct of clauses corresponding to cuts.

28.2.1 Upperbounding the number of near-minimum cuts

The first idea has its roots in the fact that one can place upper bounds on the number of minimum and near-minimum capacity cuts in an undirected graph. Let c be the capacity of a minimum cut in G. Recall that all edges in G are assumed to be of unit capacity, and that G is allowed to have parallel edges between any two vertices.

Lemma 28.8 *The number of minimum cuts in $G = (V, E)$ is bounded by $n(n-1)/2$.*

Proof: By *contracting* an edge (u, v) in a graph we mean merging the vertices u and v into a single vertex. All edges running between u and v are discarded. Those running between u or v and some other vertex w will now run between the merged vertex and w, their number being conserved.

Now consider the following *random contraction* process. Iteratively, pick a random edge (u, v) in the current graph and contract it. Terminate when exactly two vertices are left. Suppose these two vertices correspond to sets S and $V - S$, $S \subset V$, of vertices of the starting graph G. Then, the algorithm outputs the cut $(S, \overline{S})$. We will say that this cut *survives*. Clearly, a cut survives iff none of its edges is contracted during the algorithm.

Let $(C, \overline{C})$ be any minimum cut in G. We will show

$$\mathbf{Pr}[(C, \overline{C}) \text{ survives}] \geq \frac{1}{\binom{n}{2}}.$$

This statement yields the lemma via an interesting argument. Let M be the number of minimum cuts in G. The survival of each of these cuts is a mutually exclusive event, and the total probability of these events adds up to at most 1. Hence $M/(n(n-1)/2) \leq 1$, thereby giving the desired bound.

Consider an arbitrary iteration in the random contraction process, and let H be a graph at the beginning of this iteration. Since the process of contraction cannot decrease the capacity of the minimum cut, the capacity of each cut in H is at least c. This holds for cuts separating one vertex of H from the rest. Therefore, the degree of each vertex in H must be at least c. Hence, H must have at least $cm/2$ edges, where m is the number of vertices in H.

Now, the conditional probability that cut $(C, \overline{C})$ survives the current iteration, given that it has survived so far, is at least $(1 - \frac{c}{cm/2}) = (1 - 2/m)$ (this is simply the probability that the randomly chosen edge in this iteration is not picked from the cut $(C, \overline{C})$). The probability that $(C, \overline{C})$ survives the whole algorithm is simply the product of these conditional probabilities. This gives

$$\mathbf{Pr}[(C, \overline{C}) \text{ survives}] \geq \left(1 - \frac{2}{n}\right)\left(1 - \frac{2}{n-1}\right) \cdots \left(1 - \frac{2}{3}\right) = \frac{1}{\binom{n}{2}}.$$

□

For $\alpha \geq 1$, we will say that a cut is an *α-min cut* if its capacity is at most αc.

Lemma 28.9 *For any $\alpha \geq 1$, the number of α-min cuts in G is at most $n^{2\alpha}$.*

Proof: We will prove the lemma for the case that α is a half-integer. The proof for arbitrary α follows by applying the same ideas to generalized binomial coefficients. Let $2\alpha = k$.

Consider the following two-phase process: First, run the random contraction algorithm until there are k vertices remaining in the graph. Next, pick a random cut among all 2^{k-1} cuts in this graph. This will define a cut in the original graph.

Let $(C, \overline{C})$ be any α-min cut in G. We will show that the probability that it survives the two phase process is at least $1/n^{2\alpha}$, thereby proving the desired bound.

Let H be the graph at the beginning of an arbitrary iteration in the first phase. As argued in Lemma 28.8, if H has m vertices, it has at least $mc/2$ edges. Therefore, the conditional probability that $(C, \overline{C})$ survives the current iteration, given that it has survived so far, is at least $1 - \frac{\alpha c}{mc/2} = 1 - 2\alpha/m$. The probability that $(C, \overline{C})$ survives the first phase is at least

$$\left(1 - \frac{2\alpha}{n}\right)\left(1 - \frac{2\alpha}{n-1}\right) \cdots \left(1 - \frac{2\alpha}{k+1}\right) = \frac{1}{\binom{n}{k}}.$$

The conditional probability that $(C, \overline{C})$ survives the second phase, given that it has survived the first, is $1/2^{k-1}$. Therefore,

$$\mathbf{Pr}[(C,\overline{C}) \text{ survives both phases}] \geq \frac{1}{\binom{n}{k}2^{k-1}} \geq \frac{1}{n^k} = \frac{1}{n^{2\alpha}}.$$

□

28.2.2 Analysis

Recall that we are considering the case that $\text{FAIL}(p) \leq n^{-4}$. We can now justify this choice. The failure probability of a minimum cut is $p^c \leq \text{FAIL}(p) \leq n^{-4}$. Let $p^c = n^{-(2+\delta)}$, where $\delta \geq 2$. Now, by Lemma 28.9, for any $\alpha \geq 1$, the total failure probability of all cuts of capacity αc is at most $p^{\alpha c} n^{2\alpha} = n^{-\alpha\delta}$. This rapid decrease in the total failure probability of all cuts of capacity αc will enable us to bound the total failure probability of "large" capacity cuts.

Lemma 28.10 *For any* $\alpha \geq 1$,

$$\mathbf{Pr}[\textit{some cut of capacity} > \alpha c \textit{ fails}] \leq n^{-\alpha\delta}\left(1 + \frac{2}{\delta}\right).$$

Proof: Number all cuts in G by increasing capacity. Let c_k and p_k denote the capacity and failure probability of the kth cut in this numbering. Let a be the number of the first cut of capacity greater than αc. It suffices to show that

$$\sum_{k \geq a} p_k \leq n^{-\alpha\delta}\left(1 + \frac{2}{\delta}\right).$$

We will evaluate this sum in two steps. First, we will consider the first $n^{2\alpha}$ terms. Each of these terms is at most $p^{\alpha c} = n^{-\alpha(2+\delta)}$. Therefore, their sum is at most $n^{-\alpha\delta}$.

Next, let us bound the sum of the remaining terms. Clearly, this sum is bounded by $\sum_{k>n^{2\alpha}} p_k$. By Lemma 28.9, there are at most $n^{2\alpha}$ cuts of capacity bounded by αc. Therefore, $c_{n^{2\beta}} \geq \beta c$. Writing $k = n^{2\beta}$ we get $c_k \geq c \ln k / 2 \ln n$, and

$$p_k = p^{c_k} \leq (p^c)^{\frac{\ln k}{2 \ln n}} = k^{-(1+\delta/2)}.$$

Therefore,

$$\sum_{k>n^{2\alpha}} p_k \leq \int_{n^{2\alpha}}^{\infty} k^{-(1+\delta/2)} \mathrm{d}k \leq \frac{2}{\delta} n^{-\alpha\delta}.$$

This proves the lemma. □

Theorem 28.11 *There is an FPRAS for estimating network reliability.*

Proof: We will first consider the case that each edge in graph G has the same failure probability, p.

If $\mathrm{FAIL}(p) > n^{-4}$, then we will resort to Monte Carlo sampling. Flip a coin with bias p for failure of each edge, and check if G is disconnected. Repeat this experiment $O(\log n/(\varepsilon^2 \mathrm{FAIL}(p)))$ times, and output the mean number of times G is disconnected. A straightforward application of Chernoff bounds shows that the mean lies in $[(1-\varepsilon)\mathrm{FAIL}(p), (1+\varepsilon)\mathrm{FAIL}(p)]$ with high probability.

Next, assume that $\mathrm{FAIL}(p) \leq n^{-4}$. Now, for any $\varepsilon > 0$, we want to determine α such that the total failure probability of all cuts of capacity $> \alpha c$ is bounded by $\varepsilon \mathrm{FAIL}(p)$. By Lemma 28.10, it suffices to find α such that

$$n^{-\alpha\delta}\left(1+\frac{2}{\delta}\right) \leq \varepsilon \mathrm{FAIL}(p) \leq \varepsilon n^{-(2+\delta)}.$$

Solving, we get

$$\alpha = 1 + \frac{2}{\delta} - \frac{\ln(\varepsilon)}{\delta \ln n} \leq 2 - \frac{\ln(\varepsilon)}{2 \ln n}.$$

By Lemma 28.9, $c_{n^{2\alpha}} > \alpha c$. For the value of α computed,

$$\mathbf{Pr}[\text{one of the first } n^{2\alpha} \text{ cuts fails}] \geq (1-\varepsilon)\mathrm{FAIL}(p).$$

The first $n^{2\alpha} = O(n^4/\varepsilon)$ cuts can be enumerated in polynomial time (see Exercises). We will use these to construct the corresponding DNF formula, and estimate the probability that it is satisfiable, as described above.

Finally, we show how to "reduce" the case of arbitrary edge failure probabilities to the simpler case analyzed above. Suppose edge e has failure probability p_e. Choose a small parameter θ. Replace edge e with $k_e = -(\ln p_e)/\theta$ parallel edges each with failure probability $1-\theta$. Then, the probability that all k_e edges fail is

$$(1-\theta)^{-(\ln p_e)/\theta}.$$

As $\theta \to 0$, this failure probability converges to p_e. Let H be the graph obtained by doing this transformation on each edge of G. In the limit as $\theta \to 0$, each cut in H has the same failure probability as that in G.

Let us give an efficient implementation of this idea. All we really want is a listing of the "small" capacity cuts in G. Once this is done, we can apply the more general DNF counting algorithm developed in Exercise 28.5, where each variable is set to true with its own probability p_e. Observe that changing

θ scales the capacities of cuts in H without changing their relative values. Thus, it suffices to assign a weight of $-\ln p_e$ to each edge e of G, and find "small" capacity cuts in this graph. This completes the proof. □

28.3 Exercises

28.1 Given an FPRAS for a problem, show that its success probability can be improved to $1-\delta$, for any $\delta > 0$, by a multiplicative increase in the running time of only $O(\log(1/\delta))$.
Hint: Run the FPRAS $O(\log(1/\delta))$ times and output the median value.

28.2 Suppose we make the definition of an FPRAS more stringent by requiring it to have a fixed additive error α with high probability, i.e.,

$$\mathbf{Pr}[f(x) - \alpha \leq A(x) \leq f(x) + \alpha] \geq \frac{3}{4}.$$

Show that if there were such an algorithm for a #**P**-complete problem, then **P**= **NP**.

28.3 Show that if there were an FPRAS for counting the number of satisfying truth assignments to SAT then every problem in **NP** could be solved in random polynomial time. How weak an approximate counting algorithm for SAT suffices to give this consequence? What does this say for the question of approximately counting the number of solutions to other **NP**-complete problems?
Hint: Use solution amplification. Given SAT formula f, define formula f' over k new Boolean variables which is a tautology. Then the number of solutions of $\phi = f \wedge f'$ is $\#\text{f} \cdot 2^k$.

28.4 Given a DNF formula f, let Y be a random variable that on a random truth assignment τ is 2^n if τ satisfies f and 0 otherwise. Show that Y is an unbiased estimator for #f. How large can the ratio $\sigma(Y)/\mathbf{E}[Y]$ be?

28.5 (Karp and Luby [172]) You are given a DNF formula f on n Boolean variables, $x_1, \ldots, x_n$, and probabilities $p_1, \ldots, p_n$ with which these variables are (independently) set to true. Let D denote the resulting probability distribution over the 2^n truth assignments to the Boolean variables, and p denote the probability that f is satisfied by a truth assignment picked from D. Construct an FPRAS for estimating p.
Hint: Let q_i denote the probability that clause C_i is satisfied by a truth assignment picked from D, and $Q = \sum_i q_i$. Now, consider random variable X that assigns to each satisfying truth assignment τ a probability of $\mathbf{Pr}_D[\tau]c(\tau)/Q$, and define $X(\tau) = Q/c(\tau)$.

28.6 A *uniform generator* for an **NP** problem Π is a randomized polynomial time algorithm $\mathcal{A}$ that given an instance I of a problem, outputs either a solution to I, or else the special symbol "$\perp$", such that

- each solution to I is output with the same probability, i.e., there is a number $\alpha \in (0, 1]$ such that

$$\mathbf{Pr}[\mathcal{A} \text{ outputs } s] = \alpha, \text{ for each solution } s \text{ of } I, \text{ and}$$

- the probability of outputting $\perp$, i.e., failing to output a solution, is $< 1/2$.

Give a uniform generator for picking a random satisfying truth assignment to a given DNF formula.
Hint: The essential idea behind the construction of random variable X works.

28.7 (Jerrum, Valiant, and Vazirani [154]) Let Π be an **NP** problem that is self-reducible (see Section A.5 and Exercise 1.15). Show that there is an FPRAS for Π iff there is an almost uniform generator for it. An *almost uniform generator* for Π is a randomized polynomial time algorithm $\mathcal{A}$ such that for any $\mu > 0$ and instance I of Π, there is a number $\alpha \in (0, 1]$ such that

- for each solution s of I, $\mathbf{Pr}[\mathcal{A} \text{ outputs } s] \in [(1-\mu)\alpha, (1+\mu)\alpha]$,
- $\mathbf{Pr}[\mathcal{A} \text{ fails to output a solution}] < 1/2$, and
- the running time of $\mathcal{A}$ is polynomial in $|I|$ and $\log(1/\mu)$.

Observe that unlike an FPRAS, which can only achieve inverse polynomial error, a uniform generator can achieve inverse exponential error (μ), in polynomial time.
Hint: For the forward direction, first construct a uniform generator, assuming that the FPRAS makes no error. (Traverse down the self-reducibility tree for I, with biases determined by estimates on the number of solutions. Accept leaf with appropriate probability, to achieve uniform generation.) Use the fact that the error probability of the FPRAS can be made exponentially small to obtain an almost uniform generator. For the reverse direction, obtain instance I_α, with $|I_\alpha| < |I|$, and a good estimate of the ratio of the number of solutions to I and I_α.

28.8 (Jerrum, Valiant, and Vazirani [154]) This exercise leads to strong evidence that the problem of estimating the number of simple cycles in a directed graph is essentially not approximable. Show that if there is an almost uniform generator for this problem, then there is a randomized polynomial time algorithm for deciding if a given directed graph has a Hamiltonian cycle.

Hint: Obtain a graph G' from G that amplifies the number of cycles of each length. However, it amplifies bigger cycles more than it amplifies smaller cycles, so that most cycles in G' are of maximum length and correspond to the largest cycles in G.

28.9 Show that the random contraction algorithm of Lemma 28.8 can be used to obtain a randomized algorithm for finding a minimum cut in an undirected graph.

28.10 (Karger and Stein [166]) Obtain a randomized algorithm for enumerating all α-min cuts in G using the random contraction algorithm and Lemma 28.9.

In the next three exercises (from Vazirani and Yannakakis [260]), we will develop a deterministic algorithm for enumerating in an undirected graph by increasing weight, with *polynomial delay*, i.e., the algorithm spends polynomial time between successive outputs.

Assume that graph $G = (V, E)$ has n vertices besides s and t, numbered 1 to n. Every s–t cut in G can be represented as an n bit 0/1 vector. A partially specified cut, in which the sides of only vertices numbered 1 to k are decided, is represented as a k bit 0/1 vector. Consider a binary tree T of height n. Its leaves represent s–t cuts and internal nodes represent partially specified cuts. All cuts consistent with a partially specified cut lie in the subtree rooted at it. Clearly, a minimum weight cut in this subtree can be computed with one max-flow computation.

28.11 Let $\boldsymbol{a}$ be an $n-k$ bit 0/1 vector representing a partially specified cut, as well as an internal node in T. The subtree, T', rooted at this node is of height k and contains 2^k leaves (s–t cuts). Among these 2^k cuts, let $\boldsymbol{a}'$ be a minimum weight cut. Show how the remaining $2^k - 1$ cuts of T' can be partitioned into k subtrees which are of height $0, 1, \ldots, k-1$.

28.12 Using a heap, give an algorithm for enumerating s–t cuts in G by increasing weight.
Hint: The heap is initialized with a minimum cut in G. At an arbitrary point, the cuts not enumerated so far can be partitioned into subtrees (see Exercise 28.11). The heap contains a minimum cut from each subtree.

28.13 Give an algorithm for enumerating all cuts in an undirected graph by increasing weight with polynomial delay.
Hint: Assume that the graph has a special vertex s, which always goes on side 0 of the cut, and n other vertices, numbered 1 to n. A cut is specified by an n bit vector specifying the sides of vertices numbered 1 to n. The main difference arises in finding a minimum cut in the subtree rooted at the internal node $0^k, k < n$. This is done by finding a minimum cut separating the vertices $s, 1, \ldots, i$ from vertex $i+1$ for $k \leq i < n$, and picking the lightest of these cuts.

28.14 (Karger [163]) Consider the generalization of network reliability to estimating the probability that G disconnects into r or more components, where r is a fixed constant. Obtain an FPRAS for this problem.

28.4 Notes

The counting class **#P** was defined by Valiant [257]. The FPRAS for counting DNF solutions is due to Karp and Luby [172], who also gave the definition of FPRAS (see also Karp, Luby, and Madras [173]). The FPRAS for estimating network reliability is due to Karger [163].

Most algorithms for approximate counting work by constructing an almost uniform generator for the problem and appealing to the equivalence established in Exercise 28.7. Broder [34] introduced the use of rapidly mixing Markov chains for almost uniform generation (see also Mihail [213]).

Jerrum and Sinclair [152] gave the first FPRAS using this approach, for counting the number of perfect matchings in dense bipartite graphs (each vertex should have a degree $\geq n/2$; see also Section 30.3). They also showed that a crude approximate counter, with polynomial error, can be transformed into an FPRAS (with inverse polynomial error), by defining an appropriate Markov chain on the self-reducibility tree of an instance. As a result, **#P**-complete problems either admit an FPRAS or are essentially not approximable at all (see Exercise 28.8). For an example of a problem that is **NP**-complete but whose counting version is not in **#P**, unless **P** = **NP**, see Durand, Hermann, and Kolaitis [68] (however, they do establish completeness of the counting problem in a higher class: **#NP**). For Markov–chain based approximate counting algorithms, see Jerrum and Sinclair [149], Sinclair [250], and the references in Section 30.3.

29 Hardness of Approximation

A remarkable achievement of the theory of exact algorithms is that it has provided a fairly complete characterization[1] of the intrinsic complexity of natural computational problems, modulo some strongly believed conjectures. Recent impressive developments raise hopes that we will some day have a comprehensive understanding of the approximability of **NP**-hard optimization problems as well. In this chapter we will give a brief overview of these developments.

Current hardness results fall into three important classes. For minimization problems, the hardness factors for these classes are constant (> 1), $\Omega(\log n)$, and n^{ε} for a fixed constant $\varepsilon > 0$, where n is the size of the instance. For maximization problems, the factors are constant (< 1), $O(1/\log n)$, and $1/n^{\varepsilon}$ for a fixed $\varepsilon > 0$. In this chapter we will present hardness results for MAX-3SAT, vertex cover, and Steiner tree in the first class, set cover in the second class, and clique in the third class. For all these problems, we will establish hardness for their cardinality versions, i.e., the unit cost case.

29.1 Reductions, gaps, and hardness factors

Let us start by recalling the methodology for establishing hardness results for exact optimization problems. The main technical core is the Cook–Levin theorem which establishes the hardness, assuming $\mathbf{P} \neq \mathbf{NP}$, of distinguishing between instances of SAT that are satisfiable and those that are not. To show hardness of computing an optimal solution to, say the cardinality vertex cover problem, one shows, via a polynomial time reduction from SAT, that it is hard to distinguish between graphs that have covers of size at most k from graphs that don't, where k is provided as part of the input. Since an exact algorithm can make this distinction, this reduction establishes the non-existence of an efficient exact algorithm.

The main technical core of hardness of approximation results is the PCP theorem, which is stated in Section 29.2. For establishing a hardness of approximation result for, say, the vertex cover problem, this theorem is used to

[1] A few (important) exceptions, such as the graph isomorphism problem, remain uncharacterized.

show the following polynomial time reduction. It maps an instance ϕ of SAT to a graph $G = (V, E)$ such that

- if ϕ is satisfiable, G has a vertex cover of size $\leq \frac{2}{3}|V|$, and
- if ϕ is not satisfiable, the smallest vertex cover in G is of size $> \alpha \cdot \frac{2}{3}|V|$,

where $\alpha > 1$ is a fixed constant.

Claim 29.1 *As a consequence of the reduction stated above, there is no polynomial time algorithm for vertex cover that achieves an approximation guarantee of α, assuming* $\mathbf{P} \neq \mathbf{NP}$.

Proof: Essentially, this reduction establishes the hardness, assuming $\mathbf{P} \neq \mathbf{NP}$, of distinguishing graphs having a cover of size $\leq \frac{2}{3}|V|$ from those having a cover of size $> \alpha \cdot \frac{2}{3}|V|$. An approximation algorithm for vertex cover, having a guarantee of α or better, will find a cover of size $\leq \alpha \cdot \frac{2}{3}|V|$ when given a graph G from the first class. Thus, it will be able to distinguish the two classes of graphs, leading to a contradiction. □

The reduction stated above introduces a gap, of factor α, in the optimal objective function value achieved by the two classes of graphs (if $\alpha = 1$ then this is an ordinary polynomial time reduction from SAT to vertex cover). Let us formally state the central notion of a *gap-introducing reduction.* The definition is slightly different for minimization and maximization problems. For simplicity, let us assume that we are always reducing from SAT.

Let Π be a minimization problem. A gap-introducing reduction from SAT to Π comes with two parameters, functions f and α. Given an instance ϕ of SAT, it outputs, in polynomial time, an instance x of Π, such that

- if ϕ is satisfiable, $\mathrm{OPT}(x) \leq f(x)$, and
- if ϕ is not satisfiable, $\mathrm{OPT}(x) > \alpha(|x|) \cdot f(x)$.

Notice that f is a function of the instance (such as $\frac{2}{3}|V|$ in the example given above), and α is a function of the size of the instance. Since Π is a minimization problem, the function α satisfies $\alpha(|x|) \geq 1$.

If Π is a maximization problem, we want the reduction to satisfy

- if ϕ is satisfiable, $\mathrm{OPT}(x) \geq f(x)$, and
- if ϕ is not satisfiable, $\mathrm{OPT}(x) < \alpha(|x|) \cdot f(x)$.

In this case, $\alpha(|x|) \leq 1$. The gap, $\alpha(|x|)$, is precisely the hardness factor established by the gap-introducing reduction for the **NP**-hard optimization problem.

Once we have obtained a gap-introducing reduction from SAT (or any other **NP**-hard problem) to an optimization problem, say Π_1, we can prove a hardness result for another optimization problem, say Π_2, by giving a special reduction, called a *gap-preserving reduction*, from Π_1 to Π_2. Now there are four possibilities, depending on whether Π_1 and Π_2 are minimization or maximization problems. We give the definition below assuming that Π_1 is a minimization problem and Π_2 is a maximization problem. The remaining cases are similar.

A gap-preserving reduction, Γ, from Π_1 to Π_2 comes with four parameters (functions), f_1, α, f_2, and β. Given an instance x of Π_1, it computes, in polynomial time, an instance y of Π_2 such that

- $\text{OPT}(x) \le f_1(x) \Rightarrow \text{OPT}(y) \ge f_2(y)$,
- $\text{OPT}(x) > \alpha(|x|)f_1(x) \Rightarrow \text{OPT}(y) < \beta(|y|)f_2(y)$.

Observe that x and y are instances of two different problems, and so it would be more appropriate to write $\text{OPT}_{\Pi_1}(x)$ and $\text{OPT}_{\Pi_2}(y)$ instead of $\text{OPT}(x)$ and $\text{OPT}(y)$, respectively. However, we will avoid this extra notation, since the context clarifies the problems being talked about. In keeping with the fact that Π_1 is a minimization problem and Π_2 is a maximization problem, $\alpha(|x|) \ge 1$ and $\beta(|y|) \le 1$.

Composing a gap-introducing reduction with a gap-preserving reduction gives a gap-introducing reduction, provided all the parameters match up. For example, suppose that in addition to the reduction Γ defined above, we have obtained a gap-introducing reduction, Γ', from SAT to Π_1, with parameters f_1 and α. Then, composing Γ' with Γ, we get a gap-introducing reduction from SAT to Π_2, with parameters f_2 and β. This composed reduction shows that there is no $\beta(|y|)$ factor approximation algorithm for Π_2, assuming $\mathbf{P} \neq \mathbf{NP}$. In each gap-preserving reduction stated below, we will take special care to ensure that the parameters match up.

Remark 29.2

- The "gap" β can, in general, be bigger or smaller than α. In this sense, "gap-preserving" is a slight misnomer.
- We do not require any guarantee from reduction Γ if instance x of Π_1 falls in the first gap, i.e., satisfies $f_1(x) < \text{OPT}(x) \le \alpha(|x|)f_1(x)$.
- An approximation algorithm for Π_2 together with a gap-preserving reduction Γ from Π_1 to Π_2 does not necessarily yield an approximation algorithm for Π_1. Observe the contrast with an approximation factor preserving reduction (see Section A.3.1 for definition). The latter reduction additionally requires a means of transforming a near-optimal solution to the transformed instance y of Π_2 into a near-optimal solution to the given instance x of Π_1.
  On the other hand, Γ together with an appropriate gap-introducing reduction from SAT to Π_1 does suffice for proving a hardness of approximation result for Π_2. Obviously the less stringent requirement on gap-preserving reductions makes them easier to design.
- We have already presented some gap-introducing reductions, e.g., Theorems 3.6 and 5.7. The reader may wonder why these do not suffice as the starting point for further hardness results and why the PCP theorem was needed. The reason is that these reductions simply exploit the freedom to choose edge costs and not the deep combinatorial structure of the problem.

The following figure shows the gap-preserving reductions presented in this chapter:

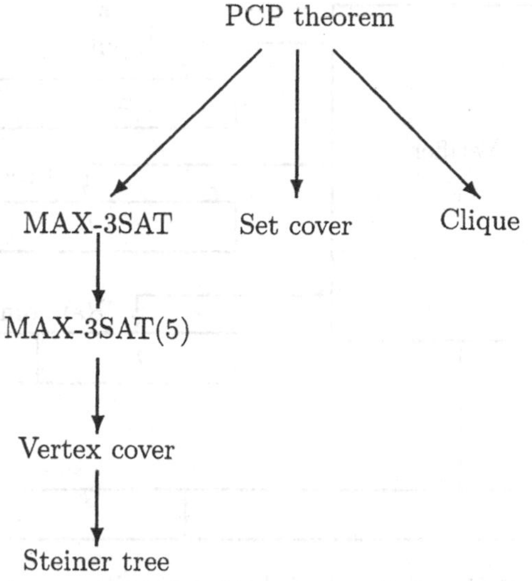

29.2 The PCP theorem

Probabilistic characterizations of the class **NP** yield a general technique for obtaining gap-introducing reductions. The most useful of these characterizations is captured in the PCP theorem. PCP stands for *probabilistically checkable proof systems.*

Recall the usual definition of **NP** (see Appendix A) as the class of languages whose yes instances support short (polynomial in the length of the input) witnesses that can be verified quickly (in polynomial time). Informally, a probabilistically checkable proof for an **NP** language encodes the witness in a special way so that it can be verified probabilistically by examining very few of its bits.

A probabilistically checkable proof system comes with two parameters, the number of random bits required by the verifier, and the number of bits of the witness that the verifier is allowed to examine. In keeping with established terminology, let us call a witness string the *proof.* The most useful setting for these parameters is $O(\log n)$ and $O(1)$, respectively. This defines the class $\mathbf{PCP}(\log n, 1)$.

The *verifier* is a polynomial time Turing machine which, besides its input tape and work tape, has a special tape that provides it with a string of random bits and another special tape on which it is provided with the proof. The machine can read any bit of the proof by simply specifying its location. Of course, the particular locations it examines are a function of the input string and the random string. At the end of its computation, the machine goes into either an accept state or a reject state.

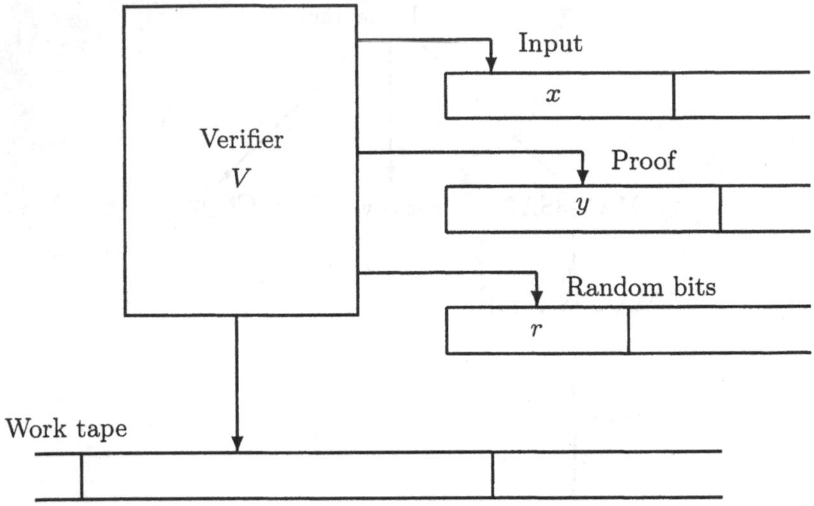

A language $L \in \mathbf{PCP}(\log n, 1)$ if there is a verifier V, and constants c and q, such that on input x, V obtains a random string, r, of length $c \log |x|$ and queries q bits of the proof. Furthermore,

- if $x \in L$, then there is a proof y that makes V accept with probability 1,
- if $x \notin L$, then for every proof y, V accepts with probability $< 1/2$,

where the probability is over the random string r. The probability of accepting in case $x \notin L$ is called the *error probability.*

In general, for two functions $r(n)$ and $q(n)$, we can define the class $\mathbf{PCP}(r(n), q(n))$, under which the verifier obtains $O(r(n))$ random bits and queries $O(q(n))$ bits of the proof. The acceptance criteria for input strings are the same as above. In this terminology, $\mathbf{NP} = \mathbf{PCP}(0, \text{poly}(n))$, where $\text{poly}(n) = \bigcup_{k \geq 0} \{n^k\}$. In this case, the verifier is not allowed any random bits. It must deterministically accept strings in the language and reject strings not in the language, as in the definition of **NP**. The PCP theorem gives another characterization of **NP**.

Theorem 29.3 $\mathbf{NP} = \mathbf{PCP}(\log n, 1)$.

One half of this theorem, that $\mathbf{PCP}(\log n, 1) \subseteq \mathbf{NP}$, is easy to prove (see Exercise 29.1). The other half, that $\mathbf{NP} \subseteq \mathbf{PCP}(\log n, 1)$, is a difficult result, and gives a useful tool for establishing hardness of approximation results. The currently known proof of this half is too complicated for exposition in this book. Fortunately, the statement of the theorem is sufficient to derive the hardness results.

In order to provide the reader with some feel for the PCP theorem, let us make an observation. It is easy to construct a verifier for 3SAT whose error probability (i.e., probability of accepting unsatisfiable formulae) is $\leq 1-1/m$, where m is the number of clauses in the input 3SAT formula, say ϕ. The verifier expects a satisfying truth assignment to ϕ as the proof. It uses the $O(\log n)$ random bits to pick a random clause of ϕ. It then reads the truth assignments for the three variables occurring in this clause. Notice that this is only a constant number of bits. It accepts iff the truth setting for these three variables satisfies the clause. Clearly, if ϕ is satisfiable, there is a proof that makes the verifier accept with probability 1, and if ϕ is not satisfiable, on every proof, the verifier accepts with probability $\leq 1 - 1/m$. The interesting and difficult part of the PCP theorem is decreasing the error probability to $< 1/2$, even though the verifier is allowed to read only a constant number of bits of the proof. It involves a complex algebraic construction that ensures that small parts of the proof depend on every bit of the input.

The PCP theorem directly gives an optimization problem – in particular, a maximization problem – for which there is no factor 1/2 approximation algorithm, assuming $\mathbf{P} \neq \mathbf{NP}$.

Problem 29.4 (Maximize accept probability) Let V be a $\mathbf{PCP}(\log n, 1)$ verifier for SAT. On input ϕ, a SAT formula, find a proof that maximizes the probability of acceptance of V.

Claim 29.5 *Assuming* $\mathbf{P} \neq \mathbf{NP}$, *there is no factor* $1/2$ *approximation algorithm for Problem 29.4.*

Proof: If ϕ is satisfiable, then there is a proof that makes V accept with probability 1, and if ϕ is not satisfiable, then on every proof, V accepts with probability $< 1/2$. Suppose there is a factor 1/2 approximation algorithm for Problem 29.4. If ϕ is satisfiable, then this algorithm must provide a proof on which V's acceptance probability is $\geq 1/2$. The acceptance probability can be computed in polynomial time, by simply simulating V for all random strings of length $O(\log n)$. Thus, this approximation algorithm can be used for deciding SAT in polynomial time, contradicting the assumption $\mathbf{P} \neq \mathbf{NP}$. □

Claim 29.5 directly gives the following corollary. In subsequent sections, we will use the PCP theorem to obtain hardness results for natural computational problems. A similar corollary follows in each case.

Corollary 29.6 *Assuming* $\mathbf{P} \neq \mathbf{NP}$, *there is no PTAS for Problem 29.4.*

29.3 Hardness of MAX-3SAT

MAX-3SAT is the restriction of MAX-SAT (see Problem 16.1) to instances in which each clause has at most three literals. This problem plays a similar role in hardness of approximation as 3SAT plays in the theory of **NP**-hardness, as a "seed" problem from which reductions to numerous other problems have been found. The main result of this section is:

Theorem 29.7 *There is a constant $\varepsilon_M > 0$ for which there is a gap-introducing reduction from SAT to MAX-3SAT that transforms a Boolean formula ϕ to ψ such that*

- *if ϕ is satisfiable,* $\text{OPT}(\psi) = m$, *and*
- *if ϕ is not satisfiable,* $\text{OPT}(\psi) < (1 - \varepsilon_M)m$,

where m is the number of clauses in ψ.

Corollary 29.8 *There is no approximation algorithm for MAX-3SAT with an approximation guarantee of $1 - \varepsilon_M$, assuming* $\mathbf{P} \neq \mathbf{NP}$, *where $\varepsilon_M > 0$ is the constant defined in Theorem 29.7.*

The exact solution of MAX-3SAT is shown hard under the assumption $\mathbf{P} \neq \mathbf{NP}$. It is interesting to note that hardness of approximate solution of MAX-3SAT is also being established under the same assumption.

For clarity, let us break the proof into two parts. We will first prove hardness for the following problem.

Problem 29.9 (MAX k-FUNCTION SAT) Given n Boolean variables $x_1, \ldots, x_n$ and m functions $f_1, \ldots, f_m$, each of which is a function of k of the Boolean variables, find a truth assignment to $x_1, \ldots, x_n$ that maximizes the number of functions satisfied. Here k is assumed to be a fixed constant. Thus, we have a class of problems, one for each value of k.

Lemma 29.10 *There is a constant k for which there is a gap-introducing reduction from SAT to MAX k-FUNCTION SAT that transforms a Boolean formula ϕ to an instance I of MAX k-FUNCTION SAT such that*

- *if ϕ is satisfiable,* $\text{OPT}(I) = m$, *and*
- *if ϕ is not satisfiable,* $\text{OPT}(I) < \frac{1}{2}m$,

where m is the number of formulae in I.

Proof: Let V be a $\mathbf{PCP}(\log n, 1)$ verifier for SAT, with associated parameters c and q. Let ϕ be an instance of SAT of length n. Corresponding to each string, r, of length $c \log n$ (the "random" string), V reads q bits of the proof. Thus, V reads a total of at most qn^c bits of the proof. We will have a Boolean variable corresponding to each of these bits. Let B be the set of Boolean variables. Thus, the relevant part of each proof corresponds to a truth assignment to the variables in B.

We will establish the lemma for $k = q$. Corresponding to each string r, we will define a Boolean function, f_r. This will be a function of q variables from B. The acceptance or rejection of V is of course a function of ϕ, r, and the q bits of the proof read by V. For fixed ϕ and r, consider the restriction of this function to the q bits of the proof. This is the function f_r.

Clearly, there is a polynomial time algorithm which, given input ϕ, outputs the $m = n^c$ functions f_r. If ϕ is satisfiable, there is a proof that makes V accept with probability 1. The corresponding truth assignment to B satisfies all n^c functions f_r. On the other hand, if ϕ is not satisfiable, then on every proof, V accepts with probability $< 1/2$. Thus, in this case every truth assignment satisfies $< \frac{1}{2}n^c$ of these functions. The lemma follows. □

Proof of Theorem 29.7: Using Lemma 29.10 we transform a SAT formula ϕ to an instance of MAX k-FUNCTION SAT. We now show how to obtain a 3SAT formula from the n^c functions.

Each Boolean function f_r constructed in Lemma 29.10 can be written as a SAT formula, say ψ_r, containing at most 2^q clauses. Each clause of ψ_r contains at most q literals. Let ψ be the SAT formula obtained by taking the conjunct of all these formulae, i.e., $\psi = \bigwedge_r \psi_r$.

If a truth assignment satisfies formula f_r, then it satisfies all clauses of ψ_r. On the other hand, if it does not satisfy f_r, then it must leave at least one clause of ψ_r unsatisfied. Therefore, if ϕ is not satisfiable, any truth assignment must leave $> \frac{1}{2}n^c$ clauses of ψ unsatisfied.

Finally, let us transform ψ into a 3SAT formula. This is done using the standard trick of introducing new variables to obtain small clauses from a big clause. Consider clause $C = (x_1 \vee x_2 \vee \ldots \vee x_k)$, with $k > 3$. Introduce $k-2$ new Boolean variables, $y_1, \ldots, y_{k-2}$, and consider the formula

$$f = (x_1 \vee x_2 \vee y_1) \wedge (\overline{y}_1 \vee x_3 \vee y_2) \wedge \ldots \wedge (\overline{y}_{k-2} \vee x_{k-1} \vee x_k).$$

Let τ be any truth assignment to $x_1, \ldots, x_k$. If τ satisfies C, then it can be extended to a truth assignment satisfying all clauses of f. On the other hand, if τ does not satisfy C, then for every way of setting $y_1, \ldots, y_{k-2}$, at least one of the clauses of f remains unsatisfied.

We apply this construction to every clause of ψ containing more than 3 literals. Let ψ' be the resulting 3SAT formula. It contains at most $n^c 2^q(q-2)$ clauses. If ϕ is satisfiable, then there is a truth assignment satisfying all clauses of ψ'. If ϕ is not satisfiable, $> \frac{1}{2}n^c$ of the clauses remain unsatisfied, under every truth assignment. Setting $\varepsilon_M = 1/(2^{q+1}(q-2))$ gives the theorem. □

29.4 Hardness of MAX-3SAT with bounded occurrence of variables

For each fixed k, define MAX-3SAT(k) to be the restriction of MAX-3SAT to Boolean formulae in which each variable occurs at most k times. This problem leads to reductions to some key optimization problems.

Theorem 29.11 *There is a gap preserving reduction from MAX-3SAT to MAX-3SAT(29) that transforms a Boolean formula ϕ to ψ such that*

- *if* $\mathrm{OPT}(\phi) = m$, *then* $\mathrm{OPT}(\psi) = m'$, *and*
- *if* $\mathrm{OPT}(\phi) < (1 - \varepsilon_M)m$, *then* $\mathrm{OPT}(\psi) < (1 - \varepsilon_b)m'$,

where m and m' are the number of clauses in ϕ and ψ, ε_M is the constant determined in Theorem 29.7, and $\varepsilon_b = \varepsilon_M/43$.

Proof: The proof critically uses expander graphs. Recall, from Section 20.3, that graph $G = (V, E)$ is an *expander* if every vertex has the same degree, and for any nonempty subset $S \subset V$,

$$|E(S, \overline{S})| > \min(|S|, |\overline{S}|),$$

where $E(S, \overline{S})$ denotes the set of edges in the cut $(S, \overline{S})$, i.e., edges that have one endpoint in S and the other in $\overline{S}$. Let us assume that such graphs are efficiently constructible in the following sense. There is an algorithm $\mathcal{A}$ and a constant N_0 such that for each $N \geq N_0$, $\mathcal{A}$ constructs a degree 14 expander graph on N vertices in time polynomial in N (Remark 29.12 clarifies this point).

Expanders enable us to construct the following device whose purpose is to ensure that in any optimal truth assignment, a given set of Boolean variables must have consistent assignment, i.e., all true or all false. Let $k \geq N_0$, and let G_x be a degree 14 expander graph on k vertices. Label the vertices with distinct Boolean variables $x_1, \ldots, x_k$. We will construct a CNF formula ψ_x on these Boolean variables. Corresponding to each edge (x_i, x_j) of G_x, we will include the clauses $(\overline{x}_i \vee x_j)$ and $(\overline{x}_j \vee x_i)$ in ψ_x. A truth assignment to $x_1, \ldots, x_k$ is said to be *consistent* if either all the variables are set to true or all are set to false. An inconsistent truth assignment partitions the vertices of G_x into two sets, say S and $\overline{S}$. Assume w.l.o.g. that S is the smaller set. Now, corresponding to each edge in the cut $(S, \overline{S})$, ψ_x will have an unsatisfied clause. Therefore, the number of unsatisfied clauses, $|E(S, \overline{S})|$, is at least $|S| + 1$. We will use this fact critically.

Next, we describe the reduction. We may assume w.l.o.g. that every variable occurs in ϕ at least N_0 times. If not, we can replicate each clause N_0 times without changing the approximability properties of the formula in any essential way.

Let B denote the set of Boolean variables occurring in ϕ. For each variable $x \in B$, we will do the following. Suppose x occurs $k \geq N_0$ times in ϕ. Let $V_x = \{x_1, \ldots, x_k\}$ be a set of completely new Boolean variables. Let G_x be a degree 14 expander graph on k vertices. Label its vertices with variables from V_x and obtain formula ψ_x as described above. Finally, replace each occurrence of x in ϕ by a distinct variable from V_x. After this process is carried out for each variable $x \in B$, every occurrence of a variable in ϕ is replaced by a distinct variable from the set of new variables

$$V = \bigcup_{x \in B} V_x.$$

Let ϕ' be the resulting formula. In addition, corresponding to each variable $x \in B$, a formula ψ_x has been constructed.

Finally, let

$$\psi = \phi' \wedge (\bigwedge_{x \in B} \psi_x).$$

Observe that for each $x \in B$, each variable of V_x occurs exactly 29 times in ψ – once in ϕ', and 28 times in ψ_x. Therefore, ψ is an instance of MAX-3SAT(29). We will say that the clauses of ϕ' are Type I clauses, and the remaining clauses of ψ are Type II clauses.

Now, the important claim is that an optimal truth assignment for ψ must satisfy all Type II clauses, and therefore must be consistent for each set $V_x, x \in B$. Suppose that this is not the case. Let τ be an optimal truth assignment that is not consistent for V_x, for some $x \in B$. τ partitions the vertices of G_x into two sets, say S and $\overline{S}$, with S being the smaller set. Now, flip the truth assignment to variables in S, keeping the rest of the assignment the same as τ. As a result, some Type I clauses that were satisfied under τ may now be unsatisfied. Each of these must contain a variable of S, and so their number is at most $|S|$. On the other hand we get at least $|S| + 1$ new satisfied clauses corresponding to the edges in the cut $(S, \overline{S})$. Thus, the flipped assignment satisfies more clauses than τ, contradicting the optimality of τ.

Let m and m' be the number of clauses in ϕ and ψ. The total number of occurrences of all variables in ϕ is at most $3m$. Each occurrence participates in 28 Type II two-literal clauses, giving a total of at most $42m$ Type II clauses. In addition, ψ has m Type I clauses. Therefore, $m' \leq 43m$.

If ϕ is satisfiable, then so is ψ. Next, consider the case that $\mathrm{OPT}(\phi) < (1 - \varepsilon_M)m$, i.e., $> \varepsilon_M m$ clauses of ϕ remain unsatisfied under any truth assignment. If so, by the above claim, $> \varepsilon_M m \geq \varepsilon_M m'/43$ of the clauses of ψ must remain unsatisfied. The theorem follows. □

Remark 29.12 The assumption about the efficient construction of expander graphs is slightly untrue. It is known that for each $N \geq N_0$, an expander of size $\leq N(1+o(1))$ can be constructed efficiently (see Section 29.9). The reader can verify that this does not change the status of Theorem 29.11.

Exercise 29.4 extends Theorem 29.11 to establishing hardness for MAX-3SAT(5).

29.5 Hardness of vertex cover and Steiner tree

In this section, we will apply the machinery developed above to some graph theoretic problems. For integer $d \geq 1$, let VC(d) denote the restriction of the cardinality vertex cover problem to instances in which each vertex has degree at most d.

Theorem 29.13 *There is a gap preserving reduction from MAX-3SAT(29) to VC(30) that transforms a Boolean formula ϕ to a graph $G = (V, E)$ such that*

- *if* $\mathrm{OPT}(\phi) = m$, *then* $\mathrm{OPT}(G) \leq \frac{2}{3}|V|$, *and*
- *if* $\mathrm{OPT}(\phi) < (1 - \varepsilon_b)m$, *then* $\mathrm{OPT}(G) > (1 + \varepsilon_v)\frac{2}{3}|V|$,

where m is the number of clauses in ϕ, ε_b is the constant determined in Theorem 29.11, and $\varepsilon_v = \varepsilon_b/2$.

Proof: Assume w.l.o.g. that each clause of ϕ has exactly 3 literals (this can be easily accomplished by repeating the literals within a clause, if necessary). We will use the standard transformation. Corresponding to each clause of ϕ, G has 3 vertices. Each of these vertices is labeled with one literal of the clause. Thus, $|V| = 3m$. G has two types of edges (see the illustration below):

- for each clause, G has 3 edges connecting its 3 vertices, and
- for each $u, v \in V$, if the literals labeling u and v are negations of each other, then (u, v) is an edge in G.

Each vertex of G has two edges of the first type and at most 28 edges of the second type. Hence, G has degree at most 30.

We claim that the size of a maximum independent set in G is precisely $\mathrm{OPT}(\phi)$. Consider an optimal truth assignment and pick one vertex, corresponding to a satisfied literal, from each satisfied clause. Clearly, the picked vertices form an independent set. Conversely, consider an independent set I in G, and set the literals corresponding to its vertices to be true. Any extension of this truth setting to all variables must satisfy at least $|I|$ clauses.

The complement of a maximum independent set in G is a minimum vertex cover. Therefore, if $\mathrm{OPT}(\phi) = m$ then $\mathrm{OPT}(G) = 2m$. If $\mathrm{OPT}(\phi) < (1-\varepsilon_b)m$, then $\mathrm{OPT}(G) > (2 + \varepsilon_b)m$. The theorem follows. □

As an illustration, consider the formula $(x_1 \vee \overline{x}_2 \vee x_3) \wedge (\overline{x}_1 \vee x_2 \vee x_3)$. The graph produced by the reduction given in Theorem 29.13 is given below:

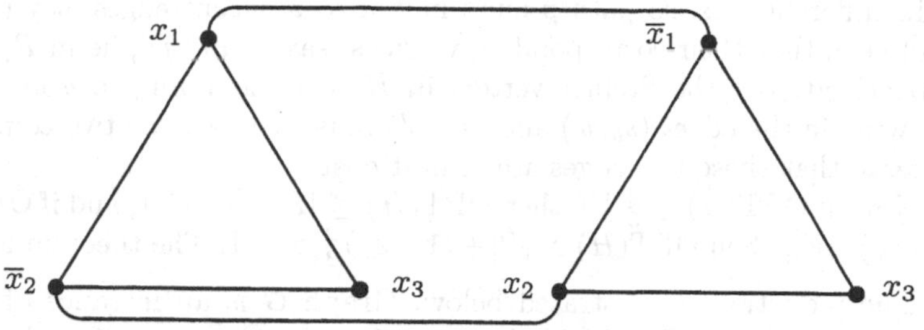

Theorem 29.14 *There is a gap preserving reduction from VC(30) to the Steiner tree problem. It transforms an instance $G = (V, E)$ of VC(30) to an instance $H = (R, S, \text{cost})$ of Steiner tree, where R and S are the required and Steiner vertices of H, and* cost *is a metric on $R \cup S$. It satisfies:*

- *if* $\text{OPT}(G) \leq \frac{2}{3}|V|$, *then* $\text{OPT}(H) \leq |R| + \frac{2}{3}|S| - 1$, *and*
- *if* $\text{OPT}(G) > (1 + \varepsilon_v)\frac{2}{3}|V|$, *then* $\text{OPT}(H) > (1 + \varepsilon_s)(|R| + \frac{2}{3}|S| - 1)$,

where $\varepsilon_s = 4\varepsilon_v/97$, and ε_v is the constant determined in Theorem 29.13.

Proof: Graph $H = (R, S, \text{cost})$ will be such that G has a vertex cover of size c iff H has a Steiner tree of cost $|R| + c - 1$. H will have a required vertex r_e corresponding to each edge $e \in E$ and a Steiner vertex s_v corresponding to each vertex $v \in V$. The edge costs are as follows. An edge between a pair of Steiner vertices is of cost 1, and an edge between a pair of required vertices is of cost 2. An edge (r_e, s_v) is of cost 1 if edge e is incident at vertex v in G, and it is of cost 2 otherwise.

Let us show that G has a vertex cover of size c iff H has a Steiner tree of cost $|R| + c - 1$. For the forward direction, let S_c be the set of Steiner vertices in H corresponding to the c vertices in the cover. Observe that there is a tree in H covering $R \cup S_c$ using cost 1 edges only (since every edge $e \in E$ must be incident at a vertex in the cover). This Steiner tree has cost $|R| + c - 1$.

For the reverse direction, let T be a Steiner tree in H of cost $|R| + c - 1$. We will show below that T can be transformed into a Steiner tree of the same cost that uses edges of cost 1 only. If so, the latter tree must contain exactly c Steiner vertices. Moreover, every required vertex of H must have a unit cost edge to one of these Steiner vertices. Therefore, the corresponding c vertices of G form a cover.

Let (u, v) be an edge of cost 2 in T. We may assume w.l.o.g. that u and v are both required. (If u is Steiner, remove (u, v) from T, getting two components. Throw in an edge from v to a required vertex to connect the two sides, and get a Steiner tree of the same cost as T.) Let e_u and e_v be the edges, in G, corresponding to these vertices. Since G is connected, there is a path, p, from one of the endpoints of e_u to one of the endpoints of e_v in G. Now, removing (u, v) from T gives two connected components. Let the set of required vertices in these two sets be R_1 and R_2. Clearly, u and v lie in different sets, so path p must have two adjacent edges, say (a, b) and (b, c) such that their corresponding vertices, say w and w', lie in R_1 and R_2, respectively. Let the Steiner vertex, in H, corresponding to b be s_b. Now, throwing in the edges (s_b, w) and (s_b, w') must connect the two components. Observe that these two edges are of unit cost.

Now, if $\mathrm{OPT}(G) \leq \frac{2}{3}|V|$, then $\mathrm{OPT}(H) \leq |R| + \frac{2}{3}|S| - 1$, and if $\mathrm{OPT}(G) > (1 + \varepsilon_v)\frac{2}{3}|V|$, then $\mathrm{OPT}(H) > |R| + (1 + \varepsilon_v)\frac{2}{3}|S| - 1$. The theorem follows. □

The reduction is illustrated below. Graph G is an instance of the vertex cover problem. The highlighted vertices form a cover. Graph H shows the Steiner tree corresponding to this cover in the reduced graph. Required vertices have been marked with squares, and the three Steiner vertices corresponding to the cover have been marked with circles (the remaining Steiner vertices have been omitted for clarity). The edge between two Steiner vertices in the tree is dotted to distinguish it from the remaining edges, which connect required and Steiner vertices.

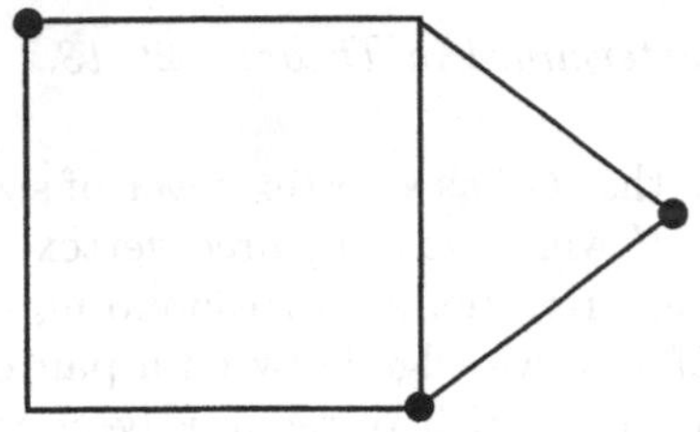

G

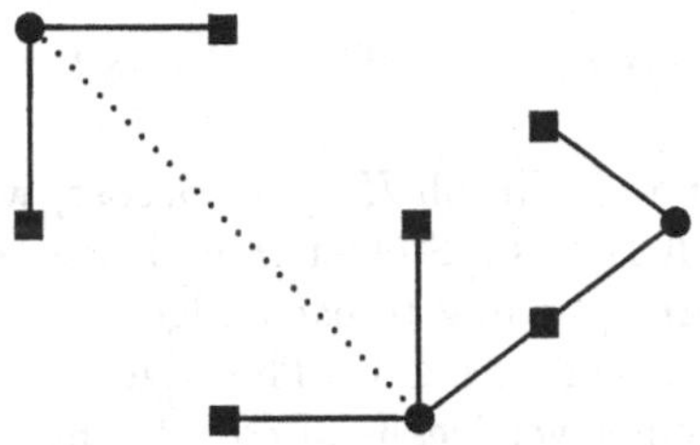

H

29.6 Hardness of clique

The best approximation algorithms known for some problems, including clique, are extremely weak – to the extent that the solution produced by the best known algorithm is only very slightly better than picking a trivial feasible solution. Recent hardness results have been invaluable in explaining why this is so: these problems are inherently inapproximable (essentially). In this section, we will establish this for clique:

Problem 29.15 (Clique) Given an undirected graph $G = (V, E)$ with nonnegative weights on vertices, find a clique of maximum weight. A *clique* in G is a subset of vertices, $S \subseteq V$, such that for each pair $u, v \in S$, $(u, v) \in E$. Its *weight* is the sum of weights of its vertices.

Consider the cardinality version of this problem, i.e., when all vertex weights are unit. In this section we will show that there is a constant $\varepsilon_q > 0$, such that there is no $1/(n^{\varepsilon_q})$ factor approximation algorithm for this problem, assuming $\mathbf{P} \neq \mathbf{NP}$. Let us first prove the following weaker result.

Lemma 29.16 *For fixed constants b and q, there is a gap-introducing reduction from SAT to clique that transforms a Boolean formula ϕ of size n to a graph $G = (V, E)$, where $|V| = 2^q n^b$, such that*

- *if ϕ is satisfiable,* $\text{OPT}(G) \geq n^b$, *and*
- *if ϕ is not satisfiable,* $\text{OPT}(G) < \frac{1}{2}n^b$.

Proof: Let F be a $\mathbf{PCP}(\log n, 1)$ verifier for SAT that requires $b \log n$ random bits and queries q bits of the proof. We will transform a SAT instance, ϕ, of size n to a graph $G = (V, E)$ as follows. For each choice of a binary string, r, of $b \log n$ bits, and each truth assignment, τ, to q Boolean variables, there is a vertex $v_{r,\tau}$ in G. Thus, $|V| = 2^q n^b$.

Let $Q(r)$ represent the q positions in the proof that F queries when it is given string r as the "random" string. We will say that vertex $v_{r,\tau}$ is *accepting* if F accepts when it is given random string r and when it reads τ in the $Q(r)$ positions of the proof; it is *rejecting* otherwise. Vertices v_{r_1,τ_1} and v_{r_2,τ_2} are *consistent* if τ_1 and τ_2 agree at each position at which $Q(r_1)$ and $Q(r_2)$ overlap. Clearly, a necessary condition for consistency is that $r_1 \neq r_2$. Two distinct vertices v_{r_1,τ_1} and v_{r_2,τ_2} are connected by an edge in G iff they are consistent and they are both accepting. Vertex $v_{r,\tau}$ is *consistent* with proof p if positions $Q(r)$ of p contain τ.

If ϕ is satisfiable, there is a proof, p, on which F accepts for each choice, r, of the random string. For each r, let $p(r)$ be the truth setting assigned by proof p to positions $Q(r)$. Now, the vertices $\{v_{r,p(r)} \mid |r| = b \log n\}$ form a clique in G of size n^b.

Next, suppose that ϕ is not satisfiable, and let C be a clique in G. Since the vertices of C are pairwise consistent, there is a proof, p, that is consistent with all vertices of C. Therefore, the probability of acceptance of F on proof p is at least $|C|/n^b$ (notice that the vertices of C must correspond to distinct random strings). Since the probability of acceptance of any proof is $< 1/2$ the largest clique in G must be of size $< \frac{1}{2}n^b$. □

As a consequence of Lemma 29.16, there is no factor 1/2 approximation algorithm for clique assuming $\mathbf{P} \neq \mathbf{NP}$. Observe that the hardness factor established is precisely the bound on the error probability of the probabilistically checkable proof for SAT. By the usual method of simulating the verifier a constant number of times, this can be made $1/k$ for any constant k, leading to a similar hardness result for clique. In order to achieve the claimed hardness, the error probability needs to be made inverse polynomial. This motivates generalizing the definition of **PCP** as follows. Let us define two additional parameters, c and s, called *completeness* and *soundness*, respectively. A language $L \in \mathbf{PCP}_{c,s}[r(n), q(n)]$ if there is a verifier V, which on input x of length n, obtains a random string of length $O(r(n))$, queries $O(q(n))$ bits of the proof, and satisfies:

- if $x \in L$, there is a proof y that makes V accept with probability $\geq c$,

• if $x \notin L$, then for every proof y, V accepts with probability $< s$.

Thus, the previously defined class $\mathbf{PCP}[r(n), q(n)] = \mathbf{PCP}_{1,\frac{1}{2}}[r(n), q(n)]$. In general, c and s may be functions of n.

We would like to obtain a PCP characterization of **NP** which has inverse polynomial soundness. An obvious way of reducing soundness is to simulate a $\mathbf{PCP}[\log n, 1]$ verifier multiple number of times and accept iff the verifier accepts each time. Simulating k times will reduce soundness to $1/2^k$; however, this will increase the number of random bits needed to $O(k \log n)$ and the number of query bits to $O(k)$. Observe that the number of vertices in the graph constructed in Lemma 29.16 is $2^{O(r(n)+q(n))}$. To achieve inverse polynomial soundness, k needs to be $\Omega(\log n)$. For this value of k, the number of bits queried is $O(\log n)$, which is not a problem. However, the number of random bits needed is $O(\log^2 n)$, which leads to a superpolynomial sized graph.

The following clever idea overcomes this difficulty. We will use a constant degree expander graph to generate $O(\log n)$ strings of $b \log n$ bits each, using only $O(\log n)$ truly random bits. The verifier will be simulated using these $O(\log n)$ strings as the "random" strings. Clearly, these are not truly random strings. Properties of expanders help show that they are "almost random" – the probability of error still drops exponentially in the number of times the verifier is simulated.

Let H be a constant degree expander on n^b vertices, each vertex having a unique $b \log n$ bit label. A random walk on H of length $O(\log n)$ can be constructed using only $O(\log n)$ bits, $b \log n$ bits to pick the starting vertex at random and a constant number of bits to pick each successive vertex. (Observe that the random walk is started in the stationary distribution, which is uniform since the graph is regular.) The precise property of expanders we will need is the following.

Theorem 29.17 *Let S be any set of vertices of H of size $< (n^b)/2$. There is a constant k such that*

$$\mathbf{Pr}[\text{ all vertices of a } k \log n \text{ length random walk lie in } S\ ] < \frac{1}{n}.$$

For intuitive justification for Theorem 29.17, observe that a constant fraction of the edges incident at vertices of S have their other end points in $\overline{S}$ – these help the walk escape from S. The following figure shows a walk on H that does not lie in S:

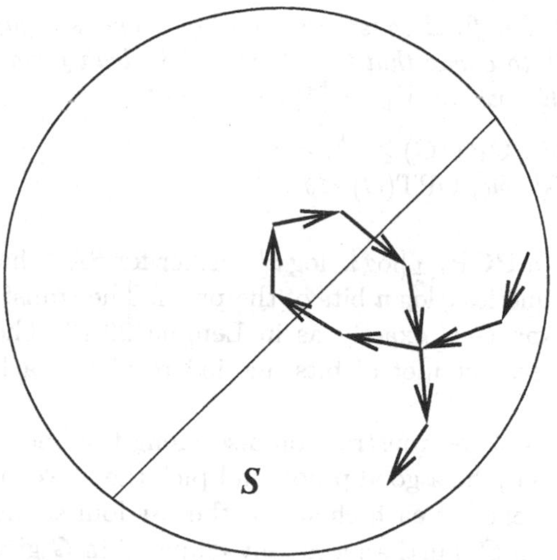

Theorem 29.18 $\mathbf{NP} = \mathbf{PCP}_{1,\frac{1}{n}}[\log n, \log n]$

Proof: We will prove the difficult half,

$$\mathbf{PCP}_{1,\frac{1}{2}}[\log n, 1] \subseteq \mathbf{PCP}_{1,\frac{1}{n}}[\log n, \log n],$$

and leave the rest as Exercise 29.5. Let $L \in \mathbf{PCP}_{1,\frac{1}{2}}[\log n, 1]$. Let F be a verifier for L which requires $b \log n$ random bits and queries q bits of the proof, where b and q are constants.

Next, we give a $\mathbf{PCP}_{1,\frac{1}{n}}[\log n, \log n]$ verifier for L, F', which constructs the expander graph H defined above. It then constructs a random walk of length $k \log n$ on H, using $O(\log n)$ random bits. Both constructions can be accomplished in polynomial time. The label of each vertex on this path specifies a $b \log n$ bit string. It uses these $k \log n + 1$ strings as the "random" strings on which it simulates verifier F. F' accepts iff F accepts on all $k \log n + 1$ runs.

Consider string $x \in L$, and let p be a proof that makes verifier F accept x with probability 1. Clearly, F', given proof p, also accepts x with probability 1. Hence the completeness of the new proof system is 1.

Next, consider string $x \notin L$, and let p be an arbitrary proof supplied to F'. When given proof p, verifier F accepts on $< (n^b)/2$ random strings of length $b \log n$. Let S denote the corresponding set of vertices of H, $|S| < (n^b)/2$. Now, F' accepts x iff the random walk remains entirely in S. Since the probability of this event is $< 1/n$, the soundness of F' is $1/n$. Finally observe that F' requires only $O(\log n)$ random bits and queries $O(\log n)$ bits of the proof. □

Theorem 29.19 *For fixed constants b and q, there is a gap-introducing reduction from SAT to clique that transforms a Boolean formula ϕ of size n to a graph $G = (V, E)$, where $|V| = n^{b+q}$, such that*

- *if ϕ is satisfiable,* $\text{OPT}(G) \geq n^b$, *and*
- *if ϕ is not satisfiable,* $\text{OPT}(G) < n^{b-1}$.

Proof: Let F be a $\mathbf{PCP}_{1,\frac{1}{n}}[\log n, \log n]$ verifier for SAT that requires $b \log n$ random bits and queries $q \log n$ bits of the proof. The transformation of SAT instance ϕ to graph G is exactly as in Lemma 29.16. The only difference is that the increased number of bits queried results in a larger number of vertices.

The correctness of the construction also along the lines of Lemma 29.16. If ϕ is satisfiable, let p be a good proof, and pick the n^b vertices of G that are consistent with p, one for each choice of the random string. These vertices will form a clique in G. Furthermore, any clique C in G gives rise to a proof that is accepted by F with probability $\geq |C|/n^b$. Since the soundness of F is $1/n$, if ϕ is not satisfiable, the largest clique in G is of size $< n^{b-1}$. □

Corollary 29.20 *There is no $1/(n^{\varepsilon_q})$ factor approximation algorithm for the cardinality clique problem, assuming* $\mathbf{P} \neq \mathbf{NP}$, *where $\varepsilon_q = 1/(b+q)$, for constants b and q defined in Theorem 29.19.*

29.7 Hardness of set cover

As stated in Chapter 2, the simple greedy algorithm for the set cover problem, which is perhaps the first algorithmic idea one would attempt, has remained essentially the best algorithm. Since set cover is perhaps the single most important problem in the theory of approximation algorithms, a lot of effort was expended on obtaining an improved algorithm.

In this section, we will present the remarkable result that the approximation factor of this algorithm is tight up to a constant multiplicative factor. Improved hardness results show that it is tight up to lower order terms as well (see Section 29.9). This should put to rest nagging doubts about the true approximability of this central problem.

29.7.1 The two-prover one-round characterization of NP

Observe that for the purpose of showing hardness of MAX-3SAT and clique (Theorems 29.7 and 29.19), we did not require a detailed description of the kinds of queries made by the verifier – we only required a bound on the number of queries made. In contrast, this time we do need a description, and moreover, we want to first establish that a particularly simple verifier suffices. For this purpose, we will introduce a new model for probabilistically checkable proofs, the *two-prover one-round proof system*. This model is best understood by thinking of the proof system as a game between the prover and the verifier. The prover is trying to cheat – it is trying to convince the verifier that a "no" instance for language L is actually in L. Is there a verifier that can ensure that the probability of getting cheated is $< 1/2$ for every "no" instance?

In the two-prover model, the verifier is allowed to query two non-communicating provers, denoted P_1 and P_2. Since the verifier can cross-check the provers' answers, the provers' ability to cheat gets restricted in this model. In turn, we will impose restrictions on the verifier as well, and thereby obtain a new characterization of **NP**. Under a one-round proof system, the verifier is allowed only one round of communication with each prover. The simplest way of formalizing this is as follows. We will assume that the two proofs are written in two alphabets, say Σ_1 and Σ_2. In general, the sizes of these alphabets may be unbounded and may depend on the size of the input. The verifier is allowed to query one position in each of the two proofs.

The two-prover one-round model comes with three parameters: completeness, soundness and the number of random bits provided to the verifier, denoted by c, s and $r(n)$, respectively. This defines the class $\mathbf{2P1R}_{c,s}(r(n))$. A language L is in $\mathbf{2P1R}_{c,s}(r(n))$ if there is a polynomial time bounded verifier V that receives $O(r(n))$ truly random bits and satisfies:

- for every input $x \in L$, there is a pair of proofs $y_1 \in \Sigma_1^*$ and $y_2 \in \Sigma_2^*$ that makes V accept with probability $\geq c$,
- for every input $x \notin L$ and every pair of proofs $y_1 \in \Sigma_1^*$ and $y_2 \in \Sigma_2^*$, V accepts with probability $< s$.

The PCP theorem implies, and in fact is equivalent to, the fact that there is a gap-introducing reduction from SAT to MAX-3SAT(5) (see Theorem 29.7 and Exercises 29.3 and 29.4). We will use this to show:

Theorem 29.21 *There is a constant* $\varepsilon_P > 0$ *such that*
$\mathbf{NP} = \mathbf{2P1R}_{1,1-\varepsilon_P}(\log(n))$.

Proof: We will establish the difficult half, i.e., $\mathbf{NP} \subseteq \mathbf{2P1R}_{1,1-\varepsilon_P}(\log(n))$, and leave the rest as Exercise 29.7. Clearly, it is sufficient to show that SAT $\in \mathbf{2P1R}_{1,1-\varepsilon_P}(\log(n))$.

As a result of Theorem 29.7 and Exercise 29.4, there is gap-introducing reduction from SAT to MAX-3SAT(5)[2]. More precisely, there is a constant $\varepsilon_5 > 0$ for which there is a reduction Γ from SAT to MAX-3SAT(5) that transforms a Boolean formula ϕ to ψ such that

- if ϕ is satisfiable, $\mathrm{OPT}(\psi) = m$, and

[2] The bounded occurrence version of MAX-3SAT is not essential for this theorem; however, we will require it in the main reduction.

- if ϕ is not satisfiable, $\mathrm{OPT}(\psi) < (1 - \varepsilon_5)m$,

where m is the number of clauses in ψ.

The two-prover one-round verifier, V, for SAT works as follows. Given a SAT formula ϕ, it uses the above stated reduction to obtain a MAX-3SAT(5) instance ψ. It assumes that P_1 contains an optimal truth assignment, τ, for ψ and P_2 contains, for each clause, the assignment to its three Boolean variables under τ (hence, $|\Sigma_1| = 2$ and $|\Sigma_2| = 2^3$). It uses the $O(\log n)$ random bits to pick a random clause, C, from ψ, and further, a random Boolean variable, x, occurring in C. V obtains the truth assignments to x and the three variables in C by querying P_1 and P_2, respectively. It accepts iff C is satisfied and the two proofs agree on their assignment for x.

If ϕ is satisfiable, then so is ψ. Clearly, there are proofs y_1 and y_2 such that V accepts with probability 1.

Next assume that ϕ is not satisfiable. Any truth assignment to ψ must leave strictly more than ε_5 fraction of the clauses unsatisfied. Consider any pair of proofs (y_1, y_2). Interpret y_1 as a truth assignment, say τ. The random clause, C, picked by V is not satisfied by τ with probability $> \varepsilon_5$. If so, and if the assignment for C contained in y_2 is satisfying, then y_1 and y_2 must be inconsistent. In the latter case, the verifier catches this with probability $\geq 1/3$. Hence overall, V must reject with probability $> \varepsilon_5/3$. □

Remark 29.22 Using standard techniques (see Exercise 29.8), Γ can be modified to ensure that the instance of MAX-3SAT(5) produced satisfies the following *uniformity conditions*: each Boolean variable occurs in exactly 5 clauses and each clause contains 3 distinct variables (negated or unnegated). This modification changes the constant ε_5 to some other constant, say $\varepsilon_5' > 0$. These uniformity conditions will be needed in the main reduction.

Remark 29.23 As a result of the uniformity conditions, if ψ has n variables, then it has $5n/3$ clauses. Therefore, the two proofs are of length n and $5n/3$, respectively. For carrying out the main reduction, it will be important to ensure that the two proofs are of equal length. This can be easily achieved by repeating the first proof 5 times and the second proof 3 times. The verifier will query a random copy of each proof. It is easy to verify that Theorem 29.21 still holds (even though the "copies" may be different).

29.7.2 The gadget

The following set system will be a basic gadget in the main reduction: $(U, C_1, \ldots, C_m, \overline{C}_1, \ldots, \overline{C}_m)$, where U is the universal set and $C_1, \ldots, C_m$ are subsets of U. Clearly, U can be covered by picking a set C_i and its complement $\overline{C}_i$. Such a cover will be called a *good cover*. A cover that does not include a set and its complement will be called a *bad cover*. The following theorem, which can be proven using the probabilistic method (see Exercise 29.9), shows the existence of such set systems for which the sizes of good and bad covers are widely different. Moreover, they can be constructed efficiently, with high probability.

Theorem 29.24 *There exists a polynomial $p(.,.)$ such that there is a randomized algorithm which generates, for each m and l, a set system*

$$(U, C_1, \ldots, C_m, \overline{C}_1, \ldots, \overline{C}_m),$$

with $|U| = p(m, 2^l)$. With probability $> 1/2$ the gadget produced satisfies that every bad cover is of size $> l$. Moreover, the running time of the algorithm is polynomial in $|U|$.

A good cover is well coordinated – it involves picking a set C_i and its complement. Acceptance in the two-prover one-round proof system also involves coordination – on random string r, the verifier queries the two proofs and accepts iff the answers are coordinated. The choice of this proof system, for establishing hardness of set cover, should be more convincing in light of this observation.

29.7.3 Reducing error probability by parallel repetition

Before presenting the reduction, we would like to improve the soundness of the two-prover one-round proof system for SAT. The usual way of accomplishing this is parallel repetition: The verifier picks k clauses randomly and independently, and a random Boolean variable from each of the clauses. It queries P_1 on the k variables and P_2 on the k clauses, and accepts iff all answers are accepting. One would expect that probability that the provers manage to cheat drops to $< (1 - \varepsilon_P)^k$.

Surprisingly enough, this is not true. Since each prover is allowed to look at all k questions before providing its k answers, it may be able to coordinate its answers and thereby cheat with a higher probability. Example 29.25 illustrates this in a simple setting. If the provers are required to answer each question before being given the next question, the probability of error drops in the usual fashion; however, this requires k rounds of communication and falls outside the two-prover one-round model.

Example 29.25 Consider the following setting in which the two non-communicating provers are attempting to agree on a random bit. The verifier gives random, independent bits r_1 and r_2 to P_1 and P_2, respectively. The protocol succeeds if the two provers manage to commit to one of the two bits, i.e., either both provers output $(1, r_1)$ or both provers output $(2, r_2)$; the first element of a pair says whose bit the provers are outputting and the second element is the bit itself. Since P_1 does not know r_2 and P_2 does not know r_1, the probability of their succeeding is $1/2$.

Now consider parallel repetitions of this protocol. The verifier gives two bits, r_1 and s_1, to P_1 and two bits, r_2 and s_2, to P_2. The four bits are random and independent. The provers succeed iff they can commit to one of the r's and one of the s's.

One would expect the probability of success to be $1/4$. However, by cleverly coordinating answers, the provers can make it $1/2$ as follows. The answers of P_1 are $(1, r_1)$ and $(2, r_1)$, and those of P_2 are $(1, s_2)$ and $(2, s_2)$. The provers succeed iff $r_1 = s_2$, which happens with probability $1/2$. □

Despite this difficulty, one can still prove that the probability of error does drop exponentially with k. However, the proof of this useful fact is not easy.

Theorem 29.26 *Let the error probability of a two-prover one-round proof system be $\delta < 1$. Then the error probability on k parallel repetitions is at most δ^{dk}, where d is a constant that depends only on the length of the answers of the original proof system.*

29.7.4 The reduction

We will prove the following.

Theorem 29.27 *There is a constant $c > 0$ for which there is a randomized gap-introducing reduction Γ, requiring time $n^{O(\log\log n)}$, from SAT to the cardinality set cover problem that transforms a Boolean formula ϕ to a set system $\mathcal{S}$ over a universal set of size $n^{O(\log\log n)}$ such that*

- *if ϕ is satisfiable, $\mathrm{OPT}(\mathcal{S}) = 2n^k$, and*
- *if ϕ is not satisfiable, $\mathbf{Pr}[\mathrm{OPT}(\mathcal{S}) > cn^k k \log n] > 1/2$,*

where n is the length of each of the two proofs for SAT under the two-prover one-round model (see Remark 29.23); n is polynomial in the size of ϕ. The parameter k is $O(\log\log n)$.

Remark 29.28 This is slight abuse of notation, since gap-introducing reductions were defined to run in polynomial time.

Proof: Let V be the two-prover one-round verifier for SAT, described in Theorem 29.21. Assume further that the MAX-3SAT(5) formula produced by V satisfies the uniformity conditions stated in Remark 29.22 and that the two proofs queried by V are of equal length, say n, as stated in Remark 29.23. Denote by ψ the MAX-3SAT(5) formula produced by V when given SAT formula ϕ.

Let V' be a two-prover one-round verifier that executes k parallel repetitions of V, as described in Section 29.7.3. Now, each of the proofs is of length n^k. Each position of P_1 contains a truth assignment to k Boolean variables (not necessarily distinct) and each position of P_2 contains a truth assignment to the $3k$ Boolean variables occurring in k clauses. Thus, proofs P_1 and P_2 are written in alphabets Σ_1 and Σ_2 whose sizes are 2^k and 2^{3k}, respectively. k will be fixed to be $O(\log\log n)$ for reasons clarified below.

Verifier V' uses random bits provided to it to pick k random clauses of ψ, and a random Boolean variable from each of these k clauses, thereby specifying a position in P_1 and a position in P_2. These involve picking from one of n^k and 3^k choices, respectively. Therefore, the total number of random strings is $(3n)^k$. Denote by $Q_1(r)$ and $Q_2(r)$ the positions in P_1 and P_2, respectively, specified by random string r.

Suppose the answers in positions $Q_1(r)$ and $Q_2(r)$ are a and b, respectively. Recall that V' accepts on random string r iff b satisfies all k clauses picked, and a and b assign the same truth values to the k chosen variables. Given r and the answer in $Q_2(r)$, say b, the "acceptable" answer in $Q_1(r)$ is uniquely specified. Let *projection function* $\pi(r, b)$ denote this answer.

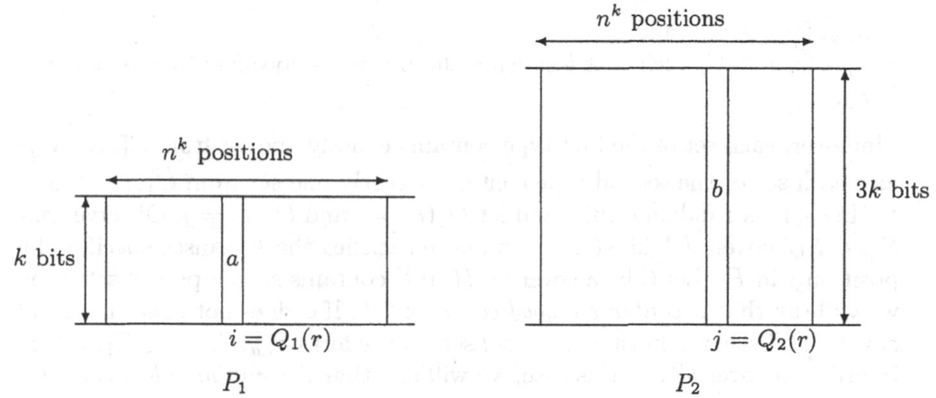

The parameters m and l for the gadget are fixed as follows. We will set $m = |\Sigma_1| = 2^k$, and $l = O(k \log n) = O(\log n \log\log n)$. Let $(U, C_1, \ldots, C_{2^k}, \overline{C}_1, \ldots, \overline{C}_{2^k})$ be the gadget with these parameters. Thus, corresponding to each answer $a \in \Sigma_1$, we have a unique set C_a. As stated in Theorem 29.24, $|U| = p(2^k, 2^l) = n^{O(\log\log n)}$, and the gadget can be constructed probabilistically in time polynomial in $|U|$.

The gadget will be constructed once, and as stated in Theorem 29.24, will satisfy the chosen parameters with probability $> 1/2$. For the rest of the proof, assume that it does. We will make $(3n)^k$ copies of the gadget over disjoint universal sets. Each copy corresponds to a random string. Denote the copy corresponding to random string r to be $(U^r, C_1^r, \ldots, C_{2^k}^r, \overline{C}_1^r, \ldots, \overline{C}_{2^k}^r)$.

The reduction Γ transforms ϕ to a set cover instance $\mathcal{S}$ as follows. The universal set

$$\mathcal{U} = \bigcup_r U^r,$$

where the union is over all $(3n)^k$ random strings. Clearly, $|\mathcal{U}| = |U|(3n)^k = n^{O(\log\log n)}$. The subsets of $\mathcal{U}$ specified by $\mathcal{S}$ are of two kinds. First, corresponding to each position i in P_1 and answer $a \in \Sigma_1$, there is a set

$$S_{i,a} = \bigcup_{r:Q_1(r)=i} C_a^r,$$

where the union is over all random strings r such that $Q_1(r) = i$. Second, corresponding to each position j in P_2 and answer $b \in \Sigma_2$, there is a set $S_{j,b}$. If b does not satisfy all k clauses of ψ, specified by position $Q_2(r)$, then $S_{j,b} = \emptyset$. Otherwise,

$$S_{j,b} = \bigcup_{r:Q_2(r)=j} \overline{C}^r_{\pi(r,b)},$$

where the union is over all random strings r such that $Q_2(r) = j$.

Let r be a random string, and let $Q_1(r) = i$ and $Q_2(r) = j$. Then, the only sets in $\mathcal{S}$ that contain elements of U^r are:

- $S_{i,a}$, for $a \in \Sigma_1$, and
- $S_{j,b}$, for $b \in \Sigma_2$ such that b satisfies the k clauses specified by position j in P_2.

Moreover, each set of the first type contains exactly one set from $C^r_1, \ldots, C^r_{2^k}$ and each set of the second type contains exactly one set from $\overline{C}^r_1, \ldots, \overline{C}^r_{2^k}$.

Let r be a random string, and let $Q_1(r) = i$ and $Q_2(r) = j$. Observe that $S_{i,a} \cup S_{j,b}$ covers U^r iff $\pi(r,b) = a$ and b satisfies the k clauses specified by position j in P_2. Let $\mathcal{C}$ be a cover for $\mathcal{U}$. If $\mathcal{C}$ contains such a pair of sets then we will say that $\mathcal{C}$ *contains a good cover* for U^r. If $\mathcal{C}$ does not contain a good cover for U^r, then it must contain $> l$ sets of the form $S_{i,a}, S_{j,b}, a \in \Sigma_1, b \in \Sigma_2$ in order to cover U^r. In this case, we will say that $\mathcal{C}$ *contains a bad cover* for U^r.

Suppose ϕ is satisfiable. Then there is a pair of proofs (y_1, y_2) on which the verifier accepts with probability 1. Let us pick a cover $\mathcal{C}$ as follows. Corresponding to each position i in P_1 and j in P_2 pick sets $S_{i,a}$ and $S_{j,b}$, where a and b are the answers for these queries in y_1 and y_2, respectively. Hence, $|\mathcal{C}| = 2n^k$. It is easy to see that $\mathcal{C}$ contains a good cover for each set U^r.

Next suppose that ϕ is not satisfiable. Now, V' will reject any pair of proofs with high probability. We have assumed that the gadget found satisfies the chosen parameters; this happens with probability $> 1/2$. Let $\mathcal{C}$ denote an optimal cover for $\mathcal{U}$. Is $\mathcal{C}$ forced to contain a bad cover for U^r, for most random strings r? Clearly, $\mathcal{C}$ is allowed to pick sets corresponding to portions of many different proofs. Using this added capability, can we not construct a cover that is only slightly larger than $2n^k$? A set from $\mathcal{S}$ helps cover elements from several different universes U^r, making the rest of the argument more involved.

Below we will give a procedure for constructing, from $\mathcal{C}$, a pair of proofs, (y_1, y_2), in such a way that if $|\mathcal{C}|$ is small, then V' must accept this pair with high probability. Hence, we will derive the desired lower bound on $|\mathcal{C}|$.

Consider the *set of answers* picked by $\mathcal{C}$ for each position of the two proofs. For each position i in P_1, define $A(i) = \{a \mid S_{i,a} \in \mathcal{C}\}$, and for each position j in P_2, define $A(j) = \{b \mid S_{j,b} \in \mathcal{C}\}$. Construct proofs y_1

and y_2 by picking for each position i in P_1 and j in P_2 a random element of $A(i)$ and $A(j)$, respectively. If any of the answer sets is empty, pick an arbitrary answer for that position. Define $B_1 = \{r \mid |A(Q_1(r))| > l/2\}$, $B_2 = \{r \mid |A(Q_2(r))| > l/2\}$ and $G = \overline{B_1 \cup B_2}$.

Thus, G is the set of random strings r for which $\mathcal{C}$ picks at most $l/2$ answers each for $Q_1(r)$ and $Q_2(r)$. Hence, $\mathcal{C}$ contains a good cover for U^r, say $S_{i,a} \cup S_{j,b}$, where $a \in A(Q_1(r))$ and $b \in A(Q_2(r))$. The pair of proofs, (y_1, y_2), contain a and b in positions $Q_1(r)$ and $Q_2(r)$, respectively, with probability $\geq (\frac{2}{l})^2$. Hence V', when given proofs (y_1, y_2), accepts on random string r with at least this probability.

Let f_G denote the fraction of random strings contained in G. Then, using Theorem 29.26,

$$f_G \left(\frac{2}{l}\right)^2 \leq \mathbf{Pr}[V' \text{accepts } \phi \text{ when given proofs } (y_1, y_2)] \leq \delta^{dk}.$$

Hence, $f_G \leq \delta^{dk} l^2/4$. Since l^2 is $O(\log^4 n)$, by picking $k = O(\log\log n)$ we can ensure that $f_G < 1/2$. As a result, $B_1 \cup B_2$ contains at least half the random strings, and therefore one of these sets contains at least a quarter. Denote this set by B_i.

Because of the uniformity property (Remark 29.22), if r is chosen at random, then $Q_1(r)$ is a random position in P_1 and $Q_2(r)$ is a random position in P_2 (although they will be correlated). Furthermore, r has probability $> 1/4$ of being in B_i. Therefore, the answer sets of $> 1/4$ of the positions of B_i are of cardinality $> l/2$. Hence the size of the cover $> ln^k/8 = \Omega(n^k k \log n)$. □

As a consequence of Theorem 29.27, inapproximability of set cover modulo **NP** not being in a one-sided-error complexity class with running time $n^{O(\log\log n)}$ follows directly. Standard techniques from complexity theory (see Exercise 1.18) lead to the following slightly stronger result.

Corollary 29.29 *There is a constant b such that if there is a $b \log n$ factor approximation algorithm for the cardinality set cover problem, where n is the size of the universal set of the set cover instance, then* $\mathbf{NP} \subseteq \mathbf{ZTIME}(n^{O(\log\log n)})$ *(see Section A.4 for definition).*

29.8 Exercises

29.1 Show that $\mathbf{PCP}(\log n, 1) \subseteq \mathbf{NP}$.
Hint: Let $L \in \mathbf{PCP}(\log n, 1)$. The **NP** machine for accepting L guesses the proof, simulates the verifier for L on all $O(\log n)$ length random strings, and accepts iff the verifier accepts on all the random strings.

29.2 Show (see Appendix A for definitions):

1. $\mathbf{PCP}(0,0) = \mathbf{PCP}(0, \log n) = \mathbf{P}$.
2. $\mathbf{PCP}(\mathrm{poly}(n), 0) = \text{co-}\mathbf{RP}$, where $\mathrm{poly}(n) = \bigcup_{k \geq 0} n^k$.
3. $\mathbf{PCP}(\log n, 1) = \mathbf{PCP}(\log n, \mathrm{poly}(n))$.
   Hint: $\mathbf{NP} \subseteq \mathbf{PCP}(\log n, 1) \subseteq \mathbf{PCP}(\log n, \mathrm{poly}(n)) \subseteq \mathbf{NP}$.

29.3 Show the converse of Theorem 29.7, i.e., if there is a gap-introducing reduction from SAT to MAX-3SAT, then $\mathbf{NP} \subseteq \mathbf{PCP}(\log n, 1)$.
Hint: Reduce the given SAT formula ϕ to an instance ψ of MAX-3SAT. The verifier expects, as proof, an optimal truth assignment to ψ. This gives an error probability of $1 - \varepsilon_M$. Repeat to decrease the error probability to $< 1/2$.

29.4 Give a gap-preserving reduction from MAX-3SAT(29) to MAX-3SAT(5), with appropriate parameters, to show hardness for the latter problem.
Hint: The reduction is similar, though easier, than that in Theorem 29.11. Instead of using an expander graph, use a cycle. Now, an inconsistent assignment can gain as many as 14 clauses corresponding to each old variable x. However, it must leave at least two clauses, corresponding to edges of the cycle of x, unsatisfied.

29.5 Complete the proof of Theorem 29.18.

29.6 (Hastad [129]) An important consideration, while obtaining a PCP characterization of **NP**, is reducing the number of bits of the proof that the verifier needs to query. The following remarkable result reduces it to just 3.

Theorem 29.30 *For every $\varepsilon > 0$,*

$$\mathbf{NP} = \mathbf{PCP}_{1-\varepsilon, \frac{1}{2}+\varepsilon}[\log n, 1].$$

Moreover, there is a particularly simple PCP verifier for SAT. It uses the $O(\log n)$ random bits to compute three positions in the proof, say i, j and k, and a bit b, and accepts iff

$$y(i) + y(j) + y(k) \equiv b \pmod 2.$$

Here $y(i)$ is the ith bit in the proof y.

1. Consider the restriction of Problem 16.12 (Exercise 16.7), linear equations over GF[2], in which each equation has exactly 3 variables. Use the characterization stated in Theorem 29.30 to give an appropriate gap-introducing reduction from SAT to this problem which shows that if, for any $\varepsilon > 0$, there is a $2 - \varepsilon$ factor approximation algorithm for the latter problem then $\mathbf{P} = \mathbf{NP}$.

2. Give an appropriate gap-preserving reduction from linear equations over GF[2] to MAX-3SAT which shows that if, for any $\varepsilon > 0$, there is a $8/7 - \varepsilon$ factor approximation algorithm for MAX-3SAT then **P** = **NP**.
Hint: The equation $x_i + x_j + x_k \equiv 0 \pmod 2$ is transformed into the clauses

$$(\overline{x}_i \vee x_j \vee x_k) \wedge (x_i \vee \overline{x}_j \vee x_k) \wedge (x_i \vee x_j \vee \overline{x}_k) \wedge (\overline{x}_i \vee \overline{x}_j \vee \overline{x}_k).$$

29.7 Complete the proof of Theorem 29.21, i.e., show that $\mathbf{2P1R}_{1,1-\varepsilon_P}(\log(n)) \subseteq \mathbf{NP}$.

29.8 Prove the uniformity conditions stated in Remark 29.22.
Hint: Use the standard technique of introducing new Boolean variables.

29.9 Prove Theorem 29.24 using the probabilistic method.
Hint: $p(m, 2^l) = O(m2^{2l})$ suffices. Pick each set C_i by including each element of U in it randomly and independently with probability 1/2.

29.10 (Feige [86]) The following stronger hardness result can be established for set cover:

Theorem 29.31 *For any constant $\delta > 0$, if there is a $(1 - \delta)\ln n$ factor approximation algorithm for the cardinality set cover problem, where n is the size of the universal set of the set cover instance, then* $\mathbf{NP} \subseteq \mathbf{DTIME}(n^{O(\log\log n)})$, *where* $\mathbf{DTIME}(t)$ *is the class of problems for which there is a deterministic algorithm running in time $O(t)$.*

Consider the maximum coverage problem, Problem 2.18 in Exercise 2.15. Using Theorem 29.31 show that if there is an $\varepsilon > 0$ for which there is a $(1 - 1/e + \varepsilon)$ factor approximation algorithm for the maximum coverage problem, then $\mathbf{NP} \subseteq \mathbf{DTIME}(n^{O(\log\log n)})$.
Hint: Use the maximum coverage algorithm to obtain a $(1 - \delta)\ln n$ factor algorithm for set cover, for some $\delta > 0$, as follows: Guess k, the optimal number of sets needed for the given instance. Run the maximum coverage algorithm, with parameter k, iteratively, until a cover is found. In each iteration, a $(1 - 1/e + \varepsilon)$ fraction of the uncovered elements is covered. Therefore, the number of iterations, l, satisfies, $(1/e - \varepsilon)^l = 1/n$.

29.11 (Jain, Mahdian, and Saberi [145]) Using Theorem 29.31 show that if there is an $\varepsilon > 0$ for which there is a $(1 + 2/e - \varepsilon)$ factor approximation algorithm for the metric k-median problem, Problem 25.1, then $\mathbf{NP} \subseteq \mathbf{DTIME}(n^{O(\log\log n)})$.

29.9 Notes

The first hardness of approximation result based on probabilistically checkable proofs was due to Feige, Goldwasser, Lovász, Safra, and Szegedy [88]. This work motivated the discovery of the PCP theorem, which additionally builds on work on interactive proof systems (Babai [19] and Goldwasser, Micali, and Rackoff [115]) and program checking (Blum and Kannan [29] and Blum, Luby, and Rubinfeld [30]), and is due to Arora and Safra [15], and Arora, Lund, Motwani, Sudan, and Szegedy [15, 13]. Theorem 29.30, which yields optimal inapproximability results for several problems, is due to Hastad [129].

Before this development, the pioneering work of Papadimitriou and Yannakakis [220] had established evidence of inapproximability of several natural problems using their notion of **Max-SNP**-completeness. Gap preserving reductions are weaker than their L-reductions. Consequently, the ideas behind their reductions carry over directly to the new development, as in the reductions given in Theorems 29.11 and 29.13. Indeed, one of the motivations for the PCP theorem was that establishing an inapproximability result for MAX SAT would directly yield inapproximability results for all **Max-SNP**-hard problems. Theorem 29.14 is from Bern and Plassmann [26].

The construction of expander graphs is due to Lubotzky, Phillips, and Sarnak [204]. Theorem 29.17 is due to Impagliazzo and Zuckerman [142]. Theorem 29.19 on hardness of clique follows from [88] and [15, 13]. The current best hardness result for clique, due to Hastad [128], states that it cannot be approximated within a factor of $n^{1-\varepsilon}$ for any $\varepsilon > 0$, unless $\mathbf{NP} = \mathbf{ZPP}$. This is quite close to the best approximation algorithm, due to Boppana and Holldórsson [31], achieving a guarantee of $O(n/(\log^2 n))$.

Lund and Yannakakis [206] gave the first hardness result for set cover, showing that it cannot be approximated within a factor of $\log n/2$ unless $\mathbf{NP} \subseteq \mathbf{ZTIME}(n^{O(\text{polylog } n)})$. The improved result, presented in Theorem 29.31, is due to Feige [86]. This enhancement comes about by using a k prover proof system. A deterministic construction of the set system gadget of Theorem 29.24, due to Naor, Schulman, and Srinivasan [219], allows replacing **ZTIME** by **DTIME** in the complexity assumption. The two-prover one-round proof system was defined by Ben-or, Goldwasser, Kilian, and Wigderson [25]. Theorem 29.26 is due to Raz [239].

Karloff and Zwick [169] give an algorithm for MAX-3SAT that achieves an approximation guarantee of 8/7 when restricted to satisfiable formulae. This complements the hardness result stated in Exercise 29.6.

For further information on this topic, see the survey by Arora and Lund [12]. For an up-to-date status of the best positive and negative results known for numerous **NP**-hard optimization problems, see the excellent compendium maintained online at
`http://www.nada.kth.se/~viggo/problemlist/compendium.html`

The compendium also appears in Ausiello, Crescenzi, Gambosi, Kann, Marchetti-Spaccamela, and Protasi [18].

30 Open Problems

This chapter is centered around problems and issues currently in vogue in the field of approximation algorithms. Important new issues are bound to arise in the future. With each of these problems two questions arise – that of obtaining the best approximation guarantee and a matching hardness result[1]

30.1 Problems having constant factor algorithms

Since a large number of important open problems in the field today involve improving the guarantee for problems for which we already know constant factor algorithms, we found it convenient to present them in a separate section. Of course, we are not looking for small improvements using incremental means. A good model is Goemans and Williamson's improvement to the MAX-CUT problem, from factor 1/2 to 0.878, which introduced semidefinite programming into the repertoire of techniques in this field. Most of the problems listed below have the potential of extending known methods in significant ways and introducing important new ideas.

Vertex cover, Problem 1.1: Improve on factor 2 (see algorithms in Chapters 1, 2, 14, and 15). Semidefinite programming may be a possible avenue, see, e.g., the attempt by Goemans and Kleinberg [109].

Set cover, Problem 2.1: This question generalizes the previous one. Consider the restriction of the set cover problem to instances in which the frequency of each element is bounded by a fixed constant f. Improve on factor f (see algorithms in Chapters 2, 14, and 15). The best hardness result known is $f^{1/19}$, assuming $\mathbf{P} \neq \mathbf{NP}$, due to Trevisan [255].

Acyclic subgraph, Problem 1.9: Improve on factor 1/2 (see Exercise 1.1). Semidefinite programming may be applicable.

Metric TSP, Problem 3.5: As stated in Exercise 23.13, the solution produced by Christofides' algorithm (Algorithm 3.10) is within a factor of

[1] For an up-to-date status of the best positive and negative results known for numerous **NP**-hard optimization problems, see the excellent compendium at `http://www.nada.kth.se/~viggo/problemlist/compendium.html`

3/2 of the subtour elimination LP-relaxation for this problem. However, the worst integrality gap example known is (essentially) 4/3. Can a 4/3 factor algorithm be obtained using this relaxation?
Christofides' algorithm consists of two steps: obtaining an MST and patching up its odd degree vertices. The above stated result follows by bounding the cost of each of these steps individually. It might be a good idea to first look for a "one-shot" factor 3/2 algorithm which compares the entire solution to the LP-relaxation. The primal–dual schema may hold the key.

Steiner tree, Problem 3.1: The best approximation guarantee known is essentially 5/3 (see Exercise 22.12). A promising avenue for obtaining an improved guarantee is to use the bidirected cut relaxation (22.7). This relaxation is exact for the minimum spanning tree problem. For the Steiner tree problem, the worst integrality gap known is (essentially) 8/7, due to Goemans (see Exercise 22.11). The best upper bound known on the integrality gap is 3/2 for quasi-bipartite graphs (graphs that do not contain edges connecting pairs of Steiner vertices), due to Rajagopalan and Vazirani [234]. Determine the integrality gap of this relaxation and obtain an algorithm achieving this guarantee[2].
Recall that in contrast, LP-relaxation (22.2) has an integrality gap of (essentially) 2, not only for this problem, but also for its special case, the minimum spanning tree problem, and its generalization, the Steiner network problem.

Steiner network, Problem 23.1: Chapter 23 gives a factor 2 algorithm. However, it uses LP-rounding and has a prohibitive running time. Obtain a factor 2 combinatorial algorithm for this problem. A corollary of Algorithm 23.7 is that the integrality gap of LP-relaxation (23.2) is bounded by 2. Therefore, this relaxation can be used as a lower bound for obtaining a factor 2 combinatorial algorithm. The primal–dual schema appears to be the most promising avenue. A starting point may be determining if the following is true:
For each instance of the Steiner forest problem (and more generally, the Steiner network problem) there is an integral primal solution $\boldsymbol{x}$ and dual feasible solution $\boldsymbol{y}$ such that each edge picked by $\boldsymbol{x}$ is tight w.r.t. the dual $\boldsymbol{y}$ and each raised dual S has degree ≤ 2 ($\leq 2f(S)$). Observe that the dual found by Algorithm 22.3 can have arbitrarily high degree.

Multiway cut, Problem 4.1: A 1.5 factor is presented in Chapter 19. As stated, this can be improved to 1.3438. However, the worst integrality gap example known for LP-relaxation (19.1) is (essentially) 8/7. Determine the integrality gap of this relaxation, and obtain an algorithm achieving

[2] A more general issue along these lines is to clarify the mysterious connection between the integrality gap of an LP-relaxation and the approximation factor achievable using it.

this guarantee. A different relaxation is presented in Exercise 19.7. How are the two relaxations related? Are they equivalent in that any feasible solution to one be converted to a solution of the other of the same cost?

Subset feedback vertex set, Problem 19.15: The best factor known is 8, via a fairly complicated algorithm (see Exercise 19.13). Is a factor 2 algorithm possible, matching several of the other related problems stated in Exercise 19.13?

30.2 Other optimization problems

Shortest vector, Problem 27.1: Obtain a polynomial factor algorithm for this problem. As shown in Chapter 27, the dual lattice helps give a factor n co-**NP** certificate for this problem. Is the dual lattice of further algorithmic use? Does it help to randomly permute the given basis vectors before running Algorithm 27.9 on them? The best hardness result known for this problem, of factor $\sqrt{2} - \varepsilon$, for any $\varepsilon > 0$, assuming **RP** $\neq$ **NP**, is due to Micciancio [212].

Sparsest cut, Problem 21.2: The best approximation factor known is $O(\log n)$ (see Chapter 21). However, no hardness of approximation results have been established for this problem – as far as we know a PTAS not yet ruled out. Is there a constant factor algorithm or a PTAS for this problem?

Minimum b-balanced cut and minimum bisection cut, Problem 21.27: An $O(\log^2 n)$ factor algorithm for both these problems was given by Feige and Krauthgamer [89]. As in the case of sparsest cut, a PTAS is not yet ruled out for these problems. Is there a constant factor algorithm or a PTAS for these problems? When restricted to planar graphs, the minimum b-balanced cut problem, for $b \leq 1/3$, can be approximated within a factor of 2, see Garg, Saran, and Vazirani [101].

Minimum multicut, Problem 18.1: An $O(\log n)$ factor algorithm is given in Chapter 20. A long standing open problem is whether there is a constant factor deterministic algorithm for this problem.

Asymmetric TSP, Problem 3.15: The best factor known is $O(\log n)$ (see Exercise 3.6). Is there a constant factor algorithm for this problem?

Vertex-connectivity network design: This variant of the Steiner network problem (Problem 23.1) asks for a minimum cost subgraph containing $r_{u,v}$ vertex-disjoint paths, instead of edge-disjoint paths, for each pair of vertices $u, v \in V$. No nontrivial approximation algorithms are known for this variant. For the special case when $r_{u,v} = k$ for each pair of vertices $u, v \in V$ and the edge costs obey the triangle inequality, a $(2 + \frac{2(k-1)}{n})$ factor algorithm is given by Khuller and Raghavachari [177].

A problem of intermediate difficulty is the element-connectivity network design problem, in which vertices are partitioned into two sets: terminals and non-terminals. Only edges and non-terminals, referred to as *elements*, can fail. Only pairs of terminals have connectivity requirements, specifying the number of element-disjoint paths required. An algorithm with an approximation guarantee of factor $2H_k$, where k is the largest requirement, is given by Jain, Măndoiu, Vazirani, and Williamson [146].

Maximum integer multicommodity flow, Problem 18.3: Example 18.8 shows that the natural LP-relaxation has an integrality gap of $\Omega(n)$. It is easy to get around this difficulty while still retaining the essence of the original problem by asking for a maximum half-integral flow. Is there an $O(\log n)$ factor algorithm for this latter problem?

Metric uncapacitated facility location and k-median, Problems 24.1 and 25.1: Determine the integrality gaps of the LP-relaxations (24.2) and (25.2).

Capacitated facility location problem, Exercise 24.8: As stated in Exercise 24.8 the modification of LP (24.2) to this problem has unbounded integrality gap. Is there some other lower bounding method that leads to a good approximation algorithm?

Directed multicut and sparsest cut: In Chapters 20 and 21 we considered two generalizations of the undirected maximum flow problem and derived approximation algorithms for the corresponding cut problems, multicut and sparsest cut. Not much is known at present about analogous problems in directed graphs.

Directed Steiner tree, Problem 3.14: As shown in Exercise 3.3 this problem is unlikely to have a better approximation guarantee than $O(\log n)$. Is a guarantee of $O(\log n)$ possible? The best guarantee known is n^{ε} for any fixed $\varepsilon > 0$, due to Charikar et. al. [38]. Generalizations of this problem to higher connectivity requirements, analogous to the Steiner network problem, also need to be studied.

Directed feedback edge (vertex) set: Given a directed graph $G = (V, E)$, a *feedback edge (vertex) set* is a set of edges (vertices) whose removal leaves an acyclic graph. The problem is to find the minimum cardinality such set. More generally, consider the weighted version in which the edges (vertices) have assigned weights, and we want to find the minimum weight such set. It is easy to see that the edge and vertex versions are inter-reducible via approximation factor preserving reductions. An $O(\log n \log \log n)$ factor approximation algorithm is known for the weighted version, due to Seymour [246]. Can this be improved to $O(\log n)$ or even a constant factor?

Cover time: Given an undirected graph $G = (V, E)$, the cover time starting at vertex $v \in V$, $C(v)$ is the expected number of steps taken by a random walk on G, which starts at v and visits all vertices. The *cover time* of G is $\max_{v \in V} C(v)$. Clearly, a randomized algorithm can estimate the cover time to any desired accuracy by empirically simulating the random walk many times and taking the average. Kahn, Kim, Lovász, and Vu [158] have given an $O((\log \log n)^2)$ factor deterministic algorithm for this problem. Is a constant factor deterministic algorithm possible?

30.3 Counting problems

For the problems presented below (other than graphs with given degree sequence and triangulations), the decision version is in **P**, the counting version is #**P**-complete, and the complexity of approximately counting the number of solutions is unresolved. The complexity of counting the number of graphs with given degree sequence and triangulations is open, though conjectured to be #**P**-complete.

Perfect matchings in general graphs: When restricted to planar graphs, this problem is polynomial time solvable using the classic algorithm of Kastelyn [175]. This result extends to $K_{3,3}$-free graphs (graphs that do not contain a subgraph homeomorphic to $K_{3,3}$) as well, see Little [198] and Vazirani [259]. A FPRAS is known for the restriction of this problem to bipartite graphs, which is the same as the problem of evaluating a 0/1 permanent, due to Jerrum, Sinclair, and Vigoda [150] (more generally, this work gives a FPRAS for evaluating the permanent of a square matrix with nonnegative integer entries).

Volume of a convex body: Given a convex body in $\mathbf{R}^n$ via an oracle, the problem is to estimate its volume. A number of other counting problems can be reduced to this fundamental problem. The first FPRAS for this problem was given by Dyer, Frieze, and Kannan [73]. Although polynomial, the running time of this algorithm was exorbitant. It required $O^*(n^{23})$ oracle calls – the "soft-O" notation of O^* suppresses factors of $\log n$ as well as ε, the error bound. The current best algorithm, due to Kannan, Lovász, and Simonovits [162] requires $O^*(n^5)$ oracle calls and $O^*(n^7)$ arithmetic operations. Can the running time be further improved?

Acyclic orientations: Count the number of acyclic orientations of a given undirected graph G. An orientation of the edges of G is acyclic if the resulting directed graph is acyclic. Several Markov chains on the set of acyclic orientations are known that asymptotically converge to the uniform distribution; however, none of them is known to be rapidly mixing. For instance, say that two orientations are adjacent if one can be obtained from the other by flipping directions of the edges incident at a source or a sink, where a source has all outgoing edges and a sink has all incoming edges. Do a random walk on this graph.

Forests: A forest in an undirected graph is a set of edges that contain no cycles. A maximal forest is a spanning tree (assume the graph is connected). Interestingly enough, the problem of counting the number of spanning trees in a graph is in **P** – this being one of the very few counting problems known to be polynomial time solvable. This follows as a consequence of the classic matrix tree theorem of Kirchhoff, see [201]. It is worth remarking that elegant polynomial time algorithms are known for generating a random spanning tree in an undirected graph using rapidly mixing Markov chains, due to Aldous [4], Broder [35], and Wilson [267]. On the other hand, the complexity of approximately counting forests in arbitrary graphs is open. The case of dense graphs (each vertex having degree at least αn, for $0 < \alpha < 1$) is handled by Annan [9]. Forests and spanning trees are the independent sets and bases, respectively, of the graphic matroid of the given graph.

Bases of a matroid: Given an arbitrary matroid via an independence oracle, count the number of bases. Define the *basis exchange graph* of a matroid as follows. Its vertices are all bases. Two bases are adjacent iff their symmetric difference is two elements. The Markov chain defined by a random walk on the basis exchange graph is conjectured to be rapidly mixing by Dagum, Luby, Mihail, and Vazirani [59]. If so, a FPRAS for approximately counting the number of bases will follow. Examples of matroids for which this conjecture has been positively resolved are graphic matroids (see previous problem) and their generalization, balanced matroids. For the latter result, see Feder and Mihail [84]. A positive resolution of this question will also resolve the question of approximately counting forests (since forests of any particular size are bases of a truncation of the graphic matroid).

Network reliability: Many versions of the network reliability problem have found practical applications and have been studied in the past. Two basic versions for undirected graphs with edge failure probabilities are s–t reliability, which asks for the probability that special vertices s and t get disconnected, and global reliability, which asks for the probability that any part of the graph gets disconnected. One can define two analogous problems in directed graphs as well. Of these four problems, only undirected global reliability is settled – a FPRAS for this version is presented in Chapter 28. In addition, for each of these four cases one can also ask for the probability that s–t or the entire graph remain connected. This version is open even for the undirected global case.

Euler tours: Count the number of Euler tours of a given undirected graph (a connected graph is Eulerian iff all vertices have even degrees). Interestingly enough, there is a polynomial time algorithm for the analogous problem for directed graphs – again following from Kirchhoff's Theorem.

Trees: Given an undirected graph G, count the number of subgraphs of G that are trees.

Antichains in a partial order: See Exercise 1.7 for the definition. For the related problem of counting the number of total orders consistent with a partial order, a FPRAS is known, due to Matthews [210], Karzanov and Khachian [174], and Bubley and Dyer [36].

Graphs with given degree sequence: Given n nonnegative integers d_1, $\ldots, d_n$, which represent the degrees of the n vertices, $v_1, \ldots, v_n$, of a simple graph, count the number of such graphs. A related problem is to count the number of connected graphs having this degree sequence. In both cases, the question of existence of one such graph can be solved in polynomial time using a matching algorithm. If G is restricted to be a bipartite graph, with the bipartition specified, then a FPRAS follows from that for 0/1 permanents [150].

Contingency tables: Given the row sums and column sums of an $m \times n$ matrix with nonnegative integer entries, count the number of such matrices. A FPRAS is known if the row sums and column sums are all sufficiently large, being at least $(m+n)mn$, due to Dyer, Kannan, and Mount [70]. Morris [216] improves this to the case where each row sum is $\Omega(n^{3/2}m \log m)$ and each column sum is $\Omega(m^{3/2}n \log n)$. If the matrices are constrained to be 0/1, this is same as the degree sequence problem restricted to bipartite graphs, for which a FPRAS follows from that for 0/1 permanents [150].

Triangulations: Compute the number of triangulations of n points on the plane, i.e., the number of ways of putting down non–intersecting line segments connecting pairs of points so that all internal faces are triangles. Consider the graph G on all possible triangulations whose edges are defined as follows: Remove an edge in a triangulation t that is not on the infinite face. If the resulting quadrilateral is convex, let t' be the triangulation obtained by adding an edge connecting the other two points of this quadrilateral. Then, G has an edge connecting t and t'. A random walk on this graph is conjectured to be rapidly mixing. If the n points form the vertices of a convex n-gon, then the number of triangulations is known to be the Catalan number C_{n-2}, and hence polynomial time computable. For this special case, the Markov chain defined above is known to be rapidly mixing, see McShine and Tetali [211].

Stable marriages: An instance of the stable marriage problem consists of n boys and n girls, together with an ordered list of the preferences of each boy and each girl (each boy orders all n girls and each girl orders all n boys). A marriage is a perfect matching of the boys and girls. Boy b and girl g who are not married to each other are said to form a *rogue couple* if b prefers g to the girl he is married to and g prefers b to the boy she is married to. The marriage is *stable* if there are no rogue couples. The complexity of approximately counting the number of stable marriages is unresolved. For numerous structural properties of the set of stable marriages, see Gusfield and Irving [126].

Colorings of a graph: Consider an undirected graph $G = (V, E)$ with maximum degree Δ. Jerrum [151] gave a FPRAS for counting the number of valid k-colorings of G for any $k > 2\Delta$, and Vigoda [262] extended this to any $k > 11\Delta/6$. Can this be improved to counting the number of valid k-colorings of G for any $k \geq \Delta + 2$? (If the number of colors is $\leq \Delta + 1$ then the natural Markov chain, that at each step picks a random vertex and recolors it with a random consistent color, may not be connected.) This quantity finds applications in statistical physics.

Hamiltonian cycles: If each vertex of an undirected graph G has degree at least $n/2$ then G must have a Hamiltonian cycles (see Dirac's condition in [201]). If the minimum degree is $(1/2+\varepsilon)n$, for $\varepsilon > 0$, Dyer, Frieze, and Jerrum [71] have given a FPRAS for this problem. Can this be extended to $\varepsilon = 0$, i.e., graphs having minimum degree $n/2$?

Independent sets: For graphs having $\Delta = 4$, a FPRAS was given by Luby and Vigoda [205], where Δ denotes the maximum degree of the graph. Dyer, Frieze, and Jerrum [72] show that the problem is not approximable for $\Delta \geq 25$, assuming **RP** $\neq$ **NP**. They also give an argument to show that the Markov chain Monte Carlo is unlikely to succeed for $\Delta \geq 6$. Besides the question of $\Delta = 5$, this leaves the question of determining whether other methods will work for $6 \leq \Delta \leq 24$ or whether these cases are also inapproximable.

Tutte polynomial: Several of the problems stated above are special cases of evaluating the Tutte polynomial of the given graph $G = (V, E)$ at a particular point of the (x, y)-plane. For $A \subseteq E$, define the *rank* of A, denoted $r(A)$, to be $|V| - k(A)$, where $k(A)$ is the number of connected components in the graph having vertex set V and edge set A. The *Tutte polynomial* of G at point (x, y) is

$$T(G; x, y) = \sum_{A \subseteq E} (x-1)^{r(E)-r(A)} (y-1)^{|A|-r(A)}.$$

Some of the natural quantities captured by this polynomial are:

- At $(1,1)$, T counts the number of spanning trees in G.
- At $(2,1)$, T counts the number of forests in G.
- At $(1,2)$, T counts the number of connected subgraphs of G.
- At $(2,0)$, T counts the number of acyclic orientations of G.
- At $(0,2)$, T counts the number of orientations of G that form a strongly connected digraph.

- The *chromatic polynomial* of G is given by
$$P(G,\lambda) = (-1)^{r(E)}\lambda^{k(E)}T(G;1-\lambda,0),$$
where $P(G,\lambda)$ is the number of colorings of G using λ colors.
- If the failure probability of each edge is p, then the probability that G remains connected is given by
$$R(G;p) = q^{|E|-r(E)}p^{r(E)}T(G;1,1/(1-p)).$$

Vertigan and Welsh [261] have shown that other than a few special points and two special hyperbolae (see next problem for definition), the exact evaluation of the Tutte polynomial is #**P**-hard. The question of designing FPRAS's is wide open. Say that a graph is α-dense if each vertex has degree $\geq \alpha n$, where $0 < \alpha < 1$. Annan [8] and Alon, Frieze, and Welsh [6] have given FPRAS's for α-dense graphs for the cases $y = 1, x \geq 1$ and $y > 1, x \geq 1$, respectively.

Partition functions of the Ising and Potts models: The hyperbolae H_α defined by
$$H_\alpha = \{(x,y) \mid (x-1)(y-1) = \alpha\}$$
play a special role in the context of the Tutte polynomial. In particular, along H_2, T gives the partition function of the Ising model for G, and along H_Q, for integer $Q \geq 2$, T gives the partition function of the Potts model for G. Both these quantities find use in statistical physics; see Welsh [264] for precise definitions and further details (the points on each hyperbola are parametrized by "temperature" and Q represents the number of "color" classes). Jerrum and Sinclair [153] gave a FPRAS for estimating, at any temperature, the partition function of the Ising model of a graph, and Randall and Wilson [236] extended this to a polynomial time sampling procedure. However, because of large exponents in the running times, these algorithms are not practical. The Swendsen-Wang process [253] provides a natural and practically used Markov chain for estimating these quantities. This leads to the question of determining, formally, whether this chain is rapidly mixing. A negative result was provided by Gore and Jerrum [118] who show that this chain is not rapidly mixing for the complete graph, K_n, for $Q \geq 3$. Positive results for certain classes of graphs were provided by Cooper and Frieze [55]. Is this chain rapidly mixing for the partition function of the Ising model for an arbitrary graph? Is there some other way of estimating the partition function of the Potts model for an arbitrary graph?

30.4 Notes

Since the appearance of this list of problems in the First Edition, the following insights have been gained. Dinur, Guruswami, Khot, and Regev [64] show that for the set cover problem, Problem 2.1, for any constant $\varepsilon > 0$, there is no approximation algorithm achieving a guarantee of $f - 1 - \varepsilon$ on instances in which the frequency of each element is bounded by f, for any fixed $f \geq 3$, assuming $\mathbf{P} \neq \mathbf{NP}$. Chekuri, Gupta, and Kumar [43] show that the relaxation presented in Exercise 19.7 has a strictly worse integrality gap than relaxation (19.1) for the multiway cut problem, Problem 4.1. For the problem of counting the number of contingency table with given row and column sums, Cryan, Dyer, Goldberg, Jerrum, and Martin [58] give a FPRAS in case the number of rows is a constant, using the MCMC method. Dyer [69] obtains a more efficient algorithm via a dynamic programming approach.

A An Overview of Complexity Theory for the Algorithm Designer

A.1 Certificates and the class NP

A *decision problem* is one whose answer is either "yes" or "no". Two examples are:

SAT: Given a Boolean formula in conjunctive normal form, f, is there a satisfying truth assignment for f?
Cardinality vertex cover: Given an undirected graph G and integer k, does G have a vertex cover of size $\leq k$?

For any positive integer k, we will denote by kSAT the restriction of SAT to instances in which each clause contains at most k literals.

It will be convenient to view a decision problem as a language, i.e., a subset of $\{0,1\}^*$. The language consists of all strings that encode "yes" instances of the decision problem. A language $L \in \mathbf{NP}$ if there is a polynomial p and a polynomial time bounded Turing machine M, called the *verifier*, such that for each string $x \in \{0,1\}^*$:

- if $x \in L$, then there is a string y (the certificate) of polynomially bounded length, i.e., $|y| \leq p(|x|)$, such that $M(x, y)$ accepts, and
- if $x \notin L$, then for any string y, such that $|y| \leq p(|x|)$, $M(x, y)$ rejects.

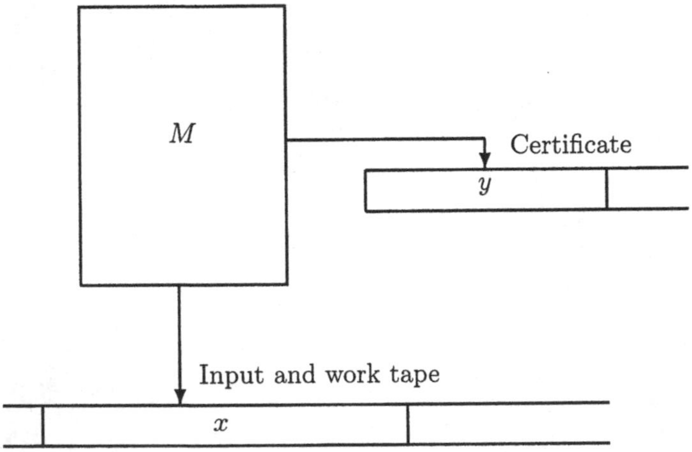

String y that helps ascertain that x is a "yes" instance will be called a *Yes certificate.* We will also refer to y as a *proof* or a *solution*; in the context of randomized computation, it is also referred to as a *witness.* Thus, **NP** is the class of languages that have "short, quickly verifiable" Yes certificates.

For example, the verifier for cardinality vertex cover assumes that y specifies a subset of the vertices. It checks whether this subset is indeed a vertex cover and is of the desired size bound. (Observe that no claim has been made about the time needed to actually *find* such a certificate.) It is also easy to see that the class **NP** defined above is precisely the class of languages that are decidable by nondeterministic polynomial time Turing machines (see Section A.6 for references), hence the name.

A language L belongs to the class co-**NP** iff $\overline{L} \in$ **NP**. Thus, co-**NP** is the class of languages that have "short, quickly verifiable" No certificates. For instance, let L be the language consisting of all prime numbers. This language allows No certificates: a factorization for number n is proof that $n \notin L$. Hence $L \in$ co-**NP**. Interestingly enough, $L \in$ **NP** as well (see Exercise 1.13), though it is not known to belong to **P**.

A.2 Reductions and NP-completeness

Next, let us introduce the crucial notion of a *polynomial time reduction.* Let L_1 and L_2 be two languages in **NP**. We will say that L_1 reduces to L_2, and write $L_1 \preceq L_2$, if there is a polynomial time Turing machine T which given a string $x \in \{0,1\}^*$, outputs string y such that $x \in L_1$ iff $y \in L_2$. In general, T does not have to decide whether x is a "yes" or a "no" instance in order to output y. Clearly, if $L_1 \preceq L_2$ and L_2 is polynomial time decidable, then so is L_1.

A language L is **NP**-*hard* if for every language $L' \in$ **NP**, $L' \preceq L$. A language L is **NP**-*complete* if $L \in$ **NP**, and L is **NP**-hard. An **NP**-complete language L is a hardest language in **NP**, in the sense that a polynomial time algorithm for L implies a polynomial time algorithm for every language in **NP**, i.e., it implies **P** = **NP**.

The central theorem of complexity theory gives a proof of **NP**-hardness for a natural problem, namely SAT. The idea of the proof is as follows. Let L be an arbitrary language in **NP**. Let M be a nondeterministic polynomial time Turing machine that decides L, and let p be the polynomial bounding the running time of M. The proof involves showing that there is a deterministic polynomial time Turing machine T, that "knows" M and p, and given a string $x \in \{0,1\}^*$, outputs a SAT formula f such that each satisfying truth assignment of f encodes an accepting computation of M on input x. Thus, f is satisfiable iff there is an accepting computation of M on input x, i.e., iff $x \in L$.

Once one problem, namely SAT, has been shown to be **NP**-hard, the hardness of other natural problems can be established by simply giving polynomial time reductions from SAT to these problems (see Exercise 1.11). Perhaps the most impressive feature of the theory of **NP**-completeness is the ease with which the latter task can be accomplished in most cases, so that with relatively little work, a lot of crucial information is obtained. Other than a handful of (important) problems, most natural problems occurring in **NP** have been classified as being either in **P** or being **NP**-complete. Indeed, it is remarkable to note that other basic complexity classes, defined using notions of time, space and nondeterminism, also tend to have natural complete problems (under suitably defined reducibilities).

Establishing **NP**-hardness for vertex cover involves giving a polynomial time algorithm that, given a SAT formula f, outputs an instance (G, k) such that G has a vertex cover of size $\leq k$ iff f is satisfiable. As a corollary, we get that under the assumption $\mathbf{P} \neq \mathbf{NP}$, there is no polynomial time algorithm that can distinguish "yes" instances of vertex cover from "no" instances. As stated above, this also shows that if $\mathbf{P} \neq \mathbf{NP}$, there is no polynomial time algorithm for solving vertex cover exactly.

Considering the large and very diverse collection of **NP**-complete problems, none of which has yielded to a polynomial time algorithm for so many years, it is widely believed that $\mathbf{P} \neq \mathbf{NP}$, i.e., that there is no polynomial time algorithm for deciding an **NP**-complete language.

The $\mathbf{P} \neq \mathbf{NP}$ conjecture has a deep philosophical point to it. The conjecture asserts that the task of finding a proof for a mathematical statement is qualitatively harder than the task of simply verifying the correctness of a given proof for the statement. To see this, observe that the language

$$L = \{(S, 1^n) \mid \text{statement } S \text{ has a proof of length} \leq n\}$$

is in **NP**, assuming any reasonable axiomatic system.

A.3 NP-optimization problems and approximation algorithms

Combinatorial optimization problems are problems of picking the "best" solution from a finite set. An **NP**-*optimization problem*, Π, consists of:

- A set of *valid instances*, D_Π, recognizable in polynomial time. We will assume that all numbers specified in an input are rationals, since our model of computation cannot handle infinite precision arithmetic. The *size* of an instance $I \in D_\Pi$, denoted by $|I|$, is defined as the number of bits needed to write I under the assumption that all numbers occurring in the instance are written in binary.
- Each instance $I \in D_\Pi$ has a set of *feasible solutions*, $S_\Pi(I)$. We require that $S_\Pi(I) \neq \emptyset$, and that every solution $s \in S_\Pi(I)$ is of length polynomially bounded in $|I|$. Furthermore, there is polynomial time algorithm that, given a pair (I, s), decides whether $s \in S_\Pi(I)$.

- There is a polynomial time computable *objective function*, obj_Π, that assigns a nonnegative rational number to each pair (I, s), where I is an instance and s is a feasible solution for I. The objective function is frequently given a physical interpretation, such as *cost, length, weight*, etc.
- Finally, Π is specified to be either a *minimization problem* or a *maximization problem.*

The restriction of Π to unit cost instances will be called the *cardinality version* of Π.

An *optimal solution* for an instance of a minimization (maximization) problem is a feasible solution that achieves the smallest (largest) objective function value. $\text{OPT}_\Pi(I)$ will denote the objective function value of an optimal solution to instance I. We will shorten this to OPT when it is clear that we are referring to a generic instance of the particular problem being studied.

With every **NP**-optimization problem, one can naturally associate a decision problem by giving a bound on the optimal solution. Thus, the decision version of **NP**-optimization problem Π consist of pairs (I, B), where I is an instance of Π and B is a rational number. If π is a minimization (maximization) problem, then the answer to the decision version is "yes" iff there is a feasible solution to I of cost $\leq B$ ($\geq B$). If so, we will say that (I, B) is a "yes" instance; we will call it a "no" instance otherwise. For example, the decision version of cardinality vertex cover is stated in Section A.1.

Clearly, a polynomial time algorithm for Π can help solve the decision version – by computing the cost of an optimal solution and comparing it with B. Conversely, hardness established for the decision version carries over to Π. Indeed hardness for an **NP**-optimization problem is established by showing that its decision version is **NP**-hard. With a slight abuse of notation, we will also say that the optimization version is **NP**-hard.

An approximation algorithm produces a feasible solution that is "close" to the optimal one, and is time efficient. The formal definition differs for minimization and maximization problems. Let Π be a minimization (maximization) problem, and let δ be a function, $\delta : \mathbf{Z}^+ \rightarrow \mathbf{Q}^+$, with $\delta \geq 1$ ($\delta \leq 1$). An algorithm $\mathcal{A}$ is said to be a *factor δ approximation algorithm for Π* if, on each instance I, $\mathcal{A}$ produces a feasible solution s for I such that $f_\Pi(I, s) \leq \delta(|I|) \cdot \text{OPT}(I)$ ($f_\Pi(I, s) \geq \delta(|I|) \cdot \text{OPT}(I)$), and the running time of $\mathcal{A}$ is bounded by a fixed polynomial in $|I|$. Clearly, the closer δ is to 1, the better is the approximation algorithm.

On occasion we will relax this definition and will allow $\mathcal{A}$ to be randomized, i.e., it will be allowed to use the flips of a fair coin. Assume we have a minimization problem. Then we will say that $\mathcal{A}$ is a *factor δ randomized approximation algorithm for Π* if, on each instance I, $\mathcal{A}$ produces a feasible solution s for I such that

$$\mathbf{Pr}[f_\Pi(I, s) \leq \delta(|I|) \cdot \text{OPT}(I)] \geq \frac{1}{2},$$

where the probability is over the coin flips. The running time of $\mathcal{A}$ is still required to be polynomial in $|I|$. The definition for a maximization problem is analogous.

Remark A.1 Even though δ has been defined to be a function of the size of the input, we will sometimes pick δ to be a function of a more convenient parameter. For instance, for the set cover problem (Chapter 2), we will pick this parameter to be the number of elements in the ground set.

A.3.1 Approximation factor preserving reductions

Typically, polynomial time reductions map optimal solutions to optimal solutions; however, they do not preserve near-optimality of solutions. Indeed, all **NP**-complete problems are equally hard from the viewpoint of obtaining exact solutions. However, from the viewpoint of obtaining near-optimal solutions, they exhibit the rich set of possibilities alluded to earlier.

In this book we will encounter pairs of problems which may look quite different superficially, but whose approximability properties are closely linked (e.g., see Exercise 19.13). Let us define a suitable reducibility in order to formally establish such connections. Several reductions have been defined that preserve constant factor approximability. The reducibility stated below is a stringent version of these, and actually preserves the constant itself. Pair of problems that are linked in this manner are either both minimization problems or both maximization problems.

Let Π_1 and Π_2 be two minimization problems (the definition for two maximization problems is quite similar). An *approximation factor preserving reduction* from Π_1 to Π_2 consists of two polynomial time algorithms, f and g, such that

- for any instance I_1 of Π_1, $I_2 = f(I_1)$ is an instance of Π_2 such that $\text{OPT}_{\Pi_2}(I_2) \leq \text{OPT}_{\Pi_1}(I_1)$, and
- for any solution t of I_2, $s = g(I_1, t)$ is a solution of I_1 such that

$$\text{obj}_{\Pi_1}(I_1, s) \leq \text{obj}_{\Pi_2}(I_2, t).$$

It is easy to see that this reduction, together with an α factor algorithm for Π_2, gives an α factor algorithm for Π_1 (see Exercise 1.16).

A.4 Randomized complexity classes

Certain **NP** languages[1] are characterized by the fact that they possess an abundance of Yes certificates, which renders them essentially tractable, assuming availability of a source of random bits. Such languages belong to the

[1] The definitions of this section will be useful in Chapter 29.

class **RP**, short for *Randomized Polynomial Time.* A language $L \in \mathbf{RP}$ if there is a polynomial p and a polynomial time bounded Turing machine M such that for each string $x \in \{0,1\}^*$:

- if $x \in L$, then $M(x,y)$ accepts for at least half the strings y of length $p(|x|)$, and
- if $x \notin L$, then for any string y of length $p(|x|)$, $M(x,y)$ rejects.

Clearly, $\mathbf{P} \subseteq \mathbf{RP} \subseteq \mathbf{NP}$. Suppose language $L \in \mathbf{RP}$. On input x, we will pick a random string, y, of length $p(|x|)$ and will run $M(x,y)$. Clearly, the entire computation takes polynomial time. We may erroneously reject x even though $x \in L$. However, the probability of this is at most $1/2$. Let us call this the error probability. By the usual trick of making repeated independent runs, we can reduce the error probability to inverse exponential in the number of runs.

A language L belongs to the class co-**RP** iff $\overline{L} \in \mathbf{RP}$. Such a language has an abundance of No certificates. The corresponding machine may make an error on inputs $x \notin L$. Finally, let us define **ZPP**, short for *Zero-error Probabilistic Polynomial Time*, to be the class of languages for which there is a randomized Turing machine (i.e., a Turing machine equipped with a source of random bits) that always terminates with the correct answer and whose *expected* running time is polynomial. It is easy to see (Exercise 1.17) that

$$L \in \mathbf{ZPP} \text{ iff } L \in (\mathbf{RP} \cap \text{co-}\mathbf{RP}).$$

DTIME(t) denotes the class of problems for which there is a deterministic algorithm running in time $O(t)$. Thus, $\mathbf{P} = \mathbf{DTIME}(poly(n))$, where $poly(n) = \bigcup_{k \geq 0} n^k$. **ZTIME**$(t)$ denotes the class of problems for which there is a randomized algorithm running in expected time $O(t)$. Thus, $\mathbf{ZPP} = \mathbf{ZTIME}(poly(n))$.

A.5 Self-reducibility

Most known problems in **NP** exhibit an interesting property, called self-reducibility, which yields a polynomial time algorithm for finding a solution (a Yes certificate), given an oracle for the decision version. A slightly more elaborate version of this property yields an exact polynomial time algorithm for an **NP**-optimization problem, again given an oracle for the decision version. In a sense this shows that the difficult core of **NP** and **NP**-optimization problems is their decision versions (see Section 16.2 and Exercise 28.7 for other fundamental uses of self-reducibility).

Perhaps the simplest setting to describe self-reducibility is SAT. Let ϕ be a SAT formula on n Boolean variables $x_1, \ldots, x_n$. We will represent a truth assignment to these n variables as n-bit 0/1 vectors (True $= 1$ and False $= 0$). Let S be the set of satisfying truth assignments, i.e., solutions, to ϕ.

The important point is that for the setting of x_1 to 0 (1), we can find, in polynomial time, a formula ϕ_0 (ϕ_1) on the remaining $n-1$ variables whose solutions, S_0 (S_1), are precisely solutions of ϕ having $x_1 = 0$ ($x_1 = 1$).

Example A.2 Suppose $\phi = (x_1 \vee x_2 \vee x_3) \wedge (\overline{x}_1 \vee x_2 \vee x_4)$. Then $\phi_0 = (x_2 \vee x_3)$ and $\phi_1 = (x_2 \vee x_4)$. □

Using this property, an oracle for the decision version of SAT can be used to find a solution to ϕ, assuming it is satisfiable, as follows. First check whether ϕ_0 is satisfiable. If so, set $x_1 = 0$, and find any solution to ϕ_0. Otherwise, set $x_1 = 1$ (in this case ϕ_1 must be satisfiable), and find a solution to ϕ_1. In each case the problem has been reduced to a smaller one, and we will be done in n iterations.

The following representation will be particularly useful. Let T be a binary tree of depth n whose leaves are all n-bit 0/1 strings, representing truth assignments to the n variables. Leaves that are solutions to ϕ are marked special. The root of T is labeled with ϕ and its internal nodes are labeled with formulae whose solutions are in one-to-one correspondence with the marked leaves in the subtree rooted at this node. Thus, the 0th child of the root is labeled with ϕ_0 and the 1st child is labeled with ϕ_1. Tree T is called the *self-reducibility tree* for instance ϕ.

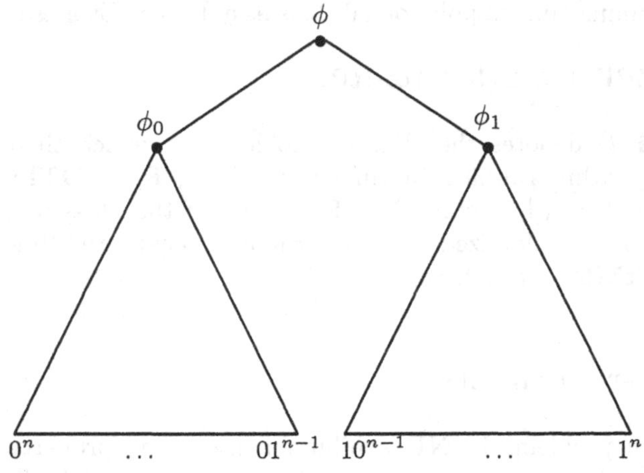

We will formalize the notion of self-reducibility for **NP**-optimization problems. Formalizing this notion for **NP** problems is an easier task and is left as Exercise 1.15.

First, let us illustrate self-reducibility for cardinality vertex cover. Observe that an oracle for the decision version enables us to compute the size of the optimal cover, OPT(G), by binary search on k. To actually find an optimal cover, remove a vertex v together with its incident edges to obtain graph G', and compute OPT(G'). Clearly, v is in an optimal cover iff OPT(G') = OPT(G) − 1. Furthermore, if v is in an optimal cover, then any optimal cover in G', together with v, is an optimal cover in G. Otherwise, any optimal cover for G must contain all neighbors, say $N(v)$, of v (in order to cover all edges incident at v). Let G'' be the graph obtained by removing v and $N(v)$ from G. Any optimal cover in G'', together with $N(v)$, is an optimal cover in G. Thus, in both cases, we are left with the problem of finding an optimal cover in a smaller graph, G' or G''. Continuing this way, an optimal cover in G can be found in polynomial time.

The above-stated reduction from the cardinality vertex cover problem to its decision version works because we could demonstrate polynomial time algorithms for

- obtaining the smaller graphs, G' and G'',
- computing the size of the best cover in G, consistent with the atomic decision, and
- constructing an optimal cover in G, given an optimal cover in the smaller instance.

The exact manner in which self-reducibility manifests itself is quite different for different problems. Below we state a fairly general definition that covers a large number of problems. In the interest of conveying the main idea behind this important concept, we will provide an intuitive, though easily formalizable, definition.

We will assume that solutions to an instance I of **NP**-optimization problem Π have *granularity*, i.e., consist of smaller pieces called *atoms* that are meaningful in the context of the problem. For instance, for cardinality vertex cover, the atoms consist of specifying whether or not a certain vertex is in the cover. Clearly, for vertex cover this can be done using $O(\log n)$ bits. Indeed, all problems considered in this book have granularity $O(\log n)$. Let us assume this for problem Π.

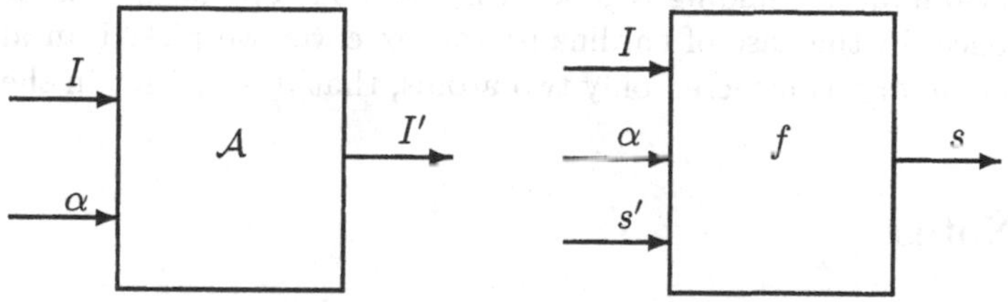

We will say that problem Π is *self-reducible* if there is a polynomial time algorithm, $\mathcal{A}$, and polynomial time computable functions, $f(\cdot,\cdot,\cdot)$ and $g(\cdot,\cdot,\cdot)$, satisfying the following conditions.

- Given instance I and an atom α of a solution to I, $\mathcal{A}$ outputs an instance I_α. We require that $|I_\alpha| < |I|$. Let $S(I \mid \alpha)$ represent the set of feasible solutions to I that are consistent with atom α. We require that the feasible solutions of I_α, $S(I_\alpha)$, are in one-to-one correspondence with $S(I \mid \alpha)$. This correspondence is given by the polynomial time computable function $f(\cdot,\cdot,\cdot)$ as follows.

$$f(I, \alpha, \cdot) : S(I_\alpha) \rightarrow S(I \mid \alpha).$$

- The correspondence $f(I, \alpha, \cdot)$ preserves order in the objective function values of solutions. Thus, if s_1' and s_2' are two feasible solutions of I_α with $\mathrm{obj}_\Pi(I_\alpha, s_1') \leq \mathrm{obj}_\Pi(I_\alpha, s_2')$, and $f(I, \alpha, s_1') = s_1$ and $f(I, \alpha, s_2') = s_2$, then $\mathrm{obj}_\Pi(I, s_1) \leq \mathrm{obj}_\Pi(I, s_2)$.
- Given the cost of an optimal solution to I_α, the cost of the best solution in $S(I \mid \alpha)$ can be computed efficiently, and is given by $g(I, \alpha, \mathrm{OPT}(I_\alpha))$.

Theorem A.3 *Let Π be an* **NP***-optimization problem that is self-reducible. There is a polynomial time (exact) algorithm for Π, given an oracle, $\mathcal{O}$, for the decision version of Π.*

Proof: As remarked earlier, via a suitable binary search we can use $\mathcal{O}$ to compute the cost of the optimal solution to an instance in polynomial time.

We will derive polynomial time algorithm $\mathcal{R}$ for solving Π exactly. Assume that $\mathcal{A}$, f, and g are defined as above for the self-reducibility of Π. Let I be an instance of Π. $\mathcal{R}$ first finds one atom of an optimal solution to I. An atom, say β, satisfies this condition iff $g(I, \beta, \mathrm{OPT}(I_\beta)) = \mathrm{OPT}(I)$, where $I_\beta = \mathcal{A}(I, \beta)$. Since atoms are only $O(\log n)$ bits long, finding such an atom involves simply searching the polynomially many possibilities. Let α be the atom found, and let $I_\alpha = \mathcal{A}(I, \alpha)$. $\mathcal{R}$ then recursively computes an optimal solution, say s', to I_α. Finally, it outputs $f(I, \alpha, s')$, which is guaranteed to be an optimal solution to I. Since $|I_\alpha| < |I|$, the recursion also takes only polynomial time. □

Remark A.4 The number of strings of length $O(\log n)$ that algorithm $\mathcal{R}$ needs to examine for finding a good atom depends on the specific problem. For instance, in the case of cardinality vertex cover we picked an arbitrary vertex, say v, and considered only two atoms, that v is or isn't in the cover.

A.6 Notes

The definition of an **NP**-optimization problem is due to Krentel [185]. Approximation factor preserving reductions are a stringent version of L-reducibility from Papadimitriou and Yannakakis [226]. Self-reducibility was first defined by Schnorr [242]. See Khuller and Vazirani [178] for a problem that is not self-reducible, assuming **P** $\neq$ **NP**. For further information on **NP**-completeness and complexity theory see Garey and Johnson [99] and Papadimitriou [224].

B Basic Facts from Probability Theory

Let us recall some useful facts from probability theory. We assume that the reader has already had a detailed exposure to this material (see Section B.4 for references).

B.1 Expectation and moments

Two quantities provide much information about a random variable: the mean, also called expectation, and variance. A key property of the expectation, which often simplifies its evaluation, is called *linearity of expectation.* It states that if $X, X_1, \ldots, X_n$ are random variables such that $X = c_1X_1 + \ldots + c_nX_n$, where $c_1, \ldots, c_n$ are constants, then

$$\mathbf{E}[X] = c_1\mathbf{E}[X_1] + \ldots + c_n\mathbf{E}[X_n].$$

(In particular, the expectation of a sum of random variables is the sum of their expectations.) The usefulness of this property arises from the fact that no assumption is made about independence between the random variables $X_1, \ldots, X_n$. Often a complex random variable can be written as the sum of indicator random variables (i.e., random variables taking on 0/1 values only), thereby simplifying the evaluation of its expectation.

The variance of random variable X measures the spread of X from its mean, and is defined as

$$\boldsymbol{V}[X] = \mathbf{E}[(X - \mathbf{E}[X])^2] = \mathbf{E}[X^2] - \mathbf{E}[X]^2.$$

Its positive square root is called the *standard deviation.* The mean and standard deviation of X are denoted by $\mu(X)$ and $\sigma(X)$, respectively.

For $k \in \mathbf{N}$, the kth *moment* and kth *central moment* of X are defined to be $\mathbf{E}[X^k]$ and $\mathbf{E}[(X - \mathbf{E}[X])^k]$, respectively. Thus the variance is the second central moment.

In general, the expectation of the product of random variables is not the product of expectations. An important exception is when the random variables are independent. Thus, if X and Y are independent random variables, then $\mathbf{E}[XY] = \mathbf{E}[X]\mathbf{E}[Y]$. This immediately implies that the variance of the sum of independent random variables is the sum of their variances, i.e., for independent random variables X and Y, $\boldsymbol{V}[X+Y] = \boldsymbol{V}[X] + \boldsymbol{V}[Y]$.

B.2 Deviations from the mean

If X is a nonnegative random variable with a known expectation, then *Markov's Inequality* helps bound the probability of deviations from the mean as follows. For $t \in \mathbf{R}^+$,

$$\mathbf{Pr}[X \geq t] \leq \frac{\mathbf{E}[X]}{t}.$$

This obvious inequality finds surprisingly many uses. For instance, it helps in obtaining a high probability statement from a bound on the expectation (e.g., see Section 14.2).

If the variance of a random variable is small, then large deviations from the mean are improbable. This intuitive statement is formalized by *Chebyshev's inequality* which states that for any random variable X and $a \in \mathbf{R}+$,

$$\mathbf{Pr}[|X - \mathbf{E}[X]| \geq a] \leq \left(\frac{\sigma(X)}{a}\right)^2.$$

See Lemma 28.5 for an application.

Poisson trials are repeated independent trials, each of which has two possible outcomes, called success and failure. In general, the success probability is allowed to change with the trials. They are called *Bernoulli trials* if the success probability is the same for each trial.

The Chernoff bounds, which provide bounds on the tail probabilities of Poisson trials, are very useful in analyzing algorithms. Let us represent n Poisson trials by indicator random variables $X_1, \ldots, X_n$, with 1 and 0 representing success and failure, respectively. Let $\mathbf{Pr}[X_i = 1] = p_i$, where $0 < p_i < 1$ for $1 \leq i \leq n$. Let random variable $X = X_1 + \ldots + X_n$ and $\mu = \mathbf{E}[X] = \sum_{i=1}^n p_i$. For the bound on the lower tail assume $0 < \delta \leq 1$. Then,

$$\mathbf{Pr}[X < (1 - \delta)\mu] < e^{(-\mu\delta^2/2)}.$$

The expression for the upper tail is more involved: for any $\delta > 0$,

$$\mathbf{Pr}[X > (1 + \delta)\mu] < \left(\frac{e^\delta}{(1+\delta)^{(1+\delta)}}\right)^\mu.$$

It can be simplified by considering two ranges for δ. For $\delta > 2e - 1$,

$$\mathbf{Pr}[X > (1 + \delta)\mu] < 2^{-(1+\delta)\mu},$$

and for $\delta \leq 2e - 1$,

$$\mathbf{Pr}[X > (1 + \delta)\mu] < e^{-\mu\delta^2/4}.$$

B.3 Basic distributions

Three distributions, of great universality, are defined below. The probability distribution of the number of successes in Bernoulli trials is called the *binomial distribution.* Consider n Bernoulli trials with probability of success p. The probability of k successes, for $0 \leq k \leq n$, is given by

$$B(k; n, p) = \binom{n}{k} p^k (1-p)^{n-k}.$$

The *Poisson distribution* with parameter $\lambda > 0$ is as follows. For each nonnegative integer k, the probability of exactly k successes is defined to be

$$p(k; \lambda) = e^{-\lambda} \frac{\lambda^k}{k!}.$$

The limit of the binomial distribution $B(k; n, p)$ as $n \to \infty$ and $np \to \lambda$, a constant, is the Poisson distribution $p(k; \lambda)$. Indeed, in many applications one comes across Bernoulli trials in which n is large, p is small, and the product $\lambda = np$ is moderate. In these situations, $p(k; np)$ is a good approximation to $B(k; n, p)$.

The *normal* density function with mean μ and standard deviation σ is

$$n(x) = \frac{1}{\sigma\sqrt{2\pi}} e^{-\frac{(x-\mu)^2}{2\sigma^2}},$$

and the *normal distribution* function is its integral,

$$N(x) = \frac{1}{\sigma\sqrt{2\pi}} \int_{-\infty}^{x} e^{-\frac{(y-\mu)^2}{2\sigma^2}} dy.$$

The normal distribution also approximates the binomial distribution. Let us state this for the case $p = 1/2$. Let n be even, $n = 2\nu$, say. For $-\nu \leq k \leq \nu$, define

$$a_k = a_{-k} = B(\nu + k; 2\nu, 1/2).$$

In the limit as $\nu \to \infty$ and k varies in the range $0 < k < \sqrt{\nu}$, a_k can be approximated by $hn(kh)$, where $h = \sqrt{\frac{2}{\nu}} = \frac{2}{\sqrt{n}}$.

B.4 Notes

For further information see the books by Feller [91], Motwani and Raghavan [217], Spencer [251], and Alon and Spencer [7].

References

1. A. Agrawal, P. Klein, and R. Ravi. When trees collide: an approximation algorithm for the generalized Steiner network problem on networks. *SIAM Journal on Computing*, 24:440–456, 1995. (Cited on pp. 129, 211)
2. R.K. Ahuja, T.L. Magnanti, and J.B. Orlin. *Network Flows.* Prentice Hall, New Jersey, 1993. (Cited on p. 107)
3. M. Ajtai. The shortest vector problem in ℓ_2 is NP-hard for randomized reductions. In *Proc. 30th ACM Symposium on the Theory of Computing*, pages 10–19, 1998. (Cited on p. 293)
4. D. Aldous. The random walk construction for spanning trees and uniform labeled trees. *SIAM Journal on Discrete Mathematics*, 3:450–465, 1990. (Cited on p. 339)
5. F. Alizadeh. Interior point methods in semidefinite programming with applications to combinatorial optimization. *SIAM Journal on Optimization*, 5:13–51, 1995. (Cited on p. 268)
6. N. Alon, A. Frieze, and D. Welsh. Polynomial time randomised approximation schemes for Tutte-Grothendieck invariants: the dense case. *Random Structures and Algorithms*, 6:459–478, 1995. (Cited on p. 342)
7. N. Alon and J. Spencer. *The Probabilistic Method.* John Wiley & Sons, New York, NY, 2000. (Cited on pp. 138, 355)
8. J. D. Annan. The complexities of the coefficients of the Tutte polynomial. *Discrete Applied Mathematics*, 57:93–103, 1995. (Cited on p. 342)
9. J.D. Annan. A randomized approximation algorithm for counting the number of forests in dense graphs. *Combinatorics, Probability and Computing*, 3:273–283, 1994. (Cited on p. 339)
10. S. Arora. Polynomial time approximation scheme for Euclidean TSP and other geometric problems. In *Proc. 37th IEEE Annual Symposium on Foundations of Computer Science*, pages 2–11, 1996. (Cited on p. 89)
11. S. Arora. Nearly linear time approximation scheme for Euclidean TSP and other geometric problems. In *Proc. 38th IEEE Annual Symposium on Foundations of Computer Science*, pages 554–563, 1997. (Cited on p. 89)
12. S. Arora and C. Lund. Hardness of approximations. In D.S. Hochbaum, editor, *Approximation Algorithms for NP-Hard Problems*, pages 46–93. PWS Publishing, Boston, MA, 1997. (Cited on p. 332)
13. S. Arora, C. Lund, R. Motwani, M. Sudan, and M. Szegedy. Proof verification and intractability of approximation problems. In *Proc. 33rd IEEE Annual Symposium on Foundations of Computer Science*, pages 13–22, 1992. (Cited on p. 332)

14. S. Arora, P. Raghavan, and S. Rao. Approximation schemes for Euclidean k-medians and related problems. In *Proc. 30th ACM Symposium on the Theory of Computing*, pages 106–113, 1998. (Cited on p. 89)
15. S. Arora and S. Safra. Probabilistic checking of proofs: a new characterization of NP. In *Proc. 33rd IEEE Annual Symposium on Foundations of Computer Science*, pages 2–13, 1992. (Cited on p. 332)
16. V. Arya, N. Garg, R. Khandekar, A. Meyerson, K. Munagala, and V. Pandit. Local search heuristics for k-median and facility location problems. In *Proc. 33rd ACM Symposium on the Theory of Computing*, 2001. (Cited on pp. 252, 253)
17. Y. Aumann and Y. Rabani. An $O(\log k)$ approximate min-cut max-flow theorem and approximation algorithms. *SIAM Journal on Computing*, 27:291–301, 1998. (Cited on p. 196)
18. G. Ausiello, P. Crescenzi, G. Gambosi, V. Kann, A. Marchetti-Spaccamela, and M. Protasi. *Complexity and Approximation. Combinatorial Optimization Problems and their Approximability Properties.* Springer-Verlag, Berlin, 1999. (Cited on pp. 11, 333)
19. L. Babai. Trading group theory for randomness. In *Proc. 17th ACM Symposium on the Theory of Computing*, pages 421–429, 1985. (Cited on p. 332)
20. V. Bafna, P. Berman, and T. Fujito. Constant ratio approximations of the weighted feedback vertex set problem for undirected graphs. In *Algorithms and Computation, 6th International Symposium, ISAAC*, volume 1004 of *Lecture Notes in Computer Science*, pages 142–151. Springer-Verlag, Berlin, 1995. (Cited on p. 60)
21. R. Bar-Yehuda and S. Even. A linear-time approximation algorithm for the weighted vertex cover problem. *Journal of Algorithms*, 2:198–203, 1981. (Cited on p. 129)
22. Y. Bartal. Probabilistic approximation of metric spaces and its algorithmic applications. In *Proc. 37th IEEE Annual Symposium on Foundations of Computer Science*, pages 184–193, 1996. (Cited on p. 253)
23. C. Bazgan, M. Santha, and Z. Tuza. Efficient approximation algorithms for the subset-sum problem. In *Proc. 25th International Colloquium on Automata, Languages, and Programming*, volume 1443 of *Lecture Notes in Computer Science*, pages 387–396. Springer-Verlag, Berlin, 1998. (Cited on p. 72)
24. A. Becker and D. Geiger. Approximation algorithms for the loop cutset problem. In *Proc. 10th Conference on Uncertainty in Artificial Intelligence*, pages 60–68, 1994. (Cited on p. 60)
25. M. Ben-or, S. Goldwasser, J. Kilian, and A. Wigderson. Multi-prover interactive proofs: How to remove intractability. In *Proc. 20th ACM Symposium on the Theory of Computing*, pages 113–131, 1988. (Cited on p. 332)
26. M. Bern and P. Plassmann. The Steiner problem with edge lengths 1 and 2. *Information Processing Letters*, 32:171–176, 1989. (Cited on p. 332)
27. S.N. Bhatt and F.T. Leighton. A framework for solving VLSI graph layout problems. *Journal of Computer and System Sciences*, 28:300–343, 1984. (Cited on p. 196)
28. A. Blum, T. Jiang, M. Li, J. Tromp, and M. Yannakakis. Linear approximation of shortest superstring. *Journal of the ACM*, 41:630–647, 1994. (Cited on p. 67)

29. M. Blum and S. Kannan. Designing programs that check their work. In *Proc. 21st ACM Symposium on the Theory of Computing*, pages 86–97, 1989. (Cited on p. 332)
30. M. Blum, M. Luby, and R. Rubinfeld. Testing/correcting with applications to numerical problems. *Journal of Computer and System Sciences*, 47:549–595, 1993. (Cited on p. 332)
31. R. Boppana and M.M. Halldórsson. Approximating maximum independent sets by excluding subgraphs. *BIT*, 32:180–196, 1992. (Cited on p. 332)
32. A. Borodin and R. El-Yaniv. *Online Computation and Competitive Analysis*. Cambridge University Press, Cambridge, UK, 1998. (Cited on p. 78)
33. J. Bourgain. On Lipschitz embedding of finite metric spaces in Hilbert spaces. *Israeli J. Math.*, 52:46–52, 1985. (Cited on p. 196)
34. A.Z. Broder. How hard is it to marry at random? In *Proc. 18th ACM Symposium on the Theory of Computing*, pages 50–58, 1986. (Cited on p. 305)
35. A.Z. Broder. Generating random spanning trees. In *Proc. 30th IEEE Annual Symposium on Foundations of Computer Science*, pages 442–447, 1989. (Cited on p. 339)
36. R. Bubley and M. Dyer. Faster random generation of linear extensions. *Discrete Mathematics*, 201:81–88, 1999. (Cited on p. 340)
37. G. Calinescu, H. Karloff, and Y. Rabani. An improved approximation algorithm for multiway cut. In *Proc. 30th ACM Symposium on the Theory of Computing*, pages 48–52, 1998. (Cited on p. 166)
38. M. Charikar, C. Chekuri, T. Cheung, Z. Dai, A. Goel, S. Guha, and M. Li. Approximation algorithms for directed Steiner tree problems. In *Proc. 9th ACM-SIAM Annual Symposium on Discrete Algorithms*, pages 192–200, 1998. (Cited on p. 337)
39. M. Charikar and S. Guha. Improved combinatorial algorithms for the facility location and k-median problems. In *Proc. 40th IEEE Annual Symposium on Foundations of Computer Science*, pages 378–388, 1999. (Cited on p. 254)
40. M. Charikar, S. Guha, É. Tardos, and D.B. Shmoys. A constant-factor approximation algorithm for the k-median problem. In *Proc. 31st ACM Symposium on the Theory of Computing*, pages 1–10, 1999. (Cited on p. 253)
41. M. Charikar, S. Khuller, D.M. Mount, and G. Narshimhan. Algorithms for facility location problems with outliers. In *Proc. 12th ACM-SIAM Annual Symposium on Discrete Algorithms*, pages 642–651, 2001. (Cited on p. 239)
42. M. Charikar, J. Kleinberg, R. Kumar, S. Rajagopalan, A. Sahai, and A. Tomkins. Minimizing wirelength in zero and bounded skew clock trees. In *Proc. 10th ACM-SIAM Annual Symposium on Discrete Algorithms*, pages 177–184, 1999. (Cited on p. 37)
43. C. Chekuri, A. Gupta, and A. Kumar. On the bidirected relaxation for the multiway cut problem. Manuscript, 2002. (Cited on p. 343)
44. J. Cheriyan and R. Thurimella. Approximating minimum-size k-connected spanning subgraphs via matching. In *Proc. 37th IEEE Annual Symposium on Foundations of Computer Science*, pages 292–301, 1996. (Cited on pp. 225, 226, 230)
45. B. Chor and M. Sudan. A geometric approach to betweenness. *SIAM Journal on Discrete Mathematics*, 11:511–523, 1998. (Cited on p. 267)
46. E.-A. Choukhmane. Une heuristique pour le problème de l'arbre de Steiner. *RAIRO Rech. Opér.*, 12:207–212, 1978. (Cited on p. 37)

47. N. Christofides. Worst-case analysis of a new heuristic for the traveling salesman problem. Technical report, Graduate School of Industrial Administration, Carnegie-Mellon University, Pittsburgh, PA, 1976. (Cited on p. 37)
48. F. Chudak, M.X. Goemans, D. Hochbaum, and D.P. Williamson. A primal–dual interpretation of two 2-approximation algorithms for the feedback vertex set problem in undirected graphs. *Operations Research Letters*, 22:111–118, 1998. (Cited on pp. 60, 129)
49. F. Chudak, T. Roughgarden, and D.P. Williamson. Approximate k-MSTs and k-Steiner trees via the primal–dual method and Lagrangian relaxation. Manuscript, 2000. (Cited on p. 251)
50. V. Chvátal. A greedy heuristic for the set covering problem. *Mathematics of Operations Research*, 4:233–235, 1979. (Cited on pp. 26, 117)
51. V. Chvátal. *Linear Programming.* W.H. Freeman and Co., New York, NY, 1983. (Cited on p. 107)
52. E.G. Coffman Jr., M.R. Garey, and D.S. Johnson. Approximation algorithms for bin backing: a survey. In D.S. Hochbaum, editor, *Approximation Algorithms for NP-Hard Problems*, pages 46–93. PWS Publishing, Boston, MA, 1997. (Cited on p. 78)
53. S.A. Cook. The complexity of theorem-proving procedures. In *Proc. 3rd ACM Symposium on the Theory of Computing*, pages 151–158, 1971. (Cited on p. 10)
54. W.J. Cook, W.H. Cunningham, W.R. Pulleyblank, and A. Schrijver. *Combinatorial Optimization.* John Wiley & Sons, New York, NY, 1998. (Cited on p. 107)
55. C. Cooper and A. Frieze. Mixing properties of the Swendsen-Wang process on classes of graphs. *Random Structures Algorithms*, 15:242–261, 1999. (Cited on p. 342)
56. T. H. Cormen, C. E. Leiserson, R. L. Rivest, and C. Stein. *Introduction to Algorithms.* Second edition. MIT Press and McGraw-Hill, 2001. (Cited on p. 11)
57. R. Courant and H. Robbins. *What Is Mathematics?* Oxford University Press, New York, NY, 1941. (Cited on p. 37)
58. M. Cryan, M. Dyer, L. Goldberg, M. Jerrum, and R. Martin. A polynomial-time alogarithm to approximately count contingency tables when the number of rows is constant. In *Proc. 43rd IEEE Annual Symposium on Foundations of Computer Science*, 2002. (Cited on p. 343)
59. P. Dagum, M. Luby, M. Mihail, and U.V. Vazirani. Polytopes, permanents and graphs with large factors. In *Proc. 29th IEEE Annual Symposium on Foundations of Computer Science*, pages 412–421, 1988. (Cited on p. 339)
60. E. Dahlhaus, D.S. Johnson, C.H. Papadimitriou, P.D. Seymour, and M. Yannakakis. The complexity of multiterminal cuts. *SIAM Journal on Computing*, 23:864–894, 1994. (Cited on p. 46)
61. G.B. Dantzig. *Linear Programming and Extensions.* Reprint of the 1968 corrected edition. Princeton University Press, Princeton, NJ, 1998. (Cited on p. 107)
62. G.B. Dantzig, L.R. Ford, and D.R. Fulkerson. Solution of a large-scale traveling-salesman problem. *Operations Research*, 2:393–410, 1954. (Cited on p. 230)

63. G.B. Dantzig, L.R. Ford, and D.R. Fulkerson. A primal–dual algorithm for linear programs. In H.W. Kuhn and A.W. Tucker, editors, *Linear Inequalities and Related Systems*, pages 171–181. Princeton University Press, Princeton, NJ, 1956. (Cited on p. 129)
64. I. Dinur, V. Guruswami, S. Khot, and O. Regev. The hardness of hypergraph vertex cover. Manuscript, 2002. (Cited on p. 343)
65. G. Dobson. Worst-case analysis of greedy heuristics for integer programming with non-negative data. *Mathematics of Operations Research*, 7:515–531, 1982. (Cited on p. 117)
66. P. Drineas, R. Kannan, A. Frieze, S. Vempala, and V. Vinay. Clustering in large graphs and matrices. In *Proc. 10th ACM-SIAM Annual Symposium on Discrete Algorithms*, pages 291–299, 1999. (Cited on p. 254)
67. D.Z. Du and F.K. Hwang. Gilbert-Pollack conjecture on Steiner ratio is true. *Algorithmica*, 7:121–135, 1992. (Cited on p. 37)
68. A. Durand, M. Hermann, and P. G. Kolaitis. Subtractive reductions and complete problems for counting complexity classes. In M. Nielsen and B. Rovan, editors, *Proceedings 25th International Symposium on Mathematical Foundations of Computer Science (MFCS 2000), Bratislava (Slovakia)*, volume 1893 of *Lecture Notes in Computer Science*, pages 323–332. Springer-Verlag, 2000. (Cited on p. 305)
69. M. Dyer. Approximate counting by dynamic programming. Manuscript, 2002. (Cited on p. 343)
70. M. Dyer, R. Kannan, and J. Mount. Sampling contingency tables. *Random Structures and Algorithms*, 10:487–506, 1997. (Cited on p. 340)
71. M.E. Dyer, A. Frieze, and M.R. Jerrum. Approximately counting hamilton cycles in dense graphs. *SIAM Journal on Computing*, 27:1262–1272, 1998. (Cited on p. 341)
72. M.E. Dyer, A. Frieze, and M.R. Jerrum. On counting independent sets in sparse graphs. In *Proc. 40th IEEE Annual Symposium on Foundations of Computer Science*, pages 210–217, 1999. (Cited on p. 341)
73. M.E. Dyer, A. Frieze, and R. Kannan. A random polynomial time algorithm for approximating the volume of convex bodies. *Journal of the ACM*, 38:1–17, 1991. (Cited on p. 338)
74. J. Edmonds. Maximum matching and a polyhedron with 0,1-vertices. *Journal of Research of the National Bureau of Standards. Section B*, 69:125–130, 1965. (Cited on p. 104)
75. J. Edmonds. Paths, trees, and flowers. *Canadian Journal of Mathematics*, 17:449–467, 1965. (Cited on pp. 10, 11)
76. J. Edmonds. Optimum branchings. *Journal of Research of the National Bureau of Standards. Section B*, 71:233–240, 1967. (Cited on p. 211)
77. J. Edmonds. Matroids and the greedy algorithm. *Mathematical Programming*, 1:127–136, 1971. (Cited on p. 105)
78. J. Edmonds. Matroid intersection. *Annals of Discrete Mathematics*, 4:185–204, 1979. (Cited on p. 227)
79. P. Erdős. Gráfok páros körüljárású részgráfjairól (On bipartite subgraphs of graphs, in Hungarian). *Mat. Lapok*, 18:283–288, 1967. (Cited on p. 10)
80. P. Erdős and J.L. Selfridge. On a combinatorial game.. *Journal of Combinatorial Theory, Series A*, 14:298–301, 1973. (Cited on p. 138)

81. G. Even, J. Naor, B. Schieber, and S. Rao. Divide-and-conquer approximation algorithms via spreading metrics. *Journal of the ACM*, 47:585–616, 2000. (Cited on p. 177)
82. G. Even, J. Naor, B. Schieber, and L. Zosin. Approximating minimum subset feedback sets in undirected graphs with applications. In *Proc. 4th Israel Symposium on Theory of Computing and Systems*, pages 78–88, 1996. (Cited on p. 166)
83. G. Even, J. Naor, and L. Zosin. An 8-approximation algorithm for the subset feedback vertex set problem. In *Proc. 37th IEEE Annual Symposium on Foundations of Computer Science*, pages 310–319, 1996. (Cited on p. 166)
84. T. Feder and M. Mihail. Balanced matroids. In *Proc. 24th ACM Symposium on the Theory of Computing*, pages 26–38, 1992. (Cited on p. 339)
85. U. Feige. Approximating the bandwidth via volume respecting embeddings. In *Proc. 30th ACM Symposium on the Theory of Computing*, pages 90–99, 1998. (Cited on p. 195)
86. U. Feige. A treshold of $\ln n$ for approximating set cover. *Journal of the ACM*, 45:634–652, 1998. (Cited on pp. 26, 331, 332)
87. U. Feige and M.X. Goemans. Approximating the value of two prover proof systems, with applications to MAX-CUT and MAX DICUT. In *Proc. 3rd Israel Symposium on Theory of Computing and Systems*, pages 182–189, 1995. (Cited on p. 269)
88. U. Feige, S. Goldwasser, L. Lovász, S. Safra, and M. Szegedy. Approximating clique is almost NP-complete. In *Proc. 32nd IEEE Annual Symposium on Foundations of Computer Science*, pages 2–12, 1991. (Cited on p. 332)
89. U. Feige and R. Krauthgamer. A polylogarithmic approximation of the minimum bisection. In *Proc. 41st IEEE Annual Symposium on Foundations of Computer Science*, pages 105–115, 2000. (Cited on pp. 196, 336)
90. U. Feige and G. Schechtman. On the optimality of the random hyperplane rounding technique for MAX-CUT. In *Proc. 33rd ACM Symposium on the Theory of Computing*, 2001. (Cited on p. 268)
91. W. Feller. *An Introduction to Probability Theory and its Applications.* John Wiley & Sons, New York, NY, 1950. (Cited on p. 355)
92. W. Fernandez de la Vega and G.S. Lueker. Bin packing can be solved within $1+\varepsilon$ in linear time. *Combinatorica*, 1:349–355, 1981. (Cited on p. 78)
93. A. Freund and H. Karloff. A lower bound of $8/(7+\frac{1}{k-1})$ on the integrality ratio of the Calinescu–Karloff–Rabani relaxation for multiway cut. *Information Processing Letters*, 75:43–50, 2000. (Cited on p. 166)
94. A. Frieze. On the Lagarias–Odlyzko algorithm for the subset sum problem. *SIAM Journal on Computing*, 15:536–539, 1986. (Cited on p. 291)
95. A. Frieze, G. Galbiati, and F. Maffioli. On the worst-case performance of some algorithms for the asymmetric traveling salesman problem. *Networks*, 12:23–39, 1982. (Cited on p. 34)
96. A. Frieze and M. Jerrum. Improved approximation algorithms for MAX k-CUT and MAX BISECTION. *Algorithmica*, 18:67–81, 1997. (Cited on p. 269)
97. M.R. Garey, R.L. Graham, and J.D. Ullman. An analysis of some packing algorithms. In *Combinatorial Algorithms (Courant Computer Science Symposium, No. 9)*, pages 39–47, 1972. (Cited on p. 10)
98. M.R. Garey and D.S. Johnson. Strong NP-completeness results: motivation, examples, and implications. *Journal of the ACM*, 25:499–508, 1978. (Cited on p. 73)

99. M.R. Garey and D.S. Johnson. *Computers and Intractability: A Guide to the Theory of NP-Completeness.* W.H. Freeman and Co., New York, NY, 1979. (Cited on p. 352)
100. N. Garg. A 3-approximation for the minimum tree spanning k vertices. In *Proc. 37th IEEE Annual Symposium on Foundations of Computer Science*, pages 302–309, 1996. (Cited on p. 251)
101. N. Garg, H. Saran, and V.V. Vazirani. Finding separator cuts in planar graphs within twice the optimal. *SIAM Journal on Computing*, 29:159–179, 1999. (Cited on p. 336)
102. N. Garg, V.V. Vazirani, and M. Yannakakis. Multiway cuts in directed and node weighted graphs. In *Proc. 21st International Colloquium on Automata, Languages, and Programming*, volume 820 of *Lecture Notes in Computer Science*, pages 487–498. Springer-Verlag, Berlin, 1994. (Cited on p. 166)
103. N. Garg, V.V. Vazirani, and M. Yannakakis. Approximate max-flow min-(multi)cut theorems and their applications. *SIAM Journal on Computing*, 25:235–251, 1996. (Cited on p. 178)
104. N. Garg, V.V. Vazirani, and M. Yannakakis. Primal–dual approximation algorithms for integral flow and multicut in trees. *Algorithmica*, 18:3–20, 1997. (Cited on pp. 151, 152, 153)
105. C.F. Gauss. *Disquisitiones Arithmeticae.* English edition translated by A.A. Clarke. Springer-Verlag, New York, NY, 1986. (Cited on p. 292)
106. E.N. Gilbert and H.O. Pollak. Steiner minimal trees. *SIAM Journal on Applied Mathematics*, 16:1–29, 1968. (Cited on p. 37)
107. M.X. Goemans and D.J. Bertsimas. Survivable networks, linear programming relaxations and the parsimonious property. *Mathematical Programming*, 60:145–166, 1993. (Cited on p. 227)
108. M.X. Goemans, A.V. Goldberg, S. Plotkin, D.B. Shmoys, É. Tardos, and D.P. Williamson. Improved approximation algorithms for network design problems. In *Proc. 5th ACM-SIAM Annual Symposium on Discrete Algorithms*, pages 223–232, 1994. (Cited on pp. 224, 230)
109. M.X. Goemans and J. Kleinberg. The Lovász theta function and a semidefinite programming relaxation of vertex cover. *SIAM Journal on Discrete Mathematics*, 11:196–204, 1998. (Cited on p. 334)
110. M.X. Goemans and D.P. Williamson. New $\frac{3}{4}$-approximation algorithms for the maximum satisfiability problem. *SIAM Journal on Discrete Mathematics*, 7:656–666, 1994. (Cited on pp. 137, 138)
111. M.X. Goemans and D.P. Williamson. A general approximation technique for constrained forest problems. *SIAM Journal on Computing*, 24:296–317, 1995. (Cited on pp. 129, 207, 211)
112. M.X. Goemans and D.P. Williamson. Improved approximation algorithms for maximum cut and satisfiability problems using semidefinite programming. *Journal of the ACM*, 42:1115–1145, 1995. (Cited on pp. 267, 268)
113. M.X. Goemans and D.P. Williamson. The primal–dual method for approximation algorithms and its applications to network design problems. In D.S. Hochbaum, editor, *Approximation Algorithms for NP-Hard Problems*, pages 144–191. PWS Publishing, Boston, MA, 1997. (Cited on pp. 129, 211)
114. O. Goldreich, D. Micciancio, S. Safra, and J.-P. Seifert. Approximating shortest lattice vectors is not harder than approximating closest lattice vectors. *Information Processing Letters*, 71, 1999. (Cited on p. 292)

115. S. Goldwasser, S. Micali, and C. Rackoff. The knowledge complexity of interactive proofs. *SIAM Journal on Computing*, 18:186–208, 1989. (Cited on p. 332)
116. R.E. Gomory and T.C. Hu. Multi-terminal network flows. *Journal of the SIAM*, 9:551–570, 1961. (Cited on p. 46)
117. T.F. Gonzalez. Clustering to minimize the maximum inter-cluster distance. *Theoretical Computer Science*, 38:293–306, 1985. (Cited on p. 52)
118. V. Gore and M. Jerrum. The Swendsen-Wang process does not always mix rapidly. In *Proc. 29th ACM Symposium on the Theory of Computing*, pages 674–681, 1997. (Cited on p. 342)
119. R.L. Graham. Bounds for certain multiprocessing anomalies. *Bell System Technical Journal*, 45:1563–1581, 1966. (Cited on pp. 10, 83)
120. R.L. Graham. Bounds on multiprocessing timing anomalies. *SIAM Journal on Applied Mathematics*, 17:416–429, 1969. (Cited on p. 83)
121. M. Grigni, E. Koutsoupias, and C. Papadimitriou. An approximation scheme for planar graph TSP. In *Proc. 36th IEEE Annual Symposium on Foundations of Computer Science*, pages 640–646, 1995. (Cited on p. 89)
122. M. Grötschel, L. Lovász, and A. Schrijver. The ellipsoid method and its consequences in combinatorial optimization. *Combinatorica*, 1:169–197, 1981. (Cited on p. 107)
123. M. Grötschel, L. Lovász, and A. Schrijver. *Geometric Algorithms and Combinatorial Optimization.* Second edition. Springer-Verlag, Berlin, 1993. (Cited on p. 107)
124. S. Guha and S. Khuller. Greedy strikes back: Improved facility location algorithms. *Journal of Algorithms*, 31:228–248, 1999. (Cited on p. 241)
125. V. Guruswami, S. Khanna, R. Rajaraman, B. Shepherd, and M. Yannakakis. Near-optimal hardness results and approximation algorithms for edge-disjoint and related problems. In *Proc. 31st ACM Symposium on the Theory of Computing*, pages 19–28, 1999. (Cited on p. 153)
126. D. Gusfield and R. W. Irving. *The Stable Marriage Problem: Structure and Algorithms.* MIT Press, Cambridge, MA, 1989. (Cited on p. 341)
127. L.A. Hall. Approximation algorithms for scheduling. In D.S. Hochbaum, editor, *Approximation Algorithms for NP-Hard Problems*, pages 1–45. PWS Publishing, Boston, MA, 1997. (Cited on p. 144)
128. J. Hastad. Clique is hard to approximate within $n^{1-\varepsilon}$. In *Proc. 37th IEEE Annual Symposium on Foundations of Computer Science*, pages 627–636, 1996. (Cited on p. 332)
129. J. Hastad. Some optimal inapproximability results. In *Proc. 29th ACM Symposium on the Theory of Computing*, pages 1–10, 1997. (Cited on pp. 330, 332)
130. M. Held and R.M. Karp. The traveling-salesman and minimum cost spanning trees. *Operations Research*, 18:1138–1162, 1970. (Cited on p. 229)
131. D. S. Hochbaum. Heuristics for the fixed cost median problem. *Mathematical Programming*, 22:148–162, 1982. (Cited on p. 241)
132. D.S. Hochbaum. Approximation algorithms for the set covering and vertex cover problems. *SIAM Journal on Computing*, 11:555–556, 1982. (Cited on pp. 25, 123)
133. D.S. Hochbaum, editor. *Approximation Algorithms for NP-Hard Problems.* PWS Publishing, Boston, MA, 1997. (Cited on p. 11)

134. D.S. Hochbaum and D.B. Shmoys. A unified approach to approximation algorithms for bottleneck problems. *Journal of the ACM*, 33:533–550, 1986. (Cited on p. 53)
135. D.S. Hochbaum and D.B. Shmoys. Using dual approximation algorithms for scheduling problems: theoretical and practical results. *Journal of the ACM*, 34:144–162, 1987. (Cited on p. 83)
136. D.S. Hochbaum and D.B. Shmoys. A polynomial approximation scheme for machine scheduling on uniform processors: using the dual approximation approach. *SIAM Journal on Computing*, 17:539–551, 1988. (Cited on p. 144)
137. J.A. Hoogeveen. Analysis of Christofides' heuristic: some paths are more difficult than cycles. *Operations Research Letters*, 10:291–295, 1991. (Cited on p. 34)
138. E. Horowitz and S.K. Sahni. Exact and approximate algorithms for scheduling nonidentical processors. *Journal of the ACM*, 23:317–327, 1976. (Cited on p. 83)
139. W.L. Hsu and G.L. Nemhauser. Easy and hard bottleneck location problems. *Discrete Applied Mathematics*, 1:209–216, 1979. (Cited on p. 53)
140. F. K. Hwang, D. S. Richards, and P. Winter. *The Steiner Tree Problem*, volume 53 of *Annals of Discrete Mathematics*. North-Holland, Amsterdam, Netherlands, 1992. (Cited on p. 37)
141. O.H. Ibarra and C.E. Kim. Fast approximation algorithms for the knapsack and sum of subset problems. *Journal of the ACM*, 22:463–468, 1975. (Cited on p. 73)
142. R. Impagliazzo and D. Zuckerman. How to recycle random bits. In *Proc. 30st IEEE Annual Symposium on Foundations of Computer Science*, pages 248–253, 1989. (Cited on p. 332)
143. A. Iwainsky, E. Canuto, O. Taraszow, and A. Villa. Network decomposition for the optimization of connection structures. *Networks*, 16:205–235, 1986. (Cited on p. 37)
144. K. Jain. A factor 2 approximation algorithm for the generalized Steiner network problem. *Combinatorica*, 1:39–60, 2001. (Cited on p. 230)
145. K. Jain, M. Mahdian, and A. Saberi. A new greedy approach for facility location problems. In *Proc. 34th ACM Symposium on the Theory of Computing*, 2002. (Cited on pp. 254, 331)
146. K. Jain, I. I. Măndoiu, V.V. Vazirani, and D. P. Williamson. Primal–dual schema based approximation algorithms for the element connectivity problem. In *Proc. 10th ACM-SIAM Annual Symposium on Discrete Algorithms*, pages 484–489, 1999. (Cited on p. 337)
147. K. Jain and V.V. Vazirani. An approximation algorithm for the fault tolerant metric facility location problem. In *Proc. 3rd International Workshop on Approximation Algorithms for Combinatorial Optimization Problems*, volume 1913 of *Lecture Notes in Computer Science*. Springer-Verlag, Berlin, 2000. (Cited on p. 239)
148. K. Jain and V.V. Vazirani. Approximation algorithms for the metric facility location and k-median problems using the primal–dual schema and Lagrangian relaxation. *Journal of the ACM*, 48:274–296, 2001. (Cited on pp. 241, 252, 253)
149. M. Jerrum and A. Sinclair. The Markov chain Monte Carlo method: an approach to approximate counting. In D.S. Hochbaum, editor, *Approximation*

Algorithms for NP-Hard Problems, pages 482–520. PWS Publishing, Boston, MA, 1997. (Cited on p. 305)

150. M. Jerrum, A. Sinclair, and E. Vigoda. A polynomial-time approximation algorithm for the permanent of a matrix with non-negative entries. *Electronic Colloquium on Computational Complexity*, pages TR00–079, 2000. (Cited on pp. 338, 340)
151. M.R. Jerrum. A very simple algorithm for estimating the number of k-colorings of a low-degree graph. *Random Structures and Algorithms*, 7, 1995. (Cited on p. 341)
152. M.R. Jerrum and A. Sinclair. Approximating the permanent. *SIAM Journal on Computing*, 18:1149–1178, 1989. (Cited on p. 305)
153. M.R. Jerrum and A. Sinclair. Polynomial time approximation algorithms for the Ising model. *SIAM Journal on Computing*, 22:1087–1116, 1993. (Cited on p. 342)
154. M.R. Jerrum, L.G. Valiant, and V.V. Vazirani. Random generation of combinatorial structures from a uniform distribution. *Theoretical Computer Science*, 43:169–188, 1986. (Cited on p. 303)
155. T. Jiang, M. Li, and D. Du. A note on shortest common superstrings with flipping. *Information Processing Letters*, 44:195–199, 1992. (Cited on p. 67)
156. D.S. Johnson. *Near-optimal bin packing algorithms*. PhD thesis, Massachusetts Institute of Technology, Department of Mathematics, Cambridge, MA, 1973. (Cited on p. 77)
157. D.S. Johnson. Approximation algorithms for combinatorial problems. *Journal of Computer and System Sciences*, 9:256–278, 1974. (Cited on pp. 10, 26, 138)
158. J. Kahn, J.H. Kim, L. Lovász, and V.H. Vu. The cover time, the blanket time, and the Matthews bound. In *Proc. 41st IEEE Annual Symposium on Foundations of Computer Science*, pages 467–475, 2000. (Cited on p. 338)
159. M. Kaib and C.-P. Schnorr. The generalized Gauss reduction algorithm. *Journal of Algorithms*, 21(3):565–578, 1996. (Cited on p. 288)
160. R. Kannan. Algorithmic geometry of numbers. In *Annual Review of Computer Science, Vol. 2*, pages 231–267. Annual Reviews, Palo Alto, CA, 1987. (Cited on p. 293)
161. R. Kannan. Minkowski's convex body theorem and integer programming. *Mathematics of Operations Research*, 12(3):415–440, 1987. (Cited on p. 293)
162. R. Kannan, L. Lovász, and M. Simonovits. Random walks and an $o^*(n^5)$ volume algorithm for convex bodies. *Random Structures and Algorithms*, 11:1–50, 1997. (Cited on p. 338)
163. D. Karger. A randomized fully polynomial time approximation scheme for the all-terminal network reliability problem. *SIAM Journal on Computing*, 29:492–514, 1999. (Cited on p. 305)
164. D. Karger, P. Klein, C. Stein, M. Thorup, and N. Young. Rounding algorithms for a geometric embedding of minimum multiway cut. In *Proc. 29th ACM Symposium on the Theory of Computing*, pages 668–678, 1999. (Cited on p. 166)
165. D. Karger, R. Motwani, and M. Sudan. Approximate graph coloring by semidefinite programming. *Journal of the ACM*, 45:246–265, 1998. (Cited on pp. 267, 269)
166. D. Karger and C. Stein. A new approach to the minimum cut problem. *Journal of the ACM*, 43(4):601–640, 1996. (Cited on p. 304)

167. H. Karloff. *Linear Programming.* Birkhäuser, Boston, MA, 1991. (Cited on p. 107)
168. H. Karloff. How good is the Goemans-Williamson MAX CUT algorithm. *SIAM Journal on Computing*, 29:336–350, 1999. (Cited on p. 268)
169. H. Karloff and U. Zwick. A 7/8-approximation algorithm for MAX-3SAT? In *Proc. 38th IEEE Annual Symposium on Foundations of Computer Science*, pages 406–415, 1997. (Cited on p. 332)
170. N. Karmakar and R.M. Karp. An efficient approximation scheme for the one-dimensional bin packing problem. In *Proc. 23rd IEEE Annual Symposium on Foundations of Computer Science*, pages 312–320, 1982. (Cited on p. 78)
171. R.M. Karp. Reducibility among combinatorial problems. In R.E. Miller and J.W. Thatcher, editors, *Complexity of Computer Computations*, pages 85–103. Plenum Press, New York, NY, 1972. (Cited on p. 10)
172. R.M. Karp and M. Luby. Monte Carlo algorithms for enumeration and reliability problems. In *Proc. 24th IEEE Annual Symposium on Foundations of Computer Science*, pages 56–64, 1983. (Cited on pp. 302, 305)
173. R.M. Karp, M. Luby, and N. Madras. Monte Carlo approximation algorithms for enumeration problems. *Journal of Algorithms*, 10:429–448, 1989. (Cited on p. 305)
174. A. Karzanov and L. Khachiyan. On the conductance of order Markov chains. Technical Report DCS 268, Rutgers University, 1990. (Cited on p. 340)
175. P.W. Kasteleyn. Graph theory and crystal physics. In F. Harary, editor, *Graph Theory and Theoretical Physics*, pages 43–110. Academic Press, New York, NY, 1967. (Cited on p. 338)
176. S. Khuller, R. Pless, and Y.J. Sussmann. Fault tolerant k-center problems. *Theoretical Computer Science*, 242:237–245, 2000. (Cited on p. 53)
177. S. Khuller and B. Raghavachari. Improved approximation algorithms for uniform connectivity problems. *Journal of Algorithms*, 21:434–450, 1996. (Cited on p. 336)
178. S. Khuller and V.V. Vazirani. Planar graph colourability is not self-reducible, assuming $P \neq NP$. *Theoretical Computer Science*, 88(1):183–190, 1991. (Cited on p. 352)
179. S. Khuller and U. Vishkin. Biconnectivity approximations and graph carvings. *Journal of the ACM*, 42, 2:214–235, 1994. (Cited on p. 227)
180. P. Klein, S. Rao, A. Agrawal, and R. Ravi. An approximate max-flow min-cut relation for undirected multicommodity flow, with applications. *Combinatorica*, 15:187–202, 1995. (Cited on pp. 178, 196)
181. D.E. Knuth. *The Art of Computer Programming. Vol. 2. Seminumerical Algorithms.* Second edition. Addison-Wesley, Reading, MA, 1981. (Cited on p. 266)
182. A. Korkine and G. Zolotareff. Sur les formes quadratiques. *Math. Annalen*, 6:366–389, 1873. (Cited on p. 290)
183. M. Korupolu, C. Plaxton, and R. Rajaraman. Analysis of a local search heuristic for facility location problems. In *Proc. 9th ACM-SIAM Annual Symposium on Discrete Algorithms*, pages 1–10, 1998. (Cited on p. 252)
184. L. Kou, G. Markowsky, and L. Berman. A fast algorithm for Steiner trees. *Acta Informatica*, 15:141–145, 1981. (Cited on p. 37)
185. M.W. Krentel. The complexity of optimization problems. *Journal of Computer and System Sciences*, 36:490–509, 1988. (Cited on p. 352)

186. H.W. Kuhn. The Hungarian method for the assignment problem. *Naval Research Logistics Quarterly*, 2:83–97, 1955. (Cited on p. 129)
187. J. Lagarias. Worst case complexity bounds for algorithms in the the theory of integral quadratic forms. *Journal of Algorithms*, 1:142–186, 1980. (Cited on p. 292)
188. J. Lagarias, H.W. Lenstra, Jr., and C.-P. Schnorr. Korkin–Zolotarev bases and successive minima of a lattice and its reciprocal lattice. *Combinatorica*, 10:333–348, 1990. (Cited on p. 293)
189. T. Leighton and S. Rao. Multicommodity max-flow min-cut theorems and their use in designing approximation algorithms. *Journal of the ACM*, 46:787–832, 1999. (Cited on p. 196)
190. A.K. Lenstra, H.W. Lenstra, Jr., and L. Lovász. Factoring polynomials with rational coefficients. *Math. Ann.*, 261:513–534, 1982. (Cited on p. 292)
191. J.K. Lenstra, D.B. Shmoys, and É. Tardos. Approximation algorithms for scheduling unrelated parallel machines. *Mathematical Programming*, 46:259–271, 1990. (Cited on p. 144)
192. H.W. Lenstra, Jr. Integer programming with a fixed number of variables. *Mathematics of Operations Research*, 8:538–548, 1983. (Cited on p. 78)
193. L.A. Levin. Universal sorting problems. *Problemy Peredaci Informacii*, 9:115–116, 1973. English translation in *Problems of Information Transmission* 9:265–266. (Cited on p. 10)
194. M. Li. Towards a DNA sequencing theory. In *Proc. 31st IEEE Annual Symposium on Foundations of Computer Science*, pages 125–134, 1990. (Cited on p. 26)
195. J. H. Lin and J. S. Vitter. Approximation algorithms for geometric median problems. *Information Processing Letters*, 44:245–249, 1992. (Cited on p. 250)
196. J. H. Lin and J. S. Vitter. ε-approximation with minimum packing constraint violation. In *Proc. 24th ACM Symposium on the Theory of Computing*, pages 771–782, 1992. (Cited on p. 253)
197. N. Linial, E. London, and Y. Rabinovich. The geometry of graphs and some of its algorithmic applications. *Combinatorica*, 15:215–245, 1995. (Cited on pp. 195, 196, 266)
198. C.H.C. Little. An extension of Kasteleyn's method of enumerating 1-factors of planar graphs. In D. Holton, editor, *Proc. 2nd Australian Conference on Combinatorial Mathematics*, volume 403 of *Lecture Notes in Computer Science*, pages 63–72. Springer-Verlag, Berlin, 1974. (Cited on p. 338)
199. L. Lovász. On the ratio of optimal integral and fractional covers. *Discrete Mathematics*, 13:383–390, 1975. (Cited on pp. 11, 26, 117)
200. L. Lovász. *An Algorithmic Theory of Numbers, Graphs and Convexity.* CBMS-NSF Regional Conference Series in Applied Mathematics, 50. SIAM, Philadelphia, PA, 1986. (Cited on p. 291)
201. L. Lovász. *Combinatorial Problems and Exercises.* Second edition. North-Holland, Amsterdam–New York, 1993. (Cited on pp. 107, 339, 341)
202. L. Lovász and M.D. Plummer. *Matching Theory.* North-Holland, Amsterdam–New York, 1986. (Cited on pp. 8, 11, 107)
203. L. Lovász and A. Schrijver. Cones of matrices and set functions, and 0-1 optimization. *SIAM Journal on Optimization*, 1:166–190, 1990. (Cited on p. 269)
204. A. Lubotzky, R. Phillips, and P. Sarnak. Ramanujan graphs. *Combinatorica*, 8:261–277, 1988. (Cited on p. 332)

205. M. Luby and E. Vigoda. Approximately counting up to four. In *Proc. 29th ACM Symposium on the Theory of Computing*, pages 682–687, 1997. (Cited on p. 341)
206. C. Lund and M. Yannakakis. On the hardness of approximating minimization problems. *Journal of the ACM*, 41:960–981, 1994. (Cited on pp. 26, 332)
207. S. Mahajan and H. Ramesh. Derandomizing semidefinite programming based approximation algoirthms. In *Proc. 36th IEEE Annual Symposium on Foundations of Computer Science*, pages 162–169, 1995. (Cited on p. 268)
208. M. Mahdian, E. Markakis, A. Saberi, and V. V. Vazirani. A greedy facility location algorithm analyzed using dual fitting. In *Proc. 4th International Workshop on Approximation Algorithms for Combinatorial Optimization Problems*, volume 2129 of *Lecture Notes in Computer Science*. Springer-Verlag, Berlin, 2001. (Cited on pp. 240, 241)
209. M. Mahdian, Y. Ye, and J. Zhang. Improved approximation algorithms for metric facility location problems. In *Proc. 4th International Workshop on Approximation Algorithms for Combinatorial Optimization Problems*, volume 2462 of *Lecture Notes in Computer Science*. Springer-Verlag, Berlin, 2002. (Cited on p. 241)
210. P. Matthews. Generating random linear extensions of a partial order. *The Annals of Probability*, 19:1367–1392, 1991. (Cited on p. 340)
211. L. McShine and P. Tetali. On the mixing time of the triangulation walk and other Catalan structures. *Randomization methods in Algorithm Design, DIMACS-AMS*, 43:147–160, 1998. (Cited on p. 340)
212. D. Micciancio. The shortest vector in a lattice is hard to approximate to within some constant. In *Proc. 39th IEEE Annual Symposium on Foundations of Computer Science*, pages 92–98, 1998. (Cited on p. 336)
213. M. Mihail. On coupling and the approximation of the permanent. *Information Processing Letters*, 30:91–95, 1989. (Cited on p. 305)
214. M. Mihail. Set cover with requirements and costs evolving over time. In *International Workshop on Randomization, Approximation and Combinatorial Optimization*, volume 1671 of *Lecture Notes in Computer Science*, pages 63–72. Springer-Verlag, Berlin, 1999. (Cited on p. 117)
215. J.S.B. Mitchell. Guillotine subdivisions approximate polygonal subdivisions: a simple polynomial-time approximation scheme for geometric TSP, k-MST, and related problems. *SIAM Journal on Computing*, 28:1298–1309, 1999. (Cited on p. 89)
216. B. Morris. Improved bounds for sampling contingency tables. In *International Workshop on Randomization, Approximation and Combinatorial Optimization*, volume 1671 of *Lecture Notes in Computer Science*, pages 121–129. Springer-Verlag, Berlin, 1999. (Cited on p. 340)
217. R. Motwani and P. Raghavan. *Randomized Algorithms*. Cambridge University Press, Cambridge, UK, 1995. (Cited on p. 355)
218. J. Naor and L. Zosin. A 2-approximation algorithm for the directed multiway cut problem. In *Proc. 38th IEEE Annual Symposium on Foundations of Computer Science*, pages 548–553, 1997. (Cited on p. 166)
219. M. Naor, L. Schulman, and A. Srinivasan. Splitters and near-optimal derandomization. In *Proc. 36th IEEE Annual Symposium on Foundations of Computer Science*, pages 182–191, 1995. (Cited on p. 332)
220. G. Nemhauser and L. Wolsey. *Integer and Combinatorial Optimization*. John Wiley & Sons, New York, NY, 1988. (Cited on p. 107)

221. G.L. Nemhauser and L.E. Trotter. Vertex packings: structural properties and algorithms. *Mathematical Programming*, 8:232–248, 1975. (Cited on p. 123)
222. Y. Nesterov and A. Nemirovskii. *Interior Point Polynomial Methods in Convex Programming.* SIAM, Philadelphia, PA, 1994. (Cited on p. 268)
223. M.L. Overton. On minimizing the maximum eigenvalue of a symmetric matrix. *SIAM J. on Matrix Analysis and Appl.*, 13:256–268, 1992. (Cited on p. 268)
224. C.H. Papadimitriou. *Computational Complexity.* Addison-Wesley, Reading, MA, 1994. (Cited on p. 352)
225. C.H. Papadimitriou and K. Steiglitz. *Combinatorial Optimization: Algorithms and Complexity.* Prentice-Hall, Englewood Cliffs, NJ, 1982. (Cited on pp. 11, 107)
226. C.H. Papadimitriou and M. Yannakakis. Optimization, approximation, and complexity classes. *Journal of Computer and System Sciences*, 43:425–440, 1991. (Cited on pp. 332, 352)
227. C.H. Papadimitriou and M. Yannakakis. The traveling salesman problem with distances one and two. *Mathematics of Operations Research*, 18:1–11, 1993. (Cited on p. 34)
228. M. Pinsker. On the complexity of a concentrator. In *Proc. 7th Annual Teletraffic Conference*, pages 318/1–318/4, 1973. (Cited on p. 178)
229. J. Plesník. A bound for the Steiner tree problem in graphs. *Math. Slovaca*, 31:155–163, 1981. (Cited on p. 37)
230. V.R. Pratt. Every prime has a succinct certificate. *SIAM Journal on Computing*, 4:214–220, 1975. (Cited on p. 9)
231. H. J. Prömel and A. Steger. RNC-approximation algorithms for the Steiner problem. In *Proc. Symposium on Theoretical Aspects of Computer Science*, volume 1200 of *Lecture Notes in Computer Science*, pages 559–570. Springer-Verlag, Berlin, 1997. (Cited on p. 211)
232. M.O. Rabin. Probabilistic algorithms. In J.F. Traub, editor, *Algorithms and Complexity, Recent Results and New Directions*, pages 21–39. Academic Press, New York, NY, 1976. (Cited on p. 11)
233. P. Raghavan. Probabilistic construction of deterministic algorithms: approximating packing integer programs. *Journal of Computer and System Sciences*, 37:130–143, 1988. (Cited on p. 138)
234. S. Rajagopalan and V.V. Vazirani. On the bidirected cut relaxation for the metric Steiner tree problem. In *Proc. 10th ACM-SIAM Annual Symposium on Discrete Algorithms*, pages 742–751, 1999. (Cited on pp. 210, 335)
235. S. Rajagopalan and V.V. Vazirani. Primal–dual RNC approximation algorithms for set cover and covering integer programs. *SIAM Journal on Computing*, 28:526–541, 1999. (Cited on p. 117)
236. D. Randall and D.B. Wilson. Sampling spin configurations of an Ising system. In *Proc. 10th ACM-SIAM Annual Symposium on Discrete Algorithms*, pages S959–960, 1999. (Cited on p. 342)
237. S. Rao and W.D. Smith. Approximating geometrical graphs via "spanners" and "banyans". In *Proc. 30th ACM Symposium on the Theory of Computing*, pages 540–550, 1998. (Cited on p. 89)
238. S.K. Rao, P. Sadayappan, F.K. Hwang, and P.W. Shor. The rectilinear Steiner arborescence problem. *Algorithmica*, 7:277–288, 1992. (Cited on p. 35)
239. R. Raz. A parallel repetition theorem. *SIAM Journal on Computing*, 27:763–803, 1998. (Cited on p. 332)

240. S.K. Sahni and T.F. Gonzalez. P-complete approximation problems. *Journal of the ACM*, 23:555–565, 1976. (Cited on p. 37)
241. H. Saran and V.V. Vazirani. Finding k-cuts within twice the optimal. *SIAM Journal on Computing*, 24:101–108, 1995. (Cited on p. 46)
242. C.P. Schnorr. Optimal algorithms for self-reducible problems. In *Proc. 3rd International Colloquium on Automata, Languages, and Programming*, pages 322–337, 1976. (Cited on p. 352)
243. C.P. Schnorr. A hierarchy of polynomial time lattice basis reduction algorithms. *Theoretical Computer Science*, 53:201–224, 1987. (Cited on p. 292)
244. P. Schreiber. On the history of the so-called Steiner Weber problem. *Wiss. Z. Ernst-Moritz-Arndt-Univ. Greifswald, Math.-nat.wiss. Reihe, 35*, 3, 1986. (Cited on p. 37)
245. A. Schrijver. *Theory of Linear and Integer Programming.* John Wiley & Sons, New York, NY, 1986. (Cited on p. 107)
246. P.D. Seymour. Packing directed circuits fractionally. *Combinatorica*, 15:281–288, 1995. (Cited on p. 337)
247. D.B. Shmoys, É. Tardos, and K.I. Aardal. Approximation algorithms for facility location problems. In *Proc. 29th ACM Symposium on the Theory of Computing*, pages 265–274, 1997. (Cited on p. 241)
248. D.B. Shmoys and D.P. Williamson. Analyzing the Held-Karp TSP bound: a monotonicity property with applications. *Information Processing Letters*, 35:281–285, 1990. (Cited on p. 230)
249. A. Sinclair. Improved bounds for mixing rates of Markov chains and multicommodity flow. *Combinatorics, Probability and Computing*, 1:351–370, 1992. (Cited on p. 196)
250. A. Sinclair. *Algorithms for Random Generation and Counting: a Markov Chain Approach.* Birkhäuser, Boston, MA, 1993. (Cited on p. 305)
251. J. Spencer. *Ten Lectures on the Probabilistic Method.* SIAM, Philadelphia, PA, 1987. (Cited on pp. 138, 355)
252. A. Srinivasan. Improved approximations of packing and covering problems. In *Proc. 27th ACM Symposium on the Theory of Computing*, pages 268–276, 1995. (Cited on p. 123)
253. R.H. Swendsen and J.S. Wang. Non-universal critical dynamics in Monte Carlo simulations. *Physics Review Letters*, 58:86–90, 1987. (Cited on p. 342)
254. R.E. Tarjan. *Data Structures and Network Algorithms.* SIAM, Philadelphia, PA, 1983. (Cited on p. 11)
255. L. Trevisan. Non-approximability results for optimization problems on bounded degree instance. In *Proc. 33rd ACM Symposium on the Theory of Computing*, 2001. (Cited on p. 334)
256. J.D. Ullman. The performance of a memory allocation algorithm. Technical Report 100, Princeton University, Princeton, NJ, 1971. (Cited on p. 78)
257. L.G. Valiant. The complexity of computing the permanent. *Theoretical Computer Science*, 8:189–201, 1979. (Cited on p. 305)
258. L. Vandenberghe and S. Boyd. Semidefinite programming. *SIAM Review*, 38:49–95, 1996. (Cited on p. 268)
259. V.V. Vazirani. NC algorithms for computing the number of perfect matchings in $K_{3,3}$-free graphs and related problems. *Information and Computation*, 80:152–164, 1989. (Cited on p. 338)

260. V.V. Vazirani and M. Yannakakis. Suboptimal cuts: their enumeration, weight and number. In *Proc. 19th International Colloquium on Automata, Languages, and Programming*, volume 623 of *Lecture Notes in Computer Science*, pages 366–377. Springer-Verlag, Berlin, 1992. (Cited on p. 304)
261. D.L. Vertigan and D.J.A. Welsh. The computational complexity of the Tutte plane. *Combinatorics, Probability and Computing*, 1:181–187, 1992. (Cited on p. 342)
262. E. Vigoda. Improved bounds for sampling colorings. In *Proc. 40th IEEE Annual Symposium on Foundations of Computer Science*, pages 51–59, 1999. (Cited on p. 341)
263. V.G. Vizing. On an estimate of the chromatic class of a p-graph. *Diskret. Analiz.*, 3:25–30, 1964 (in Russian). (Cited on p. 10)
264. D.J.A. Welsh. *Knots, Colourings and Counting.* Cambridge University Press, Cambridge, UK, 1993. (Cited on p. 342)
265. A. Wigderson. Improving the performance guarantee for approximate graph coloring. *Journal of the ACM*, 30:729–735, 1983. (Cited on p. 23)
266. D.P. Williamson, M.X. Goemans, M. Mihail, and V.V. Vazirani. A primal–dual approximation algorithm for generalized Steiner network problems. *Combinatorica*, 15:435–454, 1995. (Cited on pp. 129, 223, 230)
267. D. B. Wilson. Generating random spanning trees more quickly than the cover time. In *Proc. 30th ACM Symposium on the Theory of Computing*, pages 296–303, 1996. (Cited on p. 339)
268. H. Wolkowitz. Some applications of optimization in matrix theory. *Linear Algebra and its Applications*, 40:101–118, 1981. (Cited on p. 268)
269. L.A. Wolsey. Heuristic analysis, linear programming and branch and bound. *Mathematical Programming Study*, 13:121–134, 1980. (Cited on p. 230)
270. M. Yannakakis. On the approximation of maximum satisfiability. *Journal of Algorithms*, 3:475–502, 1994. (Cited on p. 138)
271. A.Z. Zelikovsky. An 11/6-approximation algorithm for the network Steiner problem. *Algorithmica*, 9:463–470, 1993. (Cited on p. 211)
272. A.Z. Zelikovsky and I. I. Măndoiu. Practical approximation algorithms for zero- and bounded-skew trees. In *Proc. 12th ACM-SIAM Annual Symposium on Discrete Algorithms*, pages 407–416, 2001. (Cited on p. 37)

Problem Index

2CNF≡ clause deletion **176**, 179

Acyclic subgraph **7**, 334
Antichain cover 8

Bandwidth minimization **196**
Betweenness **267**
Bin covering **77**
Bin packing **74**, 74–78, 80, 124
– with fixed number of object sizes 81

Chain cover 8
Clique 9, 306, 309, 318–322
Closest vector 292
Clustering 243
– ℓ_2^2 253, 254
– metric k-cluster **52**
Counting problems 294–305
– acyclic orientations 338
– antichains 340
– bases of a matroid 339
– colorings of a graph 341
– contingency tables 340
– DNF solutions 295, 305
– – weighted version 302
– Euler tours 339
– forests 339
– graphs with given degree sequence 340
– Hamiltonian cycles 341
– independent sets 341
– perfect matchings 305, 338
– simple cycles in a directed graph 303
– stable marriages 340
– trees 340
– triangulations 340
– volume of a convex body 338
Cover time 337
Covering integer programs 112, 116, 118
Cycle cover 35, 62

Dominating set 48, 50, 52

Edge coloring 10
Edge expansion **192**
Enumerating cuts 304

Feedback edge set
– directed 337
– subset 166, **166**, 167
Feedback vertex set 25, **54**, 54–60, 129, 166
– directed 337
– subset 166, **166**, 167, 336

Graph bipartization by edge deletion **178**

Hamiltonian cycle 30, 303

Independent set 48, 51–53
– maximal 239

Knapsack **68**, 68–73

Linear equations over GF[2] 138

Matching **3**, 104
– b-matching 152, 227
– bipartite 129
– – maximum weight 129
– maximal **3**, 8
– – minimum cardinality 8
– maximum **3**, 5, 9, 124, 152, 153
– minimum weight 107
– perfect **105**, 142, 143

-- minimum weight 32, 35, 62, 105, 230
Matroid intersection 228
Matroid parity **212**, 212
MAX k-CUT **23**, 138, 267, 269
Maximum antichain 8
Maximum coverage **25**
Maximum cut (MAX-CUT) 10, 22, 138, 255, **255**, 256, 260–263, 267, 268, 334
- directed 23, 138, 267, 269
Maximum flow 38, **97**, 97–100, 168
Maximum satisfiability (MAX-SAT) 9, **131**, 131–139, 263, 306
- MAX k-FUNCTION SAT 312
- MAX-2SAT **131**, 263, 268
- MAX-3SAT **131**, 309, 311–315, 322, 323, 326, 330, 331
-- with bounded occurrence of variables 313–316, 330
Metric k-center **47**, 47–50, 53
- fault-tolerant 52
- weighted **50**, 50–52
Metric k-median **243**, 243–254, 337
Metric k-MST **252**
Metric facility location
- capacitated 240, 337
- fault tolerant 240
- metric uncapacitated 242
- prize-collecting 240
- uncapacitated **232**, 232–239, 242, 337
Minimum k-connected subgraph
- edge **228**
- vertex **226**
Minimum k-cut **38**, 40–44
Minimum bisection **193**, 196, 197, 336
Minimum chain cover 8
Minimum cut 38, 298
- b-balanced **193**, 193–194, 196, 197, 336
- s–t 38, **98**, 97–100, 146
Minimum cut linear arrangement **194**, 194–195, 197
Minimum length linear arrangement **178**
Minimum makespan scheduling 9, 10, **79**, 79–83, 140
- uniform parallel machines 140, 145
Minimum spanning tree (MST) 28–31, 105, 206, 207, 212
Multicommodity flow 97, 147, 163
- demands 168, **180**, 180–197
- directed 165
- integer **148**, 153, 154, 337
-- in trees 146–154
-- in trees of height one 152
-- in unit capacity trees 153
- sum **168**, 168–176, 179
- uniform 192, 197
Multicut **146**, 153, 168–179, 336
- directed 337
- in trees 146–154, 166
- in trees of height one 152
Multiway cut **38**, 38–40, 155–167, 335
- bidirected integer program formulation 164
- directed **165**, 166, 167
- fractional 156
- node **160**, 160–163, 166

Network design
- element connectivity 337
- vertex connectivity 336
Network reliability **297**, 304, 305, 339
- s–t reliability 339
- global 339

Point-to-point connection **208**

Satisfiability (SAT) 9, 330, **343**, 344
- 3SAT 310, **343**
Scheduling on unrelated parallel machines **140**, 140–145
Semidefinite programming **258**, 255–269
Set cover VIII, 11, **15**, 15–26, 34, 108–122, 124–130, 239, 251, 306, 309, 322–329, 334
- constrained set multicover 112, 116, 118
- multiset multicover 112, 116, 117, 123
- set multicover 24, 112, 116, 123
- with concave costs 117
Shortest superstring 9, **20**, 19–22, 26, 61–67
- variants 25, 67

Shortest vector **273**, 273–293, 336
Sparsest cut **180**, 180–197, 336, 337
Steiner arborescence
– rectilinear **35**
Steiner forest **198**, 198–213
Steiner network **213**, 213–231, 335
Steiner tree **27**, 27–30, 33, 37, 198, 213, 306, 309, 335
– directed **34**, 337
– Euclidean 89
– prize-collecting 208, 252
Subset sum 291
Subset-sum ratio problem **72**
Survivable network design *see* Steiner network and network design

Traveling salesman problem (TSP) **30**, 229, 231
– asymmetric 34, 336
– Euclidean **84**, 84–89
– metric 30–33, 37, 229, 231, 334
– – lengths one and two 34
– – variants 34
Tutte polynomial 341

Vertex coloring **23**
– k-coloring 267, 269
Vertex cover **1**, 15, 17–19, 23, 24, 104, 122–124, 129, 146, 152, 166, 306, 307, 309, 334
– cardinality **1**, 2–5, 8, 152

Zero-skew tree
– rectilinear **36**, 37

Subject Index

α-min cut 304
#P **294**, 305
1-tree 230

Active set **200**, 209
Approximation algorithm **2**, 345–347
- approximation factor **346**
- randomized **346**
Approximation scheme **68**
- fully polynomial randomized (FPRAS) 295, **295**, 297, 300, 302, 303, 305, 338–340
- fully polynomial time (FPTAS) **68**, 69–70, 72, 77, 83
- polynomial time (PTAS) **68**, 80–89, 140, 145, 311, 336
-- asymptotic **75**, 74–78
Arborescence 228
Arithmetic-geometric mean inequality 135

Basis of a lattice **274**
- Gauss reduced **281**, 290
- KZ reduced **290**
- Lovasz reduced 283
- weakly reduced 283, 290
Bernoulli trials 190, 353

Catalan numbers 86, 340
Certificate
- co-NP 336
- Yes/No 5–7, 93, 96, 294, 343–344, 348
-- approximate 274, 288
Chebyshev's inequality 297, **353**
Chernoff bounds 9, 190, **353**
Christofides' algorithm 37, 229, 334
Chromatic polynomial 342
co-NP **344**
co-RP 10, 330, **348**
Complementary slackness conditions **97**, 100, 105, 125, 149, 161, 178, 199, 233
- relaxed **126**, 129, 130, 146, 149, 199, 234
Compression 64
Concave function 135
Convex combination 258, 259
Convex set 259
Cost-effectiveness of a set 16, 113
Counting problems VII, 294–305, 338–342
- #P-complete VII, 294, **294**, 305, 338
Covering LP 109
Crossing sets **215**, 219
Cut packing 183–191
- approximate **184**
Cut requirement function 213
Cycle space 54
- cyclomatic number 54
Cyclomatic weighted graphs 54–57

Decision problem **343**
- NP-complete **344**
- well-characterized **6**, 5–7, 10, 93
- Yes/No certificate
-- approximate 7
Deficiency of a set **226**
Demand graph 182
Derandomization 132–134, 138, 248–250, 268
Determinant of a lattice **274**
Dilworth's theorem 8
Divide-and-conquer algorithm 179, 193

DTIME 331, 332, **348**
Dual fitting 101, 108–118, 241
Dual growing
– synchronized 198
Dual lattice **284**, 284–288
Dynamic programming 69, 81, 153

Edge expansion 192
Edge-disjoint s–t paths 103, 336
Eigenvalue 257
Eigenvector 257
Ellipsoid algorithm 170, 214, 255, 259
Euclid's algorithm 273, 276–278
Euler tour 28, 32
Eulerian graph 28, 31
Expander graph **175**, 179, 192, 320, 332
Expander graphs 314
Extreme point solution **100**, 102–104, 119, **122**, 141–145, 214, 219–221

First-fit algorithm 74, 77
Flow-equivalent tree 44
Forward delete 153
Frequency of an element 15, 119
Function
– degree-weighted 17
– proper 208
Fundamental cycle 54

Game
– two-person zero-sum 106
Gauss' algorithm 273, 276–278, 288
Gomory–Hu tree 40, 44, 46
Gram–Schmidt lower bound 287, 288
Gram–Schmidt orthogonalization **278**, 278–280, 282, 285
Greedy algorithm 8, 16–17, 24, 44, 60, 64, 72, 108, 138, 241

Half-integrality 119, 122–124, 153, 160–163, 165, 213–221
Hall's theorem 144
Hamiltonian cycle 29, 214
Hardness of approximation VIII, 306–333
Hungarian method 129

Integrality gap **102**, 101–103, 111, 129, 137, 151, 164, 167, 207, 210, 211, 218, 229, 254, 262, 335, 337
Integrality ratio *see* Integrality gap
Interactive proof systems 332
Ising model 342
Isolating cut 38

Kirchhoff's theorem 339
Konig-Egervary theorem 5, 104
Kruskal's algorithm 105, 206

Lagrangian relaxation 250–252
Laminar family of sets **219**
Layering 17–19, 25, 57, 60, 129
Linearity of expectation 136, **352**
Local search 23, 253
Lower bounding OPT 2, 17, 31, 32, 39, 47, 62, 79, 89, 108, 206, 278–280
Lowest common ancestor 149
LP-duality
– theorem 6, **95**, 93–97, 100, 106, 107, 148, 183
– – weak **96**, 148, 169
– theory 6, 17, 29, 97, 101, 108, 147

Mader's theorem 227, 231
Markov chain 192, 338, 339
– conductance 192–193, 197
– Markov chain Monte Carlo method VIII, 294
– rapidly mixing 305, 339, 340
– stationary probability distribution 192
– Swendsen-Wang process 342
– transition matrix 192
Markov's inequality 88, **353**
Matroid 339
– balanced 339
– basis exchange graph 339
– graphic 339
– independent sets 212
Max-flow min-cut theorem 97, 103, 168, 207
– approximate version for demands multicommodity flow 191
– approximate version for uniform multicommodity flow 197
MAX-SNP-completeness 332

Maximum weight spanning tree 44
Menger's theorem 103
Method of conditional expectation 131–134, 138, 139, 248
Metric 183–191
– ℓ_1-embedding 183–191
– – β-distortion **185**
– – isometric **185**, 186
– ℓ_2-embedding 196
– ℓ_2^2-embedding
– – isometric 195
– – optimal distortion 197, 266
– ℓ_p **185**
Min–max relation 5–7, 11, 97–100, 168
– approximate 7, 151
Minkovski's theorem 287
Moments of a random variable **352**
– central **352**
Monte Carlo sampling 297, 301

Near-minimum cuts 298–299
Next-fit algorithm 77
Norm **185**
– ℓ_p **185**
NP **343**

Odd set cover **6**
Optimization problem 2, **345**, 351
– NP-complete 10
– NP-hard 68, **344**
– strongly NP-hard 71
Orthogonality defect **275**, 279
Overlap graph 66

P≠NP conjecture VII, 10, 68, 71, 345
Packing LP 110
Parametric pruning 47–52, 140–141, 252
Parsimonious property 229, 230
Partial ordering 8
PCP theorem VIII, 306, 308–311, 323, 332
Petersen graph **6**, 214
Poisson trials 353
Positive semidefinite matrix **257**, 257–258
Potts model 342
Prefix graph 62

Primal–dual schema VII, 101, 125–130, 149–152, 235–236, 335
– with synchronization 199–204
Primitive root 9
Primitive vector **275**, 285, 286, 290
Principal submatrix 265
Probabilistic argument 179
Probabilistic method 324
Probabilistically checkable proof system (PCP) 309, 332
– completeness 319
– parallel repetition 325–326
– soundness 319
– two-prover one round 322–324, 332
Probability distribution
– binomial **354**
– normal 261, 266, **354**
– Poisson **354**
– spherically symmetric 261
Probability theory 352–354
Pseudo-approximation algorithm 193–195, 197
Pseudo-forest **143**
Pseudo-polynomial time algorithm **69**, 69, 71–73
Pseudo-tree **143**

Quadratic forms 292
Quadratic program **255**
– strict **255**, 255–257, 267, 268

Random contraction algorithm 298, 304
Random walk 320, 338–340
Reduction
– L- 332, 351
– approximation factor preserving 24, 27, 34, 60, 152, 160, 166, 196, 242, **347**, 351
– gap-introducing **307**
– gap-preserving **307**
– polynomial time **344**
– randomized 293
Region growing 171–175
Relaxation
– convex 269
– exact **102**
– – for maximum weight bipartite matching 129

- for MST 212, 230
- LP- VII, 39, **99**, 100–106, 109, 111, 113, 119, 120, 122, 124, 125, 134, 147, 153, 155–157, 160, 164, 165, 179, 199, 206, 209, 211, 213–221, 224, 229–231, 233, 240, 244, 251, 335, 337
-- bidirected cut relaxation for Steiner tree 210, 335
-- subtour elimination relaxation for TSP **229**, 229–231
Reverse delete 149, 210
- dynamic 209
Rounding VII, 101, 119–124, 134–136, 170–175, 191
- iterated 213, 217–218
- randomized 120–122, 124, 157–160, 164, 247–248, 260–263
RP **348**

Scaling and rounding 73, 117
Self-reducibility IX, 9, 10, 303, 348–351
- tree 10, 303, 349
Semidefinite program 197, 266, 267
- duality theory 268
Separating hyperplane **259**
Separation oracle **102**, 107, 170, 179, 217
Short-cutting 29, 31, 32, 85, 241
Simplex 155
Sparsity of a cut **181**
Spread of an edge 196
Square of a graph **48**
Standard deviation **352**
Steiner tree 316–318
Sublattice 285, 290
Submodular function **215**, 224
Supermodular function
- weakly supermodular **216**

Throughput **180**, 182
Tight example IX, **4**, 8, 17, 19, 23–25, 29, 31, 33, 39, 43, 49, 51, 59, 80, 83, 120, 123, 128, 137, 144, 153, 165, 175, 206, 218, 238, 239, 249, 268
Totally unimodular matrix 104
Tournament **25**
Traveling salesman tour
- maximum weight 66
- minimum weight 62
Triangle inequality **27**, 51, 52, 178
- directed **34**

Unbiased estimator 295
Uncrossable function 224
Uniform generator 302, 303
- almost uniform 303
Unimodular matrix **274**, 274–276, 288
Unit sphere 260
Upper bounding OPT 256

Vector program **256**, 255–257, 266, 267
Verifier 309
Vertex cover 316–318
Vertex-disjoint s–t paths 103, 336
VLSI design 178
- clock routing 36
von Neumann's minimax theorem 106

Witness family 225

ZPP 10, **348**
ZTIME 329, 332, **348**